The Digital Transformation of the Automotive Industry

Uwe Winkelhake

The Digital Transformation of the Automotive Industry

Catalysts, Roadmap, Practice

MÄNGELEXEMPLAR

Uwe Winkelhake
IBM Germany GmbH
Hannover, Germany

Originally published in German language by SpringerVieweg, Wiesbaden, 2017

ISBN 978-3-319-71609-1   ISBN 978-3-319-71610-7 (eBook)
https://doi.org/10.1007/978-3-319-71610-7

Library of Congress Control Number: 2017960433

© Springer International Publishing AG 2018

This work is subject to copyright. All rights are reserved by the Publisher, whether the whole or part of the material is concerned, specifically the rights of translation, reprinting, reuse of illustrations, recitation, broadcasting, reproduction on microfilms or in any other physical way, and transmission or information storage and retrieval, electronic adaptation, computer software, or by similar or dissimilar methodology now known or hereafter developed.

The use of general descriptive names, registered names, trademarks, service marks, etc. in this publication does not imply, even in the absence of a specific statement, that such names are exempt from the relevant protective laws and regulations and therefore free for general use.

The publisher, the authors and the editors are safe to assume that the advice and information in this book are believed to be true and accurate at the date of publication. Neither the publisher nor the authors or the editors give a warranty, express or implied, with respect to the material contained herein or for any errors or omissions that may have been made. The publisher remains neutral with regard to jurisdictional claims in published maps and institutional affiliations.

Printed on acid-free paper

This Springer imprint is published by Springer Nature
The registered company is Springer International Publishing AG
The registered company address is: Gewerbestrasse 11, 6330 Cham, Switzerland

Contents

1 **Introduction** 1
   1.1 Digitisation – A Hot Topic 1
   1.2 IT Development – The Exponential Function Explodes 2
   1.3 Transformation of the Automotive Industry 3
   1.4 Structure of the Book 5
   1.5 Definition of Focus and Readership 6
   References 7

2 **Information Technology as Driver of Digitisation** 9
   2.1 Moore's Law 9
   2.2 Exponential Growth Also for Digitisation 11
   2.3 Energy Requirement of IT 14
   2.4 IT Security 16
   2.5 Handling of Personal Data 18
   2.6 Powerful Networks 19
   2.7 Technology Outlook 19
         2.7.1 3D Chip Architectures 20
         2.7.2 Flow Batteries 21
         2.7.3 Carbon Nanotubes 22
         2.7.4 Neuronal Networks 22
         2.7.5 Quantum Computer 23
   2.8 Technological Singularity 24
   Annex A1 25
   References 27

3 **"Digital Lifestyle" – Future Employees and Customers** 29
   3.1 Always On 30
   3.2 Mobile Economy 32
   3.3 "Real-Time" Expectation in the Mobile Ecosystem 34
   3.4 Sharing Economy 35
   3.5 Start Up Mentality 37
   3.6 Innovative Work Models 38

		3.6.1	Digital Nomads	39
		3.6.2	Crowdsourcing and Liquid Workforce	39
		3.6.3	Wikinomics	41
	3.7	Google – The Goal of the Digital Natives		41
	References			44
4	**Technologies for Digitisation Solutions**			**45**
	4.1	IT Solutions		47
		4.1.1	Cloud Services	48
		4.1.2	Big Data	49
		4.1.3	Mobile Applications and Apps	52
		4.1.4	Collaboration Tools	55
		4.1.5	Cognitive Computing and Machine Learning	56
	4.2	Internet of Things		58
	4.3	Industry 4.0		59
	4.4	3D Printing		62
	4.5	Virtual and Augmented Reality		64
	4.6	Wearables		67
	4.7	Blockchain		68
	4.8	Robotics		69
	4.9	Drones		71
	4.10	Nanotechnology		72
	4.11	Gamification		73
	References			74
5	**Vision Digitised Automotive Industry 2030**			**77**
	5.1	Development of the Automotive Market		78
	5.2	Future Customer Expectations in the Passenger Car Area		80
	5.3	Digitisation Situation in the Automotive Industry		83
	5.4	Vision Digitised Automotive Industry		86
		5.4.1	Mobility Services Instead of Vehicle Ownership	89
		5.4.2	Connected Services	90
		5.4.3	Autonomous Driving	92
		5.4.4	Electromobility	96
		5.4.5	Centralised Embedded IT Architecture	98
		5.4.6	Prototype-Free Process-Based Development	103
		5.4.7	Internet Based Multichannel Distribution	105
		5.4.8	Digitised Automotive Banks	109
		5.4.9	Flexible Production Structures/Open Networks/Industry 4.0	110
		5.4.10	Automated Business Processes	113
		5.4.11	Cloud-Based IT Services	115
	5.5	General Electric – An Example of Sustainable Digitisation		117
	Annex A2			121
	References			123

6	**Roadmap for Sustainable Digitisation**		127
	6.1	Digitisation Roadmap as Part of Company Planning	127
		6.1.1 Assessment of Market Potential and Customer Requirements	128
		6.1.2 Adaptation of the Company Strategy	132
		6.1.3 Business Model and Lean Enterprise	136
		6.1.4 Frame for Digitisation	139
	6.2	Roadmap for Digitisation	143
		6.2.1 Roadmap Connected Services and Digital Products	144
		6.2.2 Roadmap Mobility Services and Autonomous Driving	151
		6.2.3 Roadmap Processes and Automation	157
		6.2.4 Roadmap Customer Experience, Sales and After-Sales	170
	6.3	Overview Roadmap and KPIs	174
	References		175
7	**Corporate Culture and Organisation**		179
	7.1	Communication and Leadership	180
	7.2	Agile Project Management Methods	184
		7.2.1 Design Thinking	186
		7.2.2 Scrum	188
	7.3	Entrepreneurship	190
	7.4	Resourcing for Digitisation	191
		7.4.1 E-Learning as Basis for Digital Education	192
		7.4.2 New Ways of Learning	193
		7.4.3 Knowledge Management	195
		7.4.4 Hiring	196
	7.5	Cooperation Forms	198
	7.6	Open Innovation	200
	7.7	Organisational Aspects of Digitisation	203
		7.7.1 Chief Digital Officer (CDO)	204
		7.7.2 Adaptation of the IT Organisation	206
		7.7.3 New Occupational Profiles and Career Models	210
		7.7.4 Change Management	211
	7.8	Case Study: Transformation IBM	214
	References		218
8	**Information Technology as an Enabler of Digitisation**		223
	8.1	IT Transformation Strategy	224
	8.2	Building Blocks of an IT Strategy	225
	8.3	Cost and Benefit Transparency	227
	8.4	Transformation Projects	229
		8.4.1 Development in the Status Quo	229
		8.4.2 Microservice-Based Application Development	232
		8.4.3 Data Lakes	235

		8.4.4	Mobile Strategy	238
		8.4.5	Infrastructure Flexibilisation Through Software Defined Environment	240
		8.4.6	Computing Centre Consolidation	243
		8.4.7	Business-Oriented Security Strategy	245
		8.4.8	Security of the Factory IT and Embedded IT	248
	8.5	Case Studies on IT Transformations		251
		8.5.1	Transformation Netflix	251
		8.5.2	Transformation General Motors	254
	8.6	Conclusion		257
	References			258
9	**Examples of Innovative Digitisation Projects**			263
	9.1	Framework		263
	9.2	Connected Services/Digital Products		264
	9.3	Mobility Services and Autonomous Driving		272
	9.4	Efficient Processes and Automation		275
	9.5	Customer Experience – Marketing, Sales, After-Sales		283
	9.6	Corporate Culture and Change Management		288
	References			293
10	**Car Mobility 2040**			297
	10.1	Environment		297
	10.2	Electric Drive and Autonomous Driving		298
	10.3	Market Shift		299
	10.4	Mobility Services and Vehicle Equipment		299
	10.5	Innovative Process- and Production Structures		300
	10.6	Day-in-a-Life of a "Liquid Workforce"		301
	10.7	Conclusion		304
	References			304
Glossary				307

Chapter 1
Introduction

1.1 Digitisation – A Hot Topic

The subject of "digitisation" is a key issue in all companies – often driven by the fear of overseeing a potential attack on the own traditionally established business by new entrants from Silicon Valley. These attacks are leveraging the potential of platform economy principles. Who would want to be left behind like Kodak, Nokia, or the many video stores? Such attacks – bringing "disruption" to established business models – are to be repulsed. The challenge is to recognise the potential of new digitised business models as early as possible and to integrate them into the own company via a transformation process. Responsiveness and creativity are of the essence. Even if it does not involve a completely new business model right away, digitisation should at least though achieve a noticeable increase in process management efficiency and help to sell more products, for example through deepened customer insight and comprehensive evaluation of social media data, or also availing of new digitised distribution channels.

The need to survive, as well as the prospect of higher profits and revenue, rightly put the subject of digitisation at the top of the agenda in today's businesses. This is underpinned by Fig. 1.1, which shows the results of a survey [Sto16].

In addition to these objectives, many companies consider the potential of digitisation as a means of improving customer satisfaction and thus increasing sales opportunities, opening up new markets and implementing product innovations. Thus it is obvious to all parties and stakeholders in the companies that something is to be done – yet what is it? Many engage with the issue of digitisation and launch initiatives and projects. However, there is a lot of uncertainty about how to proceed, what to do, and how deeply and comprehensively to implement changes. Occasionally, at this stage already doubts are cast which suspect "new wine in old wineskins" behind the keyword of digitisation and recommend calm in the form of compact projects. To demonstrate at least actionism, for example, the replacement of paper-based order documents is addressed by iPad-based visualisation.

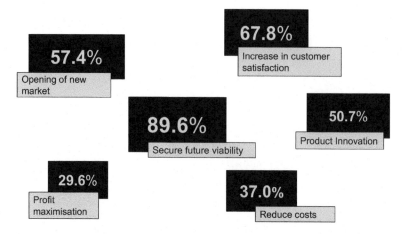

Fig. 1.1 Cross-departmental corporate objectives for digital transformation [Sto16]

Based on the author's wide experience, it is profoundly wrong to just garnish existing processes "as is" with some IT and aim to get on top of the subject of digitisation this way. We are at the beginning of a "tsunami" which will strike all industrial enterprises – associated with high risks, but also with immense new opportunities. It is safe to assume that everything which can be networked and automated with the help of automation, will in fact be involved – it is just a matter of time. Therefore, with any idea of digitisation it is imperative to begin by fully scrutinising present and so far well-proven business models, processes and the organisation. The topic of digitisation is then to be approached in-depth and in a sustainable way based on a compelling vision and business strategy derived from this – not as a single project, but as a continuous process of transformation.

1.2 IT Development – The Exponential Function Explodes

The digitisation will affect all companies vehemently simply due to the fact that more and more powerful and inexpensive information technologies become available as drivers. This kind of explosion can best be explained by a brief review on the development of IT. The performance and thus the penetration of business and private processes with information technology (IT) solutions follows an exponential function [Kur05]. Reminder: an exponential function goes first in a gradual, almost linear increase and then after a curve assumes a massive surge within a short time, the exponential growth.

During the first linear climb, after the Second World War until the 1970s, company-specific software programs written by specialists in FORTRAN or COBOL were implemented via punch cards on the computer systems in the corporate data centres. Selected users, specialists of their business function, were trained for the operation of the programmes. In a first increase in growth, which saw the spread of computers in the 1980s and '90s, almost all jobs in business management and administration were equipped with IT solutions, and typewriters were replaced by word processors. Standard software solutions to support processes spread as well. Originally, IBM had dominated the market for the production industry with COPICS, then SAP evolved into the de facto standard in the area of ERP solutions. Almost all private households used Windows PCs for word processing or spreadsheet programmes for private administration tasks.

In the late 1990s, the use of the Internet expanded, eBay became a platform for private and increasingly also the professional trade, and Amazon in the last decade within a short time first became the world's largest bookseller and also the dominant retailer. Solutions were used widely to book overnight stays or theatre seats, and the term of the so-called platform economy became popular. The development of digitisation, from the author's point of view, at that time already entered the "curve" of the exponential function, in enterprises as well as in private households.

With Apple's introduction of the iPhone in 2007 and its extremely rapid worldwide penetration and acceptance, the above mentioned exponential function of IT enters the phase of massive surge. This is underpinned by the introduction and high acceptance of other mobile devices such as Android smartphones and the success of tablets which are gradually replacing PC's and notebooks as full-fledged computers.

1.3 Transformation of the Automotive Industry

This development which is just briefly outlined here, will continue at an accelerated pace and cause substantial upheavals in all companies and also private processes. The automotive industry is being affected in particular and is facing several changes at the same time:

- Electric drive technology
- Autonomous driving
- Transformation of the business model from a vehicle manufacturer to becoming a mobility provider
- Digitisation of the vehicle – Connected Services; Software-oriented
- configuration
- Multiple distribution channels – from the centric importer/distributor to
- the customer-centric direct sales partner
- Use of the digitisation for process automation
- Overarching value chains: intermodal transport – electricity supplier – service provider
- Change of customer needs from vehicle ownership to mobility on demand

These foreseeable changes, according to the author's observations and experiences, highlight that the automotive industry needs to reinvent itself. Namely the established manufacturers are being challenged and under intense time pressure in the transformation because new competitors push into the market, which are free of any "inherited burdens" and can have a fully digitised flying start with new structures "born on the web". The aggressive new entrants often focus on new technologies, such as the electric drive.

The established companies find it particularly hard to aggressively implement the new requirements because this often comes at the expense of existing products [Wes12]. The initial success and the market response of Tesla Motors which was founded in 2007 are quite impressive; other companies already being formed are Faraday in California, but also in the Chinese region with the online retailer Alibaba and search engine provider Baidu as well entering the automotive business. Both companies have announced to offer driverless cars. It certainly remains to be seen of how these newcomers will develop, however these challengers definitely represent a threat to established car manufacturers with their current business model. In addition, new competitors cavort in the future business focus of mobility providers, which will make it extra difficult for producers to make a difference in the market and still dominate it. All automotive providers will be aware of this challenging situation so that the results of a KPMG survey as shown in Fig. 1.2 do not surprise [KPM16].

The subject of digitisation, along with connected services and alternative drive technologies, is evaluated as a key trend. For the automotive industry there is no alternative to vigorously embracing the impending fundamental changes and to turning potential threats into opportunities through proactive action. In doing so,

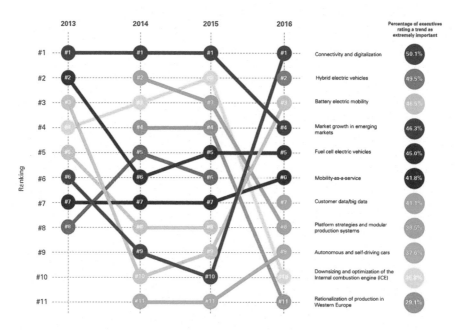

Fig. 1.2 Key trends of the automotive industry 2016 (KPMG)

current and aggressively prognosticated technological possibilities are to be involved, operating hand in hand with modern, highly flexible and efficient IT structures. Right in this synergetic approach are the highest potentials for optimisation. Nevertheless, digitisation projects are often approached just one-dimensionally as islands. Further obstacles are in the traditional project management and budgeting methods, and in many cases also in a lack of change culture and the incomplete expertise of the team.

In the future, Industry 4.0 for instance will be part of the digitisation strategy with the aim of highly automated production processes, in which robots directly cooperate with employees. The requirements for modern IT structures to be derived from these thematic fields lead to hybrid cloud architectures in order to achieve digitisation in a targeted and cost-effective manner.

The topic of digitisation must also be looked at from the product angle of view. What does autonomous driving or the conversion of vehicles into "driving IP addresses" mean as part of a global "Internet of Things"? How can you master the masses of data from the vehicles, business processes and customer activities in order to turn these into business benefits and competitive advantage? How should you protect yourselves against new market participants from the IT environment, use long-standing traditional experiences and thus emerge stronger from the transformation?

1.4 Structure of the Book

Against this background, this book addresses many of the common shortcomings in the implementation and the related problems. A methodically sound and well-tried guide to the implementation of digitisation in the automotive industry is developed, thus ensuring the competitiveness of this key industry in the long run. Comprehensive and pragmatic recommendations for action in the automotive and supplier industries are pointed out to shape the transition from the discrete vehicle-focused business model to a continuous and mobility-oriented model. The path to the automatic, highly efficient execution of lean, integrated business processes is also discussed, as are the collection of significant changes in sales, aftersales and marketing structures and the new design of customer relationships. Under this objective the book is divided into 4 blocks:

Block 1 with Chapters 2, 3 and 4: Drivers IT Technology, Digital Natives, Technology for Digitisation

To understand why there is no alternative to dealing intensively with the topic of digitisation, and also to assess future potentials, starting from the Moore's Law and Nanotechnology up to Singularity, an outlook is given on the future development of IT technology. It is important to understand future customers and at the same time future employees in their behaviour, their expectations and interaction. This topic is explained in one chapter as well as in the following chapter the technologies important for future consideration, both on the IT side and complementary such as 3D printing, Wearables, or new concepts such as additive manufacturing and Blockchain for instance.

Block 2 with Chapters 5 and 6: Vision Automotive 2030; Roadmap digitisation

In this block, a vision or rather an outlook on the automotive industry in the year 2030 is developed. For this purpose, "software defined vehicles", Internet-based sales and also service platforms for administrative services are highlighted. This way a comprehensive basis is provided to give recommendations for the development of a concrete roadmap for the implementation of a goal-oriented digitisation strategy. The recommendations are derived from concrete project experience and case studies.

Block 3 with Chapters 7 and 8: Corporate culture; Flexible IT structures

Prerequisite for a successful implementation is a transformation culture with leadership exemplarily communicated by the board, accompanied by appropriate incentives as well as the necessary basic training of employees and the use of innovative, agile implementation methods in the projects. Another important prerequisite for successful implementation of digitisation strategies are efficient and flexible IT structures. These have to be designed in such a way that they meet the needs of the business appropriately and react swiftly. Hybrid cloud architectures as well as the consideration of open standards along with effective security concepts and requirements for data storage are the basis for successful digitisation projects.

Block 4 with Chapters 9 and 10: Implementation Examples, Outlook 2040, Conclusion

In the fourth and final block of the book, current implementation examples are presented, challenges of the implementation pointed out, and a short outlook on the automotive industry in the year 2040 is given.

1.5 Definition of Focus and Readership

The book provides recommendations for the development and implementation of digitisation strategies for the automotive industry with a focus on manufacturers and distributors of cars and vans. This means that the largest market or rather enterprise sector of this industry is addressed. With some limitations, the advice is also interesting to other manufacturers (trucks, commercial vehicles, special machines) and suppliers. Within the addressed segment, both make-to-stock producers, which are predominantly found in the USA and Japan, as well as contract manufacturers, are addressed. Especially the second field will grow in the course of finer customer segmentation and increasing individualisation.

The book is aimed at executives from all business sectors of the automotive and supplier industries, as well as research institutes and consultancies, plus students of production and business science who are interested in taking up the topic of digitisation.

References

[KPM16] KPMG: KPMG's Global Automotive Executive Survey 2016. https://home.kpmg.com/xx/en/home/. Drawn: 10.06.2016

[Kur05] Kurzweil, R.: The singularity is near: when humans transcend biology. Viking Books, New York (2005)

[Sto16] Stoll, I, Buhse, W. (eds.): Transformationswerkreport 2016. www.transformationswerk.de/studie. Drawn: 10.06.2016

[Wes12] Wessel, M., Christensen, C.M.: Surviving disruption. Harvard Business Review, 12/2012

Chapter 2
Information Technology as Driver of Digitisation

Driven by the extreme increase in the efficiency of information technology (IT), the digitisation wave keeps approaching us unstoppably and ever faster. The so-called Moore's Law, which more than 50 years ago has already described a doubling of the capacity of integrated circuits over a period of 12 months [Moo65], has been synonymous with the ongoing massive increase in the performance of IT. If the basic technology had not changed, this law would no longer be valid. However, due to technological leaps, the principle of exponential growth in performance still applies. There don't appear to be technological limits, and it is only a matter of time until the human intelligence will be overtaken by "machine intelligence" and so-called singularity is reached.

In order to understand this situation and to prove why digitisation is proceeding at an unstoppable pace and will dramatically change our private and business processes, this chapter first explains the fundamentals of IT development. Subsequently, the subjects of IT security and energy demand are highlighted as potential developmental impediments. The conclusion of the chapter is the concept of technological singularity thereby being a more visionary outlook.

2.1 Moore's Law

In April 1965 Gordon Moore described in an expert article an observation on integrated circuits [Moo65]. He noted that the number of transistors on a silicon chip regularly doubled with minimal component costs at a fixed time interval. As a result, the computer performance increases exponentially without the costs increasing as well. The underlying period has since then been adjusted several times due to changes in the conditions of the technological framework. The basic explanation for exponential growth, however, continues to apply – nowadays usually within a period of 18 to 24 months. The basic context is illustrated in Fig. 2.1, which shows the number of transistors in different process types in logarithmic representation and

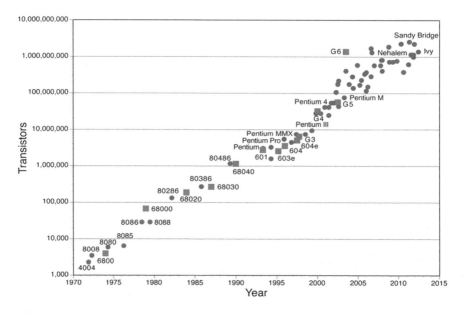

Fig. 2.1 Packing density of processor types in the course of time (Schau et al.)

thus as a straight line over time [Cha03]. Causes for this immense increase in packing density are the continuous reduction of component sizes and improved manufacturing processes.

The component size and density on the chips is directly related to their performance – the smaller the size and the denser the pack, the greater the performance. When in 2005 the mass production of chips with structures from 130 down to 90 nm was established, the 65 nm technology was already in pre-production. Laboratories began to work on first prototypes with even smaller structural sizes down to 10 nm. This technology is expected to become the standard in mass production by about 2017 and thus to further confirm the Moore's Law [Boh15].

Moore's Law is based on observations and is not scientifically substantiated. Nevertheless, it has established itself as the standard of the digital revolution in industry, and the industry is defining milestones by it for its planning. This is why it is also referred to as a self-fulfilling prophecy, which is quasi the drive to the IT power increase. To link the performance of a processor directly to the number of transistors, is a simplification which is however sufficient for a basic understanding of the strongly growing IT performance which is the focus here. In today's high-performance chips, not all transistors directly serve the computing power, but also, for example, the temporary data storage (so-called Cache). The aspect of multi-processor architectures and their influence on computer performance is only outlined here. Elaboration or deeper understanding of these details is not required for the purpose of this book.

2.2 Exponential Growth Also for Digitisation

It is much more interesting that the basic context of the exponential growth as observed and identified by Moore for integrated circuits, has already been applied to IT technologies used before the arrival of chips; see Fig. 2.2 [Cha03].The computer power per second or per $1000 of value was already subject to an exponential course at the times of punch card technology as well as in the following technology phases of mechanical relays, electron tubes and individual transistors.

Further analysis shows that this development applies to all information technology parameters, such as bandwidth, storage capacity, clock rate and the prices of the corresponding technology components [Cha03]. In this regard, the discussion is obsolete in which temporal distance a doubling of the respective performance parameter may take place. Be it 12, 18 or 24 months – in any case there will be large-scale increases, even across technological frontiers. The resulting dynamic developments are shown in Fig. 2.3 for smartphone users [Mee16] and Fig. 2.4 for the number of network nodes in an automobile [Reg16].

Both the increase in smartphone users in the dynamic Asia/Pacific market and the performance of the network in automobiles are subject to exponential growth in

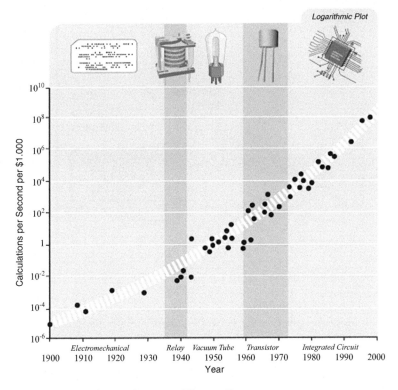

Fig. 2.2 Development of processing power (Chau et al.)

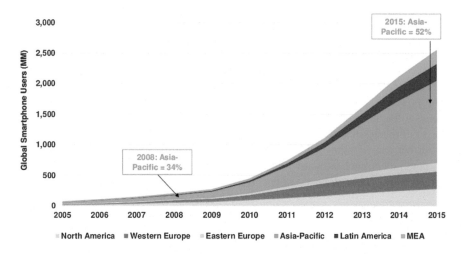

Fig. 2.3 Global development of smartphone users 2005–2015 (Meeker)

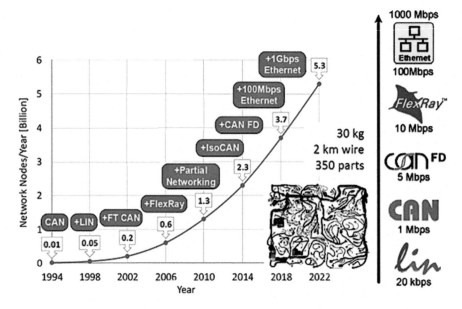

Fig. 2.4 Development of internal network power when adapting bus technology in vehicles (Reger)

analogy to Moore's Law. For the networks in automobiles the bus technology is being developed further to achieve growth, extending from Lin via CAN to Ethernet. As supplementary information, the figure shows as an example the extent of a car network. It consists of over 300 components, the wiring has a length of 2 km and overall the network weighs 30 kg (as of 2017).

2.2 Exponential Growth Also for Digitisation

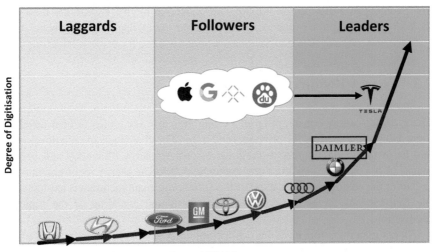

Fig. 2.5 Digitisation degree of automotive manufacturers as of 2016 (U. Winkelhake, see Annex A1)

Generalising this trend, it can be assumed that the penetration of digitisation into enterprises is also subject to exponential growth and will therefore take up speed significantly. This raises the question of how the major automotive manufacturers are comparable in terms of their digitisation activities. In order to answer this question, a digitisation degree has been estimated by the author using the following parameters: the provision of digital services, partnerships in the digital field, the offering or timely announcement of vehicles with electric drive or even autonomous cars, organisational adjustments and, last but not least, "Google hits" for the term combination manufacturer name/digitisation. The parameters are derived from annual reports or net searches for instance. A ranking was determined for each parameter in a comparison, and these were evaluated by points. In summary, this results in Fig. 2.5, the basis of which can be found in Annex A1. The three groups of Leaders, Followers and Laggards are distinguished.

Based on the exclusively offered electric drive, with "update over the air" of the vehicle software and an innovative distribution channel, Tesla Motors has established itself as a digitisation benchmark in the industry. Amongst the German manufacturers, Mercedes and BMW are nearly even, followed by AUDI. A little further behind, Volkswagen leads the group of laggards ahead of Toyota, GM and Ford. This snapshot (as of August 2016) demonstrates that many manufacturers are in a considerable backlog. Taking into account the exponential growth, the catching up of even small distances means an immense effort and investment. Additional transformation pressure on the established OEM's is exerted by those new industry entrant companies which have from the start already a high degree of digitisation and thus high process efficiency and strong customer orientation.

The derivation of the necessary measures for this transformation is the goal and main part of this book. First, potential obstacles to continuous IT performance improvement or a digitisation initiative are looked at briefly. Essentially these are the energy consumption of the IT, security and the lawful handling of personal data.

2.3 Energy Requirement of IT

The growing energy consumption of IT, and thus also the topics of heat development and pollution, affect IT providers and users. The challenges of chip developers and hardware manufacturers to continue to ensure energy efficiency and performance improvements, is being discussed in Sect. 2.5. The energy requirement of IT as part of the ecological assessment of the automotive manufacturers is explained in the following as far as required for the overall understanding of the topic of digitisation.

All car manufacturers have their own computing centres (CC) as the heart of the required information technology. Comprehensive consolidations of the server and storage systems into a "global mega-CC" are not established so far, yet "regional CC's" distributed across the continents are common due to latencies in the network. These are located, for example, in North and South America, Europe, Asia, China and possibly the ASEAN countries and are often linked in order to guarantee availability.

The demand of the automotive industry for computing capacity and thus for CC space is growing continuously. Drivers are growing business volumes and namely the digitisation trend with more and more mobile devices or rather smartphones, the massive growth of structured and unstructured data, for example by simulations in product development and videos in the marketing sector, as well as through automated process flows. Also the Internet of Things (IoT) and the increasing digitisation of the production as a result of the Industry 4.0 implementation require much computing performance. This increases the need for IT hardware and the energy required for operation, data networks, air conditioning systems, emergency power units and transformers.

The power consumption for the technical building services of a computing centre is nowadays about 50% of its total power requirement, so currently only half of the energy is used to operate the actual IT infrastructure. The ratio of CC total energy demand to the power requirement of IT constitutes an industry standard for the energy efficiency of a CC. While the installed CC's operate with a characteristic value of 2 on average, new mainframe computing centres achieve values of significantly less than 1.5 [Hin16]. This is achieved on the one hand by the higher efficiency of the technical building equipment and on the other by improved organisation, methods and air-conditioning technology concepts. For example, hot water cooling, shifts in room temperatures and also the housing of servers and storage are common improvement measures.

2.3 Energy Requirement of IT

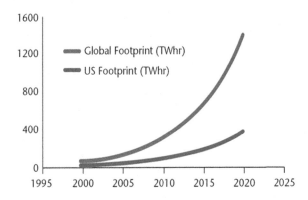

Fig. 2.6 Forecast of the energy consumption of computing centres (Burger)

At the same time, the energy efficiency of the IT infrastructure is being improved continually. As early as at the beginning of the 2010s, standard PC's consumed more than 100 Watt of power, while today's systems consume less than 30 W and smartphones less than 3 W. This is certainly a pleasing development which will continue. However, two aspects are contrary to this improvement: The power consumption per computing transaction remains almost the same, since the computing speed has increased massively. The number of end devices (PC's, notebooks, smartphones) however sharply increases. The applications on the end devices are connected to central systems and cause increasing power consumption in the networks and the CC's. It is therefore not surprising that the energy demand in the CC's continues to grow, despite improved efficiency. As a result, the capacity of CC's often is not determined by the footprint for server and storage systems, yet rather by the necessary energy supply and cooling. A forecast of the energy demand of computing centres worldwide and for the USA, measured in terawatt hours, is shown in Fig. 2.6 [Bur16].

Consequently, the IT energy demand is growing exponentially. Jonathan Koomey of Stanford University estimates the share of data centres in global power consumption at 1.1–1.5% [Koo10]. For the US, he estimates this share to be as high as 2.2%, while in Germany the CC's cause approx. 2% of the total power consumption [Hin16]. This underlines the importance of measures to increase the efficiency of the CC's.

In addition to the outlined technological aspects, conceptual, organisational and business policy options play a role as well. The utilisation factor of the servers employed is still at a relatively low level. As an example, the results of a comprehensive study are being mentioned here [Koo15]. According to it, servers operate on average at a 6% utilisation factor, and 30% of servers in the US in both virtualised and non-virtualised environments are in a "comatose state", i.e. they are fully installed and consume power however did deliver neither processing performance nor data in the past 6 month. These values point to a considerable potential for improvement and energy savings, which must be consistently realised. Extensive consolidations and cross-segment virtualisation as well as the shutdown of obsolete applications and servers are recommended. Virtualisation means the combination of different servers under a consolidating software layer, which optimises the distribution of the performance requirements to the individual servers, thus improving the respective utilisation. To support these projects, tools are available on the web [Koo15].

The optimisation of the utilisation of storage systems is also to be promoted with every emphasis. On the one hand, as the data volume is increasing rapidly in the enterprises – rates of increase of 60% per year are quite common – and on the other hand, since the virtualisation in the storage area, compared to these approaches in the server area, only began to be used later. The concept of the so-called Software Defined Storage (SDS) offers considerable utilisation and performance benefits. Here, a software layer is laid over existing storage systems by even different manufacturers, so that free memory can be recognised quickly and used by several different systems. This results in the advantage of a common efficient use of existing hardware, freedom of choice in the use of additional storage units, and the common management of the connected system as a whole. The improved utilisation in turn leads to savings in energy consumption for data storage.

These were just some notes on energy saving for computing centres. Further flexible usage concepts using so-called hybrid cloud architectures as a platform for digitisation projects will be discussed in Chap. 8.

The energy consumption of their computing centres is important for automotive manufacturers, in addition to the aspects of operational safety and profitability, also from an ecological point of view. Of course, the environment is always about consuming as little energy as possible and preferably getting the required energy from environmentally friendly sources. The energy consumption of a computing centre is at the same time part of the overall eco-balance of the automotive manufacturers. Many companies have embedded environmental objectives in their strategy, covering the entire life cycle of the vehicles. The established parameter for this is the "CO_2-Footprint per Vehicle". This has to include the proportionate CC energy consumption – a further reason to pay attention to the energy efficiency of the IT.

2.4 IT Security

In the context of digitisation, IT security and proper handling of personal data are to be addressed similarly to the subject of energy consumption. Traditionally, Germany is dealing intensively and particularly sensitively with these topics. This certainly is appropriate indeed. However, this should not lead to any obstacles to meaningful digitisation projects, as the author repeatedly experienced in practice. Both topics are challenging, comprehensive and complex and are dealt with in detail in the relevant specialist literature. For this reason, they are not discussed in-depth here; rather it follows an overview to provide a basic understanding and problem-consciousness as the basis for planned digitisation projects.

2.4 IT Security

Basic principles for the proper execution and the corresponding audits of IT security are laid down in numerous laws, standards and instructions. The most important and comprehensive regulations are provided by the ISO 2700x series of standards. These cover, for example, identity management, authentification, encryption incl. key management and monitoring as well as implementation instructions for the detection and reporting of intruders. In addition, there are numerous special standards such as DIN EN 50600 for computing centre setup and infrastructure, and the IEC62443 for the certification of IT security in industrial automation and control systems.

These standards offer a suitable basis and a coherent framework of action. A full discourse or even an enumeration of all standards and guidelines relevant to the subject of safety would be beyond the scope of this book. For this reason, please kindly refer to the relevant technical literature. A very good technical overview and a compilation of many further sources is given, for instance, in the study "IT Security for Industry 4.0", commissioned by the German Federal Ministry of Economics and Energy (Bundesministerium für Wirtschaft und Energie) [Bac16].

A broad overview on legal requirements, current research activities and funding programs can also be found there. The study focuses not only on technological aspects, yet also on organisational and legal questions with regard to digitisation in production and Industry 4.0, and pragmatic proposals for action are also, amongst others, given to the automotive industry. A comprehensive implementation guide, also in the sense of "best practices", related to more than 70 fields of action in IT security, is contained in a further recommendable study of the German Federal Office for Information Technology Security (Bundesamt für Sicherheit in der IT) [BSI13]. This compendium is supposed to be continually adapted and expanded with regard to upcoming challenges, especially with regard to digitisation.

The recommendations for action listed in the studies will not be delved into here. The importance is to understand the relevance of IT security from a digitisation point of view. The Internet of Things, the integration of processes, along with the comprehensive networking of all partners involved in the value-added process across national borders, as well as the automation of processes and the growing number of mobile devices, Big Data and Cloud all increase potential IT security risks and thus the importance of this topic.

The scale and complexity of threats are growing heavily with the increasing dissemination and relevance of IT. The main threats are the infiltration and the infection with malicious software via the Internet or rather via storage media and external hardware, increasingly also via smartphones. Human wrongdoing and sabotage continue to be among the greatest risks. According to a survey conducted by the German Federal Office for Information Technology Security in 2015, more than 58% of the over 400 companies surveyed were the target of a cyber-attack, and more than 40% of these attacks were considered successful, i.e. did damage to these enterprises [BSI15]. The topic thus is to be included in every digitisation roadmap and must be implemented in careful and close cooperation with the projects.

2.5 Handling of Personal Data

The protection of personal data is as important as IT security, especially in Germany. The handling of such data, i.e. to change, transfer, lock and delete data, as well as their use, has to be done in Germany according to the provisions of the Federal Data Protection Act (BDSG). Any information is personal if the data have a reference to a person. The aim of the legislation is to protect citizens from disadvantages resulting of handling their data. Basically, personal data may only be collected, processed and used if it is permitted by special laws or if the person concerned explicitly consents to it voluntarily. Prior to this consent, information must be given on the intended use and the type of processing. The consent shall apply to the specific agreed application only and needs to be renewed upon any further or different use. If the intended use is no longer to be pursued, the data must be deleted. There is certainly room for interpretation in the implementation of this specification, as the following example may show.

A customer configures his or her new car online with individual features such as sunroof, metallic paint and special finish of the steering wheel. This configuration is further processed in the manufacturer's back-end systems, for example, for material disposition, order control and logistics, and detailed information on the order is then sent electronically to the component suppliers. Body-in-white manufacturing and painting are carried out according to the configuration, and the components go exactly to the final assembly place. After completion, the vehicle is delivered to the customer in accordance with the specification. In this example, from configuration to delivery several times the processing of customer or personal data is performed. With respect to all work steps consent must have been given, as otherwise it may be a breach of the Federal Data Protection Act and trigger a fine.

This simplified example highlights the relevance of the topic for digitisation projects. This becomes even more fascinating in regard to cross-border logistics chains or the transmission and storage of these personal data, for example in Cloud computing centres abroad. There is considerable legal certainty in the transfer and processing within the EU, however the United States, or so-called third countries such as Japan, India or China, which are subject to less stringent protection laws, raise complex legal questions. Similar to IT security, it is important not to postpone this issue, but rather to involve the relevant experts, such as the data protection officer of the company, right from the outset in the digitisation projects in order to establish regulations and security at an early stage. This should happen in tandem outside of the specialist projects in order to avoid uncertainty from this discussion or to probably lose time in special discussions. Clear, pragmatic and timely guidelines for IT security and the handling of personal data assist in successful project implementation.

2.6 Powerful Networks

In addition to IT security and the adequate handling of personal data, powerful networks are an important prerequisite for the implementation of digitisation. For example, Industry 4.0 programs require reliable communications within the plant, comprehensive integration into the enterprise IT, and connections across enterprise boundaries as well. The volume and intensity of communication are increasing considerably.

The available bandwidths of the underlying network infrastructure will no longer be able to satisfy this demand, and measures must be taken at an early stage to ensure that the subject of communication does not become the "bottleneck" of digitisation. Currently, 10 Gbit networks are installed in the enterprises, while computing centres already use 40 Gbit lines and even have plans for 100 Gbit bandwidth.

Within this infrastructure, the manufacturers use the so-called Multiprotocol Label Switching (MPLS) technology to optimise the communication, which can use different protocols to send the information packets through the network. Low-cost Internet connections are being used more and more for access. In this area, so-called "All-IP" technologies will prevail in future. This refers to the bundling of different transmission techniques based on the Internet Protocol (IP). Thus, various services such as telephony, multimedia mails and data are routed through just one technology. This service from a single source results in cost and service benefits, for instance through uniform access for users from any location [NN15].The next development step is called Next Generation Network (NGN), which also provides for bundling, however not on the basis of IP technology, yet using manufacturer-specific protocols.

Further developments in the area of networking provide for the use of virtualisation. The future technologies "Software Defined Networking" (SDN) and "Network Functions Virtualisation" (NFV) are designed for this purpose. The methods decouple the infrastructure from the communication requirements by means of a software layer and optimise the resource utilisation by coordinating this layer. These technologies have already proved themselves in computing centres (see Sect. 8.4.5), hence the technical details are not discussed further however referred to studies available. The future implementation of the methods in the wide area network (WAN) offers future potential for securing communication for digitisation.

2.7 Technology Outlook

The following is a look at the future with the question as to whether and how the continuous growth of IT performance can be continued and whether the Moore's Law is still valid in the context of alternative technologies. The benchmark for this and the immense progress in IT performance are illustrated by the following comparisons: Today, a standard smartphone has 120 million times the computing power

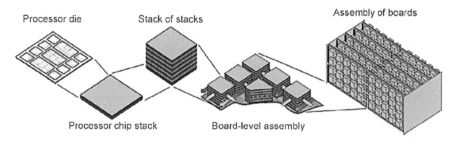

Fig. 2.7 Construction of 3D chips (Ruch et al)

of the NASA Apollo programme, and the iPad would have made it into the supercomputer rankings in 1994 [Gru16].

2.7.1 3D Chip Architectures

The basis for the previous performance increase of the chips was a continuous reduction of the chip structures and an increase in the packing density of the transistors. Limitations are looming here though. Current research is carried out on chip structures in the size of 5–7 nm. There is still some effort needed to achieve economic production. Nevertheless, it can be assumed that these will be in mass production within a few years' time. However, this path of miniaturisation is likely to encounter physical limitations, since a silicon atom has a diameter of approx. 0.3–0.5 nm, and thus only a few atoms fit in the structures next to each other. In these dimensions, safe atomic or line actions are no longer possible, and so-called quantum effects [Ruc11] occur. In addition to these physical challenges, there are further limitations in energy supply and control of heat problems. If the development of chip technology were to continue with the same basic conditions of energy requirement, systems with 30,000 times bigger computing power would be achieved within some decades, although this would require today's total world production of electricity.

In this way, an end to achieving further performance increases through continuous miniaturisation based on silicon technology seems to be near. One option now is to build the chips in multiple layers in 3D architectures instead of the existing two-dimensional 2D structure. In doing so, the electronic components are located on a number of wafer plates which are arranged one above the other. Figure 2.7 shows the principle, scalable from structures of single boards up to board groups [Ruc11].

The high spatial packing density of these 3D chip architectures allows further performance increases since with smaller chip bases the distances between the modules can be shortened and the data transmission can be optimised. However, there are two new challenges with these compact architectures: There is an extreme heat development of some kWh per cm^3 in the chip stack which by far exceeds the heat

2.7 Technology Outlook

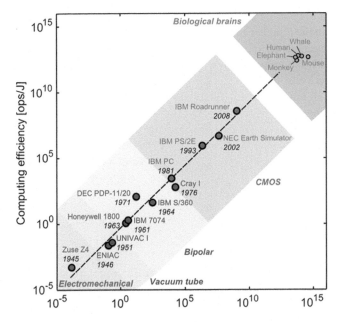

Fig. 2.8 Computational efficiency and computational density of computers compared with mammalian brains [Ruch et al.]

generation of a combustion engine, and also the required energy supply via the connection pins is not ensured sufficiently [Ruc11].

2.7.2 Flow Batteries

To control the heat problems in the chips or boards, a special liquid runs through subtle cooling channels, which pass through the chip stacks. IBM researchers from the Zurich laboratory had already made great progress with the use of hot water cooling at the beginning of the 2010s. Their use in supercomputers resulted in significant energy savings. The IBM lab is now pushing ahead with a further innovative approach. The basis for this are so-called Redox flow batteries, which are already used in practice, for instance with renewable energies. With this technology, energy is no longer transported via conductor tracks, yet by means of an electrochemically active liquid. Electrodes are used for the energy accumulation and also the energy extraction from the liquid at the point of use. The liquid is thus used for the purposive transport of energy and, at the same time, for cooling. This fluid could also be called electronic blood.

For the envisaged IT deployment, it is now important to continue to further drive performance and miniaturisation. This results in an analogy with the human brain which, as Fig. 2.8 shows, is clearly superior to today's technologies in terms of efficiency and power density [Ruc11]. Our brain is currently 10,000 times more powerful than the established IT technology.

In analogy to our brain with its energy supply and cooling through blood, researchers see similar potential in the flow battery approach. Until now, the human brain is superior to today's technology in terms of efficiency and power density [Ruc11]. As of now our brain is ten thousand times more powerful than the established IT-Technology.

Therefore, they are convinced that in a few years a supercomputer with a computing power of 1 PetaFlop/s (this means 10^{15} floating-point operations per second) can be built and the size of be reduced to the size of a conventional PC, keeping its performance [Ruc11].

2.7.3 Carbon Nanotubes

In order to exploit further options for the increase in performance, research is also carried out on material alternatives to silicon, for instance on so-called carbon nanotubes for some years. These are tiny tubes whose wall consists of a single layer of carbon atoms, connected in a honeycomb structure. These tubes can be used to conduct electrons with minimal resistance. The tube diameter is between 0.5 and 50 nm. The problem is that bundles of several intertwined fibres develop during production [Shu13].

IBM scientists have started to arrange the nanotubes side by side on a silicon wafer. Atop the established chip structures, the tubes can then be used as transistors. Scientists predict that in a few years it will be possible to create compounds on this basis in mass production, in which the metallic conductor tracks and the nanotubes have to be only 28 atoms of thickness [Quin16]. This opens up a new way to increase chip performance.

2.7.4 Neuronal Networks

Another way to increase the performance of chips is the architecture. Today's computers are all based on the so-called von-Neumann principle, in which transistors generate binary switching states specifically for the binary data processing. Furthermore, the processor and storage are separated since transistors cannot store information. Via the processors, the switching current is conducted back and forth between the arithmetic unit, the logic unit, and a buffer store. This process is very complex and energy-intensive.

An alternative to this, again in analogy to the human brain, are neuronal networks. These consist of nerve cells (neurons) which are connected to one another via communication channels (synapses). Information is processed within the network by neurons via non-linear functions, taking into account further neurons or rather switching points. This linkage results in a highly parallel processing of the input information, and very complex non-linear dependencies in the input information can be imaged quickly. Neuronal networks are able learn these dependencies and further extend their experiences [Rey10, Smh15].

In a new chip architecture, the so-called neuromorphic chips, neuronal networks are imitated in silicon circuits, and the storage and processor are combined. In this way, the human brain is mimicked with its nerve cells and thus makes it possible to solve certain problems, such as pattern recognition or prediction and recognition of contexts, more quickly and with considerably greater energy efficiency than with today's computer systems.

With these capabilities, the chips and computers that are based on them are also of interest to the automotive industry, for example for rapid pattern or image recognition in autonomous driving. With regard to the promising potential, intensive research is being carried out in this area, the first prototypes have already shown considerable results, and the path to production readiness begins to show [Dön14].

2.7.5 Quantum Computer

Finally, a brief look at the topic of quantum computers is given for the consideration of future technologies or possible approaches to achieve further massive IT performance increases. This idea has been researched for years [Mat13]. Instead of the current binary system with the two clearly defined states of a bit, these computers use quantum mechanical effects. Similar to the bit, there are so-called Qubits (derived from quantum bits), which, however, can assume any intermediate states. One can combine several Qubits, in quantum physics this is called interlacing, whereby the common state again superimposes all the individual states. If several Qubits are interlaced into so-called quantum registers and the information is distributed on these registers, a very high number of values can be processed simultaneously, thus making it possible to solve very complex problems [Sch15]. High performance is achieved by the parallelism of the calculations. However, quantum computers are not universal computers yet, but are rather particularly suitable for problems that can make good use of quantum mechanical effects. These are, for example, the simulation of superimposed magnetic fields, the search of unstructured databases or for decoding problems based on prime concepts.

Numerous large research and development organisations are working on the further development of quantum computers. The challenges are the miniaturisation and the reproducible generation and interlacing of registers with very many Qubits. Promising approaches and first prototypes have already been presented. Under laboratory conditions, current systems are at suitable tasks already 100 million times faster than conventional computers [Sch15].

This may suffice as a brief look at future technologies, which will enable further growth in IT performance. In addition, many other ideas are pursued, from photonics (light effects), spintronics (electrons as medium of two bits), and biologic

DNA computers. From the point of view of these additional options, it is obvious that Moore's Law, possibly applied to other technologies, will continue to apply. Alongside this effect, the growth of digitisation will also develop.

2.8 Technological Singularity

The following is a short overview on a futuristic topic which the now 20-year-olds will be able to experience in their lifetime: The so-called Technological Singularity. Originally a term from mathematics, this refers to a point at which a function is not defined, such as 1/x at the point of x = 0. At this point, all the curves for X go towards infinity. In physics, singularity refers to a situation in which no scientific laws are applicable, as presumably in a black hole [Rie11]. With these rather gloomy definitions, the transfer of the concept to information technology is exciting.

The so-called technological singularity is understood as a point in time at which the world-wide processing power of machines or rather high performance computers will overtake the summed performance of all human brains. From this point onwards, computers can continue to further improve independently [Kur05]. This situation is illustrated in Fig. 2.9, which shows the computing power of all today's computers, as well as of all mouse, insect and human brains.

The point of intersection, and ergo the point of the singularity according to this will be in the year 2050 approximately [Rie11]. If one assumes exponential growth here as well, two exponential processes come together in this moment, namely the development of IT technology and the independence of computers. Taking the assumption of this mind-boggling acceleration, it is exciting to assess the future

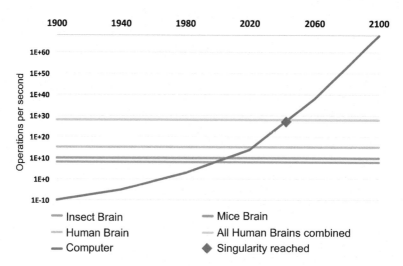

Fig. 2.9 Development of computer performance in comparison to the brain of living beings [Kur06]

consequences for all spheres of life. The digitisation of companies will presumably gain a momentum of its own as well.

As a result, revolutions are taking place in medical technology, nanotechnology, and robotics, and these areas will change lastingly. In nanotechnology, manufacturing is carried out at the atomic level. In combination with robotics, this leads to miniature robots, whose intelligence is superhuman and which, for example, are constantly circling our blood channels and monitoring health parameters. As a result, these robots can automatically take corrective action in the development of diseases. Similar applications are conceivable with miniature service robots in automobiles.

Overall, the massively growing overall intelligence will drive the exponential growth of information technology at an unprecedented pace. There are exciting visions in conjunction with many questions: How do you possibly link human and technical intelligence? Who directs and controls who? How to we prevent unwanted sprawl into unplanned areas? What is the human role? Many questions that are obvious, but do not fit into the context of this book. Thus the interested reader is again referred to the specialist literature [Kur06].

It is important to the author to underline from a technical perspective as well the necessity of confronting digitisation in a targeted and constructive way. Digitisation comes with massive momentum and is unstoppable – so it is important for every company to control these new forces and be able to use them purposefully.

Annex A1

In order to determine the general status and positioning of the largest automobile manufacturers with regard to digitisation, the author conducted a qualitative Internet search. The following evaluation parameters were used to assess the situation and compare the manufacturers:

- "Digitisation < manufacturer name>"
- Vacancies digitisation
- Cooperations in the field of digitisation
- Autonomous Driving
- Electric vehicles
- Organisational announcements in the field of digitisation
- Number of available Apps

Essentially, the evaluation was based on the number of hits in the Google search, and points were awarded in the resulting ranking order. Electric vehicles and organisational measures were put in a ranking by quality and then also awarded points. This pragmatic approach resulted in the ranking shown in "Figure 2.5: Digitisation degree of automotive manufacturers". Details of the classification are shown in the following table:

	BMW	Daimler	Volkswagen	Toyota	Audi	GM	Ford	Hyundai	Honda
Score hits	3	2	4	4	3	2	5	3	4
Score cooperations	5	3	2	4	2	2	4	2	1
Score autonomous driving	4	5	5	5	5	5	5	5	5
Score electric vehicles (planned)	2	4,5	4	0	2	2,5	1,5	1,5	1
Organisational initiatives	1	1	1	1	1	1	1	0	0
Number of apps	5	4	3	4	3	1	1	1	1
Total	**20**	**19,5**	**19**	**18**	**16**	**13,5**	**17,5**	**12,5**	**12**
Average	3,3	3,3	3,2	3,0	2,7	2,3	2,9	2,1	2,0
Ranking	**1**	**2**	**3**	**4**	**6**	**7**	**5**	**8**	**9**

References

[Bac16] Bachlechner, D., Behling, T., Holthöfer, E.: IT-Sicherheit für die Industrie 4.0 (IT Security for Industry 4.0). BMWiStudie; Final report 01/2016. http://www.bmwi.de/BMWi/Redaktion/PDF/. Drawn: 18.06.2016

[Boh15] Bohr, M.: Moore's Law will continue through 7nm chips; ISSCC Conference 2015

[BSI13] NN: ICS Security Compendium 2013. Federal Office for Security in Information Technology. https://www.bsi.bund.de/SharedDocs/Downloads. Drawn: 18.06.2016

[BSI15] NN: Cyber-Sicherheits-Umfrage 2015 (Cyber Security Survey 2015). Federal Office for Security in Information Technology. https://www.allianz-fuer-cybersicherheit.de/ACS/DEdownloads/cybersicherheitslage/umfrage2015. Drawn: 18.06.2016

[Bur16] Burger, A.: Data center emissions, coal power use much higher than thought. http://globalwarmingisreal.com/2016/02/09/. Drawn: 17.06.2016

[Cha03] Chau, R., Doyle, B., Doczy, M.: Silicon nano-transistors are breaking the 10 nm physical gate length barrier. Device Research Conference (2003)

[Dön14] Dönges, J.: Neuromorphe Computer – Der 1-Million-Neuronen-Computerchip (Neuromorphe Computers – The 1 million-neuron computer chip). Spektrum der Wissenschaft, 07.08.2014. http://www.spektrum.de/news. Drawn: 21.06.2016

[Gru16] Gruber, A.: Physikalische Grenze der Chip-Entwicklung: Kleiner geht's nicht (Physical limitation to the chip development: Smaller is not possible). Article Spiegel online: 26.03.2016. Drawn: 20.06.2016

[Hin16] Hintemann, R.: Rechenzentren – Energiefresser oder Effizienzwunder (Computing Centres – Energy Guzzlers or Efficiency Wonder) Informatik aktuell of 26.01.2016. https://www.informatik-aktuell.de/betrieb/server/rechenzentren-energiefresser-oder effizienzzwunder.html. Drawn: 20.06.2016

[Koo10] Koomey, J.G.: New study of data center electricity use 2010. http://www.koomey.com/post/8323374335. Drawn: 17.06.2016

[Koo15] Koomey, J.G., Taylor, J.: New data supports finding 30 percent of servers are 'Comatose'. Study June 2015. http://anthesisgroup.com/wpcontent/uploads/2015/06/. Drawn: 17.06.2016

[Kur05] Kurzweil, R.: The singularity is near: when humans transcend biology. Penguin Books, New York (2006)

[Mat13] Matting, M.: Die faszinierende Welt der Quanten (The fascinating world of quanta). AO edition. 21.01.2013

[Mee16] Meeker, M.: Internet Trend Trends 2016 – Code Conference KPCB Menlo Park, 1.6.2016. http://www.kpcb.com/internet-trends. Drawn: 03.07.2016

[Moo65] Moore, G.: Cramming more components onto integrated circuits. In: Electronics. vol. 8, No. 8. (1965)

[NN15] All-IP-Netze: Final Document Project group All-IP networks; Plattform "Digitale Netze und Mobilität" (Platform „Digital Network and Mobility"), National IT Summit Berlin; 27.10.2015. http://webspecial.intelligente-welt.de/app/uploads/2015/11/151030_pf1_007_fg1_abschlussdokument_pg_all_ip.pdf. Drawn: 27.02.2017

[Quin16] Qing, C., Shu, J.H., Tersoff, J.: End-bonded contacts for carbon nanotube transistors with low, size-independent resistance. Science, 17 June 2016; vol. 352, Issue 6292

[Reg16] Reger, L.: Baukasten zu autonomen Fahren (Kit for autonomous driving) Key Note ISSCC Conference 2016, San Francisco, article/127176/. Drawn: 14.06.2016

[Rey10] Rey, G.D., Wender, K.F.: Neuronale Netze – Eine Einführung in die Grundlagen, Anwendung und Datenauswertung (Neuronal Networks – An Introduction to the Basics, Application and Data Evaluation), 2. Edition Publisher Hans Huber, Bern 2011

[Rie11] Riegler, A.: Singularität: Ist die Ära der Menschen zu Ende? (Singularity: Is the era of mankind ending?) futurezone, 11.04.2011. Drawn: 22.06.2016

[Ruc11] Ruch, P., Brunschwiler, T., Escher, W.: Towards five-dimensional scaling: how density improves efficiency in future computers. IBM J.Res. & Dev.; vol. 55, No.5; Sept./Oct. 2011

[Smh15] Schmidhuber, J.: Deep learning in neuronal networks: an overview, Neural networks, vol. 61, January 2015. Drawn: 14.06.2016

[Shu13] Shulaker, M.M., Hills, G., Patil, N.: Carbon nanotube computer. Nature 501, 26.09.2013

[Sch15] Schulz, T.: Computerrevolution: Google und NASA präsentieren Quantencomputer (Google and NASA present quantum computers). Spiegel online, 09.12.2015. Drawn: 22.06.2016

Chapter 3
"Digital Lifestyle" – Future Employees and Customers

In the previous chapter it was shown that information technology will drive digitisation with exponential performance growth. Digitisation will penetrate the society and businesses extensively and will deeply change processes and organisations. These changes will encounter a very heterogeneous population of buyers and employees, with different training and experience in digitisation. More and more customers and all of today's career starters in the companies are part of the so-called Digital Natives generation. These are people who have grown up with IT-based services such as computer games, the Internet and Facebook as well as smartphones. The handling of these digital offers is natural to them and has shaped their behaviour.

In addition, so-called Digital Immigrants work in the companies. Many of them have learned how to deal with these new topics only in the adult age after their training or study. This group is also characterised by a certain behaviour as well as value systems and habits. Today's world of work and its forms of organisation, collaboration models, workplace design and established communication methods are often still oriented towards the "Immigrants". In a few years, however, the "Natives" will be the majority of the employees in the companies and as future customers. It is now up to the companies to recognise this situation in their workforce and to implement measures that will lead to the success of digitisation efforts involving all employees.

Before going into details on the subject of this chapter, the author would like to describe a personal experience which authentically illustrates the environment of digital natives. The 25-year-old son has completed a master's degree abroad. His entire studies, especially the labs and seminars, were based on the university's collaboration tools. These are software tools available on a secure platform which simplify collaboration between groups over the Internet, e.g. through audio and video conferencing systems, instant messenger services, project management tools, etc. Collaboration with his fellow students from different countries worked smoothly and flexibly in different tasks with different teams. After completing his studies, his first step in searching for a job was the research in international online platforms. The comments on social networks were decisive in evaluating the offers.

In contrast to many of his fellow students, who were looking for the risk of young startups, his career start was in a consultancy company in order to gain insight into numerous companies with different tasks. As a place of residence, Frankfurt was chosen based on the main argument of good transport links. The room in a shared flat with two other career beginners of about the same age, who he did not know previously, was found quickly via social media platforms. Although every member of this community is on an above-average salary, none of them owns their own car and does not plan to buy one. For short distances they use the offers of mobility platforms, and public transport offers for longer distances, preferably low-cost long-distance buses – also because of the stable Internet connection on board. Necessary overnight stays are booked flexibly via sharing platforms, for example via Airbnb.

In addition, a small anecdote: While having a coffee together at the first visit a few months after moving in, unknown ringtones disturbed the conversation. After some search the source was found. At the internet connection in the box room actually was a traditional telephone, which was probably ringing due to an error call. None of the residents had used it so far or would know the landline number.

This personal experience outlines the challenges for companies to properly address Digital Natives as potential employees as well as potential customers. Therefore, the following chapter presents the background and the life setting of the natives in more detail and provides recommendations on how companies should set up and organise themselves today in order to attract, long-term motivate and develop natives as new employees. Also ideas and suggestions are presented on how to win them as customers. At the same time, the immigrants of course must be kept motivated and in the company.

3.1 Always On

In 2015, 3.2 billion people worldwide had access to the Internet, of which 2 billion live in developing countries. The global number of users had also grown exponentially from 400 million in the year 2000 in this case [ICT15] (see Sect. 2.2). Much of it are teenagers, so digital natives. As a comprehensive study on usage patterns in Germany illustrates, young people between 9 and 24 years of age state that they were surfing the Internet daily (85%), but at least several times a week (28%). As shown in Fig. 3.1, the "online rate" increases with the age of the adolescents, and the Internet use of the 18 to 24-year-olds is even at 94%. The same study shows that this age group most commonly uses Facebook, Google and WhatsApp [Bor14]. As expected, these offers address the three main areas of use: social networks, search

3.1 Always On

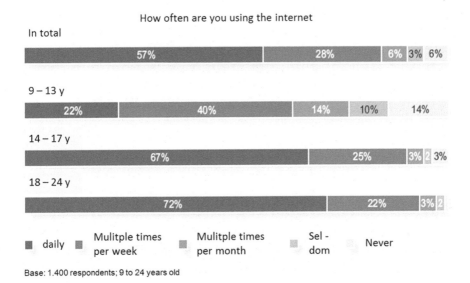

Fig. 3.1 Use of the Internet by 9 to 25-year-olds (Boasted et al.)

and communication. In addition, commerce platforms are established, and video and sharing services are growing steadily. From the author's point of view, this is a very representative statement, which also applies to other countries in a similar way. In Brazil, USA or China, Internet use is even higher, in China with the corresponding providers Baidu and WeChat.

An interesting differentiating background to the user behaviour of young people, which is also relevant to companies, is shown in Fig. 3.2, based on the age group of 14 to 24-year-olds [Bor14].

The vertical axis divides the level of education of the group under investigation to low, medium and high grades, and the horizontal axis its basic orientation into the three classes of traditional, modern and postmodern. In this field, seven characteristic patterns of behaviour could be distinguished. The pragmatic and sovereign user groups which are operating on the Internet in a targeted and risk-aware manner, are assigned to the middle and higher education level, accounting for 54% of all users. The group is also ready for transformation, combined with a curious basic attitude and the willingness to change and to set off. This is followed by the group of the carefrees (18% of the population), rather from the lower education level and with less risk awareness while surfing the Internet. The groups of the confused, cautious and skeptical make up a total of 20% of the interviewees, which come from all education levels and are rather the traditional basic attitude. Here is also the group of the conscientious, yet rather from higher education levels.

This deeper analysis of the "always on"-mentality with the presented groupings on behavioural patterns is a viable approach that companies can project on to their

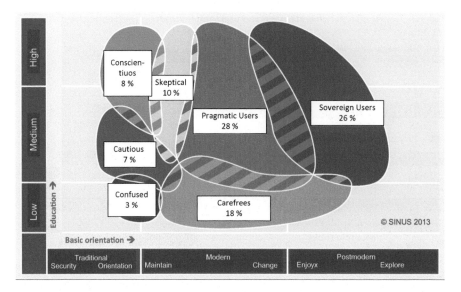

Fig. 3.2 User groups of the 14–24 year olds (Borchers et al.)

employees to derive, for example, communication and training programmes for digitisation projects. With these programs, it is important to integrate the digital natives with their creativity and with their knowledge of up-to-date communication and creation of up-to-date software solutions, thus making them the driving force behind the transformations.

3.2 Mobile Economy

Not only in the case of digital natives, are mobile devices increasingly becoming the standard tool for IT use. Forecasts predict that the global number of smartphones will grow to 3.2 billion devices in 2017 (number as of 2012: 1.6 billion phones), which means that twice as many smartphones as PC's are used. While the number of PC's will rather stagnate at the current volume or rather decrease in the long term, the number of smartphones will continue to grow significantly, thus further increasing the gap [Eva16].

It is important to understand that this smartphone hype not only replaces the PC as an access device to the Internet and IT solutions, but that these devices will establish a completely new usage culture. "Always on" is with the smartphones no conscious usage decision any longer, but rather the standard. A completely new economic system opens up through this with immense opportunities. The departure into this new business generation is also illustrated by Fig. 3.3 [Eva16].

The shown computer generations with their installed device volume, validated also in further individual studies and especially in the field of mobile devices with

3.2 Mobile Economy

Fig. 3.3 Computer generations with numbers of devices worldwide (Evans)

dynamic growth, also each represent a typical type of business. The time of mainframes, workstations and PC's represents traditional corporate structures. Companies were and still today are structured hierarchically, initially locally oriented and later also on a global scale. Mainframes and workstations are the driving force behind comprehensive application software, such as ERP and engineering solutions, which are accessed by employees from their fixed workstations during office hours.

With the arrival of smartphones the digitisation arrived in all areas of the enterprise as well as in daily personal life. They have the potential that literally every person in the world owns a smartphone. These mobile devices enable the so-called back-end systems to be accessed at any time and from different mobile workplaces. This leads to an intertwining of the private life and the work life. There is a growing number of special smartphone solutions (Apps) in the work environment which operate complementarily as a front-end to the corporate solutions, as well as in the private world, such as the popular weather, travel or stock market apps. These applications are easily and flexibly downloaded from shop platforms at low cost, or often even free of charge, financed through advertising. If these apps then stand the first quick test after the initial download, they usually remain on the device for further use.

At the same time, more and more digital natives are working in the companies, and new company cultures are evolving, frequently changing organisational forms, collaboration models and business models as well. This transformation towards a digitisation culture is discussed in Chap. 7 in detail since it is a key success factor for digitisation projects.

With the so-called "always on, mobile first"-culture, the usage behaviour changes drastically. The Internet, with its countless offerings, becomes an integral part of life. There is no longer a distinction between on/off-times. This is obvious when you look around in streets, cafes or even restaurants. Everywhere there are young people, however also more and more older digital immigrants, looking at their smartphone and repeatedly briefly interacting with the device. Various studies confirm this observation. As an example, the findings of an analysis are summarised and interpreted here [Mar15]. Eighty-eight times a day, users reach for the smartphone, which means that on average it is "checked" every 10 min, at an assumed wake-up time of 16 h per day. The total time of usage is approx. 2.5 h per day. An interaction with the smartphone takes less than 2 min on average.

On the basis of this insight, the question arises as to how, despite continuous short-term interruptions, a productive concentration can be maintained in order to achieve targeted work results. However, this is what young people and digital natives are increasingly training. They will bring these skills to "speed multitasking" to the work in the enterprise, but not only the abilities, yet the derived expectations as well. Somebody who is constantly online and communicates in a highly reactive manner also expects this from colleagues in the company, his suppliers and business partners, and also as a customer. Answers in a dialogue are expected not only within a day, but rather within hours, and even up to "real-time".

3.3 "Real-Time" Expectation in the Mobile Ecosystem

When working on the web, real-time is not required actually, but loading times of websites should be below 3 s to meet user expectations. This value will reduce even further with increasing IT performance. Short loading times are an important criterion for acceptance to continue the work from a screen or to accept the apps as a solution on the smartphones. This also applies to the loading times of the user screens of application software in the companies.

This high communication dynamics, coupled with the expectation that dialogues are answered directly, but at least soon, is also transferred to other areas in the so-called mobile ecosystem. One example is the "sameday delivery" initiative by the online retailer Amazon, i.e. the delivery of ordered goods on the day of ordering. For this service, customers are willing to accept a surcharge. Thus, the delivery acceleration may at least at the start of the initiative be a differentiation feature in the market competition and also enables an additional profit margins. As a result, the supply chain is under pressure in the entire online retail sector, and a transfer to the supply chains in the automotive industry can also be observed. New ideas such as the use of drones are already being tested, and new service providers are being established with innovative approaches such as the mobility service provider Uber, which is already offering in the US, in competition to Amazon Now, deliveries which can be flexibly demanded via apps and logistics platforms.

This example from the field of the mobile Internet and the mobile ecosystem clearly shows how strongly the technology drives the upheaval of business models. It is important that all companies recognise these risks, but also the associated potentials, at an early stage and take advantage of these. It is also indisputable that further technologies, from Big Data and analytics to cognitive computing, will be the basis of the business models to take account of customer wishes and customer history. An example is the integration of location-based services in order to optimise delivery routes and the utilisation of the means of transport. The solutions just outlined at this point are discussed in detail in Chap. 9.

As a further example of the importance of high communication dynamics, the entire area of "business to consumer" is to be mentioned. Customers expect rapid reactions in their online communication with companies, but also that the customer information already available in the company is, for instance, known at follow orders or complaints and is taken into account in the replies. Dialogues in hourly action with solutions within a day characterise the expectation of digital natives. These expectations must be fulfilled as a basis for successful customer relationships and, consequently, the necessary ability has to be organised in the companies. In addition, the customer interface, as part of the sales and aftersales processes, is subject to extensive changes and challenges on which Chap. 6 expands.

3.4 Sharing Economy

Sharing Economy is another interesting business model, which is becoming increasingly important, driven by the rapid spread of the smartphone. The basic approach of sharing or shared use has long been known with high-value investment goods. Examples are the sharing of holiday houses in the timesharing concept, the use of harvesting machines via cooperatives or also of machine tools. These models were already established before Internet times. However, special arrangements and elaborate coordination were required for handling, so that the entire Sharing business model so far only generated a modest sales volume. This changes with the availability of apps for simple, fast and extremely cost-efficient processing of the transactions. On the basis of platforms, completely new markets are emerging. New market participants, both suppliers and customers, can join existing platforms that are almost free of charge. The same applies to the spread of the platforms in new markets, resulting in impressive scale effects with exponential growth. Rifkin refers to it as the Zero-Marginal-Costs Society [Rif14].

Renowned Sharing providers are the market leaders in their respective segment, companies Uber with mobility services, and Airbnb with accommodation facilities, each offered by private individuals. Founded in 2007, Airbnb was represented in over 34,000 cities and over 190 countries, with a total of more than 1.5 million overnight stays per year. The company turnover in 2013 was $250 million and in 2015 approx. $900 million. Although the company did not make a profit in 2015, it is valued at $26 billion. Airbnb predicts for the year 2020 a profit before taxes and depreciation of $3 billion [Eic15]. The development of Uber is similarly dynamic. Founded in 2009, the company has a business valuation of $62.5 billion. Every day more than 2 million trips are provided, of which 1 million are in China alone. In the US, the company has 450,000 drivers with a total of 6700 employees. Sales in 2015 were $1.9 billion, with a loss of approx. $2.2 billion [Fre14].

Just these two examples highlight the dynamics of the shared business model, driven by smartphones, the "always on"-mentality and the corresponding lifestyle of certain population groups, a considerable proportion of which are also digital natives. The high company valuations are essentially driven by the assessment of the market potential. Sales of this Sharing business model worldwide are estimated to go up from the current $15 billion to around $335 billion by 2025 [Eic15].

Given the high growth expectations of the shared business model, the question arises as to how the areas relevant to the automotive industry are developing.

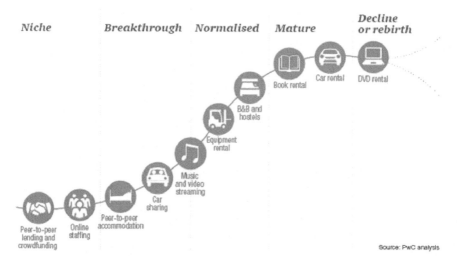

Fig. 3.4 Development of sharing offers (PWC)

Extensive studies have already been carried out. The results of a study by consulting firm Roland Berger are briefly discussed here as an example [Fre14]. The study estimates a growth in car sharing by 2020 to a market volume of €5.6 billion. This attractive business potential explains why there are now many providers in Germany, e.g. car2go (Daimler), DriveNow (BMW) and also Flinkster (Deutsche Bahn). Also for ridesharing or ridehailing, which are organised ride services, with a market volume of €5.2 billion in 2020, a strongly growing business potential is expected. Suppliers for ridesharing or ridehailing are, for example, Uber, Lyft (GM), Gett (Volkswagen) or the rideshare agency BlaBlaCar, and in China Didi Chixing. The shared parking, i.e. the sharing of parking space, is to grow by 25% annually until the year of 2020 to a volume of around €2 billion.

This means that the three fields of car sharing, ridesharing and shared parking are the business areas, and possibly also in potential offshoots, such as so-called peer-to-peer sharing, in which automobile manufacturers want to strategically position themselves. This is confirmed in a study by PricewaterhouseCoopers [Mil15]. It analysed the development status and the prospects of various sharing offerings. Figure 3.4 shows the summary in an S-curve representation and assigns the respective status of the offers to the phases of the economic life curve.

The models of peer-to-peer lending and car sharing, which are of interest to the automotive industry, are still at an early stage of development, while the traditional car rental business is already saturated. In the study, average annual growth rates of 63% are estimated for peer-to-peer sharing by 2025, and 23% for car sharing. This study also confirms the interesting business sectors as a possible part of the strategy of a mobility service provider.

The picture also shows that there are other promising sharing opportunities, for example, in online staffing or in the joint use of special machines or test equipment. The last-mentioned fields are not to be considered within this book. The mobility-relevant models are dealt with in more detail in the development of a digitisation strategy in Chap. 5. The subject of online-staffing is part of Chap. 7 "Corporate Culture".

3.5 Start Up Mentality

A remarkable behavioural pattern of digital natives derives from their frequent use of computer games. Different studies have found that teenagers are gaming for fun more than 2 h per day, older teenagers even longer. From these experiences, skills and patterns of thinking are adapted which the digital natives also show in their professional life [Sti15]. The willingness to expand borders and take on higher risks are typical behavioural patterns derived from this – for example "restart at failure". To this extent, digital natives are ready to try new ideas without reservation, to quickly assess the chances of success, to adjust them if necessary, or to completely reject them ("reset"). This mentality also plays an important role in the acceptance of new solutions in companies. Digital natives are ready to accept major innovation leaps, but expect the immediate removal of errors, or adjustments.

This new setting is cumulated in a so-called start-up mentality. This is characterised by boldly tackling subjects with having so far only very few experiences. The work is then approached with great commitment and the flexible integration of further resources on innovative paths. In doing this, working together through open communication in project-related structures, ideally without hierarchies and standards. This approach does not mean though to start frantically, aimlessly and hastily. Obviously there are agreements and objectives. In comparison to the traditional process, these are however dynamically and flexibly adjusted within the team in order to quickly realise value contributions. Fields of work with these agile approaches can be found particularly with startups in the Silicon Valley, so that a lot of digital natives and founders go there, especially as the international environment and flexible financing possibilities are a further motivator. Meanwhile, start-up centres have also established themselves in Haifa, Berlin, London and also Bangalore, which attract digital natives as well.

In order to be interesting as a possible employer to digital natives in competition with these startups, the companies have to offer a work environment that addresses this mentality. Working independently and with entrepreneurial thinking must be the standard. Then, the natives commit with high work input, creativity and also leadership. Problems are solved quickly, unconventionally and inventively. A creative work environment and a new organisational culture develop in which digital immigrants sometimes like to be pulled by the natives. Digital natives in this role in the enterprises are called Intrapreneurs – derived from "Intracorporate" and

"Entrepreneurship". For companies it is necessary to develop this new type of employee in their organisation and to establish Intrapreneurs as so-called change agents. This creates a transformation culture as the prerequisite for successful digitisation projects. This subject is discussed comprehensively in Chap. 7.

3.6 Innovative Work Models

According to their nature and behaviour patterns, digital natives have a different relation to work than had the digital immigrants at the time of their professional career start. Previously, job security, the image of the employer and the level of income were important criteria at the career start and later for promotion opportunities in the company hierarchies. Today, it is much more about the task, an international, open work environment and also flexible work models with the possibility to create a suitable "work/life balance". This assessment is also supported in a study by the Fraunhofer Institute IAO, which differentiates seven trends in this context as motivators [Kor16].

1. Competitive orientation
2. Manage changing tasks
3. Work internationally
4. Demand-oriented presence
5. Consistently acquire new knowledge
6. Work in self-organised teams
7. Work under atypical contractual relationships

The study took the opinion of 1400 students in Germany on these trends with respect to their future professional activities. As expected, digital natives welcome the opportunity to adapt flexibly to changing tasks, to learn new knowledge and to work internationally in self-organised teams. However, the latter two trends are controversial. Virtual presence on demand is preferred to sporadic changes of location, and permanent positions are preferred to temporary employment or freelance work.

The immigrants were characterised by loyalty to an employer, natives rather by loyalty to their work content and environment. As explained in the above-mentioned study and in the previous section, if given the freedom of an intrapreneur, digital natives are also increasingly willing to accept permanent positions in established companies. Nevertheless, flexible working models have to be established, which will become more and more important in the future due to the change in traditional work structures. These are discussed below.

3.6.1 Digital Nomads

For a work model with the highest degree of freedom, the term of digital nomads is established. This describes people who are involved in IT and Digitisation, for instance as programmers, web designers, authors, bloggers and also software testers. As a "one man show" they are independent and free in the choice of their place of work and their working hours. The most important thing for them and for the exercise of their activities are powerful network connections. The digital nomads earn their livelihood by marketing their product, for example, apps or blogs, independently via appropriate Internet platforms. Another avenue is to collaborate on major projects, for example when designing a web site or when programming software solutions.

Namely in innovative IT projects, many companies rely on the integration of external resources, not least due to the lack of own specialist knowledge and lack of experience, but also of scarce internal capacities. Service providers thus work on tasks in the form of defined scopes of work. In addition, entire work fields, in particular outside the core business of the companies, are given to contractors on long term bases for 3–5 years in the form of outtasking (without transfer of employees) or outsourcing contracts (with employee transfer). In this environment, digital nomads with special knowledge are often involved as subcontractors. This type of flexible, so-called staffing of projects, often under a general contractor as a contract partner of the companies, is a collaboration mode which is currently used frequently. From the author's point of view, this is just an intermediate step though towards even more flexible structures.

3.6.2 Crowdsourcing and Liquid Workforce

The labour market which traditionally is locally orientated and the traditional work organisation in the companies are undergoing a radical change. On the one hand, digitisation projects often require very innovative and specific knowledge for a manageable project duration, on the other hand, the digital natives are ready to work in new working models. A further important prerequisite is the fact that with the WEB 2.0 technologies and high-performance computer networks tools are available to co-operate almost seamlessly across countries and continents. This creates highly flexible work structures, described by terms such as Liquid Workforce, Crowdsourcing and Cloudsourcing.

Especially in the IT industry, the concept of the Liquid Workforce assumes that the core workforce of companies is concentrated in core fields with reduced staff numbers, and this is then very flexibly ("liquid") strengthened in a project-related manner with experts [Acc16], [Boe14]. The employees are selected from a worldwide pool of the Crowd or the Cloud. The sourcing or the integration of the employees is handled through specialised Internet platforms, which are increasingly available. This form of so-called online-staffing will increase considerably. Here, the already cited PwC study shows an average annual growth rate of 35%. Relevant criteria for the staffing decision are, apart from economic considerations, the applicant's knowledge and expertise.

However, some factors have to be taken into account in order to successfully implement crowdsourcing or online staffing during the project work. First, a brief summary on the most important aspects from view of the company is given based on the author's professional experience. To begin with, the technological basis of the project, such as tools, work equipment, communication and test procedures for the entire project needs to be defined, binding for all employees and sourcing partners. In doing so, one should largely rely on established standards in order to achieve low operating and adaptation costs in the subsequent use of the project results. Based on this, the project scope has to be divided into concrete work packages and described in detail. Namely the interfaces to the adjacent work packages and to existing company solutions must be specified and performance parameters be defined. On the part of the sourcing partner, it is recommended that they prove their qualification in the form of standardised certifications, ideally supported by references.

The individual work packages are then often advertised by the companies on online platforms in the form of auctions, whereby the companies besides the detailed descriptions also specify the required qualification level of the sourcing partners. Requirement for the release of payment usually is the provision of quality and timely work as per the tender to fulfil the contracted scope. A central task of the companies and the respective project manager is to ensure the reliable interplay of all part projects and, if necessary, the integration into adjacent existing systems. Despite these challenges in terms of structured project preparation and integration risks, it can be assumed that the flexibility of surfing in the form of staffing via internet-based processing platforms will further increase in future, especially as this could also be a potential answer to the growing shortage of specialists, namely in Germany – catchphrase "war for talents".

In this way, the concept of crowdsourcings is first described in terms of flexible work organisation and staffing procedures. This term is however used in many further aspects as well. Further fields of application for crowdsourcing are described, for example, in [Arn14]:

- Innovation ... e.g. the joint development of the automobile of the future
- Funding and investment ... e.g. acquisition of various investors for start ups
- Preparation and management of knowledge ... e.g. offshoots of Wikipedia
- Charity/social projects ... e.g. collection of in-kind donations or famine relief
- Creative marketplaces ... e.g. platform for digital photography

Some of these aspects are also interesting for companies. Thus the provision of knowledge leads the way out from the mass of others, with the well-known example of Wikipedia, to a further general digital lifestyle feature. This is briefly illuminated in the following as it has to be considered in the necessary transformation of the entrepreneurial cultures in the course of digitisation.

3.6.3 Wikinomics

The term of Wikinomics refers to a new culture of work organisation and cooperation – also in companies as part of a new corporate culture [Tap09]. In this, people work freely and without hierarchical structures on different tasks. Wikipedia not only is the godfather of naming for this model yet also a good example to explain the principle. Many interested people with a qualified background are working without guidelines, pressure and compensation, using a WEB 2.0 platform to gather knowledge, keep it up-to-date and make it available as a knowledge database readily, flexibly and free of charge. This model of open cooperation of many, intrinsically motivated participants under a common goal can also be transferred to tasks in companies. Four basic principles have to be considered for successful implementation [Tap09]:

- Peering ... Voluntary cooperation of individuals (including outsiders)
- Open ... Openness
- Sharing ... Culture of sharing
- Act globally ... Global action

These principles are the basis of the success of Wikipedia and also of LINUX and YouTube. There is nothing wrong with using these principles internally in order to work across the entire company successfully on development tasks for instance, to bundle knowledge in the form of so-called Wikis, and to organise the communication with contributions from many employees via internal social media platforms like Facebook. The success of such initiatives certainly depends on the participation of as many interested employees as possible, as well as motivated by the active cooperation and informal example of superiors. This type of initiative is particularly appealing to digital natives and will motivate them to become involved in corporate transformation, which is a key success factor for digitisation. The subject of transformation is dealt with extensively in Chap. 7.

3.7 Google – The Goal of the Digital Natives

As already explained, digital natives do not distinguish between on/off times, they like to think and work cross-border. In their projects they act globally and appreciate fast communication behaviour almost in the real-time mode. They work in self-organised teams and do not need hierarchical structures. They respond to changes directly and flexibly. The work is possible from anywhere in the world and after a challenging workload peak, which is accepted with high motivation for the sake of the team's success, a time-out is appreciated for the purpose of work/life balance. Working in innovative work models with changing team members is thus becoming the norm.

When companies succeed in creating a work environment that allows to live in this pattern, young professionals are highly motivated to work as intrapreneurs and change agents in traditional businesses as well. In order to attract and retain well-educated digital natives as employees, despite the massive competition in the "war for talents", the question arises as to how companies can make the work environment attractive.

As a suggestion let us have a quick look at Google which is considered the "best of breed" model for such a work environment. The appraisal of Google by graduates in the field of IT in Germany is documented in a comprehensive study on the ranking of companies by their attractiveness for a career start [Tre16]. It resulted in the following order, incl. approval rates:

1.	Google	23.7%
2.	BMW	9.1%
3.	Apple	8.6%
...		
7.	AUDI	6.9%
...		
9.	Daimler	5.5%
...		
12.	Porsche	4.5%
...		
18.	Volkswagen	3.9%

Google is therefore voted by IT graduates the most attractive employer by far for a career start. The German automakers are behind by a considerable distance, with BMW as second, then followed by the other manufacturers. This results in a clear need for action in the automotive industry to increase its attractiveness. This is all the more pressing if one takes into account that the demand for IT specialists in this industry is growing at an ever-increasing pace and will as early as in 2020 exceed the demand for engineers [Pwc16]. Hence there is a considerable resource problem ahead that needs to be addressed. Adapting the listed flexible sourcing models can only be part of the solution. It is imperative that the automotive manufacturers as potential employers become more attractive and must win IT experts as internal employees. Just relatively high remuneration will not bring improvement in the ranking, namely since it is other criteria that are more important to natives.

What are the reasons for Google being this far ahead in the assessment? Sure, the image plays an important role. Among young people, the automotive industry has the image of being rather traditional, slow and not particularly innovative. This assessment is made by the industry product itself, without knowing any details on the products, the high engineering performance required and on the companies. No doubt the Google products are closer to the lifestyle of natives and used by most of them. The image is also strongly influenced by the work environment though.

Another personal anecdote by the author: During vacation in San Francisco, his student offsprings propose a trip to the Google Headquarters in Mountain View

3.7 Google – The Goal of the Digital Natives

instead of visiting the Yosemite National Park. Without having organised anything in advance, entering the site was easily possible as well as totally freely moving on the campus, even using the well-known Google bicycles. This freedom alone is a lasting impression. The Google employees who enjoy their lunch break are much more impressive though: some of them play basketball or football, some have their dogs with them, and they are all relatively young and appear to come from different nations. This illustrates the open, inspiring environment of the Google culture.

In order to make this culture even more attractive for digital natives, this is a (translated) quote on the Google culture straight from the Google-Germany homepage:

> Our employees make Google what it is. We bring smart and purposeful people to our team. Abilities are more important to us than experience. While all Googlers share common goals and visions for the company, we have very diverse backgrounds and speak many languages, as our users come from all over the world as well. Our leisure interests range from cycling and beekeeping to Frisbee and Foxtrott. We wish to maintain an open culture like it is typical of start-ups. Everyone can actively contribute and share their ideas and opinions with others. In our weekly TGIF meetings ("Thank God it's Friday") and, of course, also by e-mail or in our cafe, the Googlers can with company-relevant questions turn directly to Larry Page, Sergey Brin and other members of the management teams. Our offices and cafes are designed to promote the interaction between Googlers within teams as well as cross-team, be it at work or at a round of football.

Overall, Google creates a lively and flexible work environment, which is skillfully published with many reports and photos in blogs, IT reports and on YouTube [Goo16]. In personnel portals are numerous ratings on Google as an employer. The tenor is consistent and confirms the independent way of working in motivated, international teams and interesting projects in an innovative environment. Overall, Google has more than 62,000 employees with a relatively low mean age of 30 years, compared to Facebook (29), Apple, HP (38) and IBM (36). Fluctuation on Google is, however, relatively high and job entrants leave the company after less than 2 years to launch their own start-ups or to find fresh challenges in their areas of interest in new positions. Employee satisfaction is very high in the top rank of 89%, together with Facebook (96%) and Salesforce (89%). As regards the payment for job starters, Google is just right behind Facebook also in the top field of the high tech companies [Pay16].

This brief overview on Google as an attractive and preferred employer can give the automotive industry an indication of how to become an employer of interest to digital natives. First of all, the products should be brought closer to the digital lifestyle of natives. To this end, future vehicle generations with connected services and autonomous driving concepts, quasi driving IP addresses, may have good opportunities on offer. Attractive mobility services which are offered through apps and platforms on the smartphones will help to improve the image of automotive manufacturers. Nevertheless, it is also important that they transform themselves into start-up-like organisations along with the digitisation in order to offer interesting tasks and innovative work environments to the digital natives.

References

[Acc16] N.N.: Liquid workforce: Building the workforce for today's digital demand. Accenture Technology Vision, 2016. https://www.accenture.com/fr-fr/acnmedia/. Drawn: 01.07.2016

[Arn14] Arns, T., Aydin, VU, Beck, M., et al.: Crowdsourcing für Unternehmen; Leitfaden (Crowdsourcing for Enterprises; Guide). BITKOM (eds.), Berlin (2014). https://www.bitkom.org/Publikationen. Drawn: 05.07.2016

[Boe14] Boes, A., Kämpf, T., Langes, B., et al.: Cloudworking und die Zukunft der Arbeit (Cloudworking and the Future of Work). BTQ Kassel inputconsulting (2014). https://www.researchgate.net/. Drawn: 05.07.2016

[Bor14] Borstedt, S., Roden, I., Borchard, I: DIVSIU25 Study; 02/2014 German Institute for Trust and Security in the Internet. https://www.divsi.de/wp-content/uploads/. Drawn: 25.06.2016

[Eic15] Eichhorst, W., Spermann, A.: Sharing Economy – Chancen, Risiken und Gestaltungsoptionen für den Arbeitsmarkt (Sharing Economy – Opportunities, Risks and Design Options for the Labour Market). IZA Research Report No. 69, 12/2015. Research Institute for the Future of Labour, Bonn. http://www.iza.org/en/. Drawn: 27.06.2016

[Eva16] Evans, B., Andreessen, H.: Mobile is eating the world. Presentation 03/2016. http://ben-evans.com/benedictevans/2016/3/29/presentation-mobile-ate-the-world. Drawn: 26.06.2016

[Fre14] Freese, C., Schönberg, T.: A Shared Mobility. How new businesses are rewriting the rules of the private transportation game. Roland Berger study (2014). https://www.rolandberger.com/. Drawn: 29.06.2016

[Goo16] N.N.: Google: Our Culture. https://www.google.com/intl/de_de/about/company/facts/culture/. Drawn: 05.07.2016

[ICT15] N.N.: ICTFact and Figures – The world in 2015. ICT Data and Statistics Division, Geneva Switzerland. http://www.itu.int/en/ITU-D/Statistics. Drawn: 25.06.2016

[Kor16] Korge, G., Buck, S., Stolze, D.: Die "Digital Natives" grenzenlos agil? ("The 'Digital Natives' endlessly agile?") Study Fraunhofer-Institut für Arbeitswirtschaft und Organisation IAO (2016). http://www.iao.fraunhofer.de/images/. Drawn 01.07.2016

[Mar15] Markowetz, A.: Digitaler Burnout. Warum unsere permanente Smartphonenutzung gefährlich ist (Why Our Permanent Use of Smartphones Is Dangerous). DroemerKnaur, Munich (2015)

[Mil15] Miller, M.J.: PwC: Americans subscribe to the sharing economy brandchannel, 21.04.2015. http://www.brandchannel.com/. Drawn: 02.07.2016

[Pay16] N.N.: Spot Check: How Do Tech Employers Compare; PayScaleStudy 2016. http://www.payscale.com. Drawn: 05.07.2016

[Pwc16] N.N.: Bis 2020 entfallen 60 Prozent aller neuen F&E-Jobs in Auto-mobilindustrie auf IT-Spzialisten ("By 2020, 60 Per cent of All New R&D Jobs in the Automotive Industry Will Go to IT Specialists"); PwC Analysis 02/2016. http://www.pwc.de/de/pressemitteilungen/2016/bis-2020-entfallen-60-prozent-aller-neuen-f-e-jobs-in-automobili.html. Drawn: 05.07.2016

[Rif14] Rifkin, J.: Die Null-Grenzkosten-Gesellschaft (The Zero-Marginal-Cost Society). Campus Verlag, Frankfurt a.M. (2014)

[Sti15] Stiegler, C., Breitenbach, P., Zorbach, T. (eds.): New Media culture: Mediale Phänomene der Netzkultur (Medial Phenomena of Network Culture). transcript Verlag, Bielefeld (2015)

[Tap09] Tapscott, D., Williams, A.D.: Wikinomics: Die Revolution im Netz (The Revolution in the Net). Deutscher Taschenbuchverlag, Munich (2009)

[Tre16] N.N.: Deutschlands 100 Top-Arbeitgeber ("Germany's top 100 employers"). trendence Institute (2016). https://www.deutschlands100.de. Drawn: 10/08/2016

Chapter 4
Technologies for Digitisation Solutions

This chapter presents technologies and innovative solutions which are available today or in the foreseeable future for digitisation projects in the automotive industry. The purpose of the discussion is to understand their potential applications and potential benefits in order to assess their relevance for current and soon upcoming projects.

The so-called Hype Cycle for innovative technologies, published by Gartner, a leading technology analyst, provides a good first overview on the technologies to come in the industry, Fig. 4.1 [Lev15].

The diagram assigns the technologies to their phases of life, from recognition of the technology, through the phase of excessive expectations to deep disillusionment, followed by initial pilot projects up to the breakthrough. Furthermore, for each technology, the period of time is indicated in which their maturity is reached.

Not all the technologies shown are already relevant to the automotive industry, and some are not yet at all. For this reason, only the fields that have demonstrated a practically relevant degree of maturity and applications at least in the first references are presented below, without entering into the details of the respective technology and thus probably confusing the reader. In addition, solutions and technologies will not be commented on here which are still in the research stage and therefore only available to the industry in the medium to long term. However, when setting up a strategic digitisation roadmap, it is important to also identify and understand these future potentials. That is why Sect. 7.6 looks at these from an innovation management point of view as part of a respective roadmap.

The selection of the technologies detailed here is based on literary research and the results of a large number of current studies. Example sources to mention are [GfK16], [Köh14], [Dum16] and [Man15]. Based on this analysis and taking into

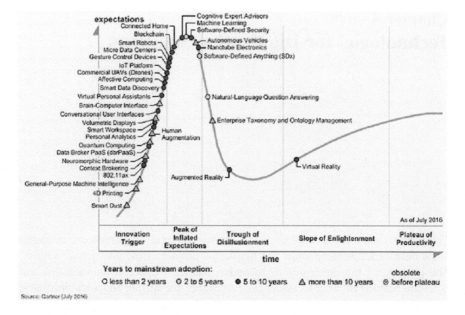

Fig. 4.1 Hype cycle of innovative technologies (Gartner)

account the author's own project and industry experience, the following topics were selected:

- IT solutions
  - Cloud Service
  - Big Data/Analytics
  - Mobility Solutions/Apps
  - Collaboration
  - Machine Learning/Cognitive Computing
- Internet of Things
- Industry 4.0/Edge Computing
- 3D Printing/Additive Manufacturing Processes
- Virtual/Augmented Reality
- Wearables/Beacon
- Blockchain
- Robotics
- Drones
- Nanotechnology
- Gamification

Vehicle-related innovations such as new materials, battery technology and also embedded software, for example for autonomous driving, are excluded here as they do not affect the topic of digitisation directly.

4.1 IT Solutions

As already explained in detail, the driver of digitisation is the information technology, which enables with ever more powerful hardware increasingly efficient software and solutions and expands into many adjacent technology areas. The provision of IT infrastructure in the automotive industry is still to a large extent in its own computing centre. The evolution steps of IT up to the end of the millennium are shown in the lower part of Fig. 4.2.

On the basis of central concepts around the Mainframe, decentralised Client/Server solutions were added in the 1980s and 1990s. Core applications of the industry, for example, from finance, development and logistics, are still running on mainframe systems, i.e. powerful central computers, while newer systems, such as ERP or CAD applications, are implemented on Client server architectures. In this, a Client programme interacts with a server programme installed elsewhere in the processing of transactions. Since the 2010s, a clear trend towards Cloud services has developed along with powerful networks and WEB 2.0 services. This is, in simplified terms, the provision of computing power in any size and of unlimited storage systems for enterprises via a network connection.

The further developments are the relatively new technologies, such as Big Data, Collaboration Tools and Cognitive Computing, which are symbolically indicated in

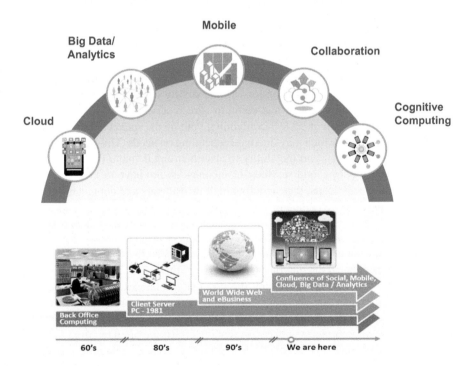

Fig. 4.2 Evolution of information technology

the upper part of Fig. 4.2. The following sections are devoted to these developments which are crucial to the automotive industry.

4.1.1 Cloud Services

Service delivery from the Cloud distinguishes three models. "Infrastructure as a Service" (IaaS) provides the hardware (server, storage, network) including operating software up to the middleware based on service level agreememts [Kav14]. The users install their own software applications on the provided technology and operate them independently. In "Platform as a Service" (PaaS), a development environment is also offered as an additional service, and "Software as a Service" (SaaS) also includes software or application programs, SAP modules for instance, as a Cloud service.

The service models are available on demand in various forms of organisation and security. In the case of the public model (public Cloud), the services are available anonymously from a computing centre network based on the software environment of the Cloud provider. In each case, free capacities are used so that the computing power and data storage are provided from different locations. The service provider can utilise its Cloud environment to capacity due to the flexible usage, so that the prices of this option are relatively low. By contrast, the private model (private Cloud) permanently assigns a certain infrastructure to the customer for the duration of the service provision, and also the software environment may be installed in a customer-specific manner. In this model, the location of the data storage can be determined, for example for the keeping of personal data, in a computing centre in Germany.

This brief overview shows the options and flexibility of today's service models. The IT performance is delivered according to agreed service quality, for example continuously 7 days/24 h at 99.8% availability, "out of the socket". The billing is made by consumption, which may fluctuate depending on needs. This flexibility, the speed of service delivery, and the ability to absorb strong demand fluctuations are the main advantages of Cloud services. Companies do not have to invest in large systems to cover peak loads, thus avoiding high investments and depreciation. The alternative of procuring a company's own infrastructure and building it at the company's computing centre, may well take several months. Then again, Cloud services are provided via internet-based platforms within hours.

Especially for the new development of applications in a fast-paced agile world, even with strong demand fluctuations, Cloud services thus offer considerable advantages. Obstacles are, on the one hand, the necessary bandwidths for an efficient, secure network connection (becoming available more and more readily and less expensively) and, on the other hand, the "Cloud capability" of the existing applications. Transformation projects with corresponding efforts are often required for this. Another issue is security. Personal data should be kept legally secure in private or dedicated Cloud environments in their own country. A further option is to organise the data maintenance in a way that sensitive data is stored in the company's own computing centre. This leads to so-called hybrid Cloud architectures, the most common type of IT architecture. These are discussed in detail in Chap. 8.

In conclusion, Cloud services offer great opportunities to make the provision of IT services more flexible. Instead of considerable time and investment, the necessary services are available quickly and geared to the needs of the business. The supply models together with the agreed service levels determine the costs. With a correct overall cost analysis, companies usually achieve significant saving potentials – in addition to the agility gained.

Nevertheless, in the automotive industry there is a heavy backlog in the transition to Cloud services. This is often due to a conservative behaviour of inertia and undepreciated investments. Potential counter-arguments such as data security and availability of power supply are often overstated. Further causes for the delayed Cloud expansion may also be in the existing organisation. Transformation projects to get applications "Cloud-ready" usually require the involvement of several areas of responsibility. In addition to the infrastructure areas, the persons responsible for the usage are required to change the applications and then have these tested by specialist departments.

Such projects bring effort and burdens to the respective teams, while the potential savings may benefit the company but not be adequately advantageous to the project participants. This Gordian knot needs to be cut by all divisional managers agreeing on Cloud conversions as a common objective, and establishing a "Cloud accelerator" as a matrix manager who is responsible for and driving this transformation integrally and across the organisation.

4.1.2 Big Data

A further topic with cross-functional potential is Big Data. This term has experienced a kind of hype which often led to unfulfilled hopes that had been put into the usage. As a result, initiatives were assigned a low priority again. From the author's point of view this is wrong as the topic of Big Data, combined with the corresponding software tools for processing large amounts of data, due to new findings in data evaluation for instance holds significant potential for savings and improvements in process flows through to the development of new business models [Win14].

Big Data is the basic term for large data volumes from different sources and in different structures, but also for their storage, processing, purposeful analysis and evaluation. This term was created in connection with the exponentially growing flood of data. Especially the Internet of Things, WEB 2.0, Smartphone and Apps account for this growth, as Fig. 4.3 demonstrates.

Over the next few years, the worldwide data volume is expected to double annually, reaching a value of 44 zetabytes by 2020 (Zeta equals a 1 with 21 zeros). More vividly: The data volume 2020 is more than 57 times the number of sand grains of all the beaches on earth [Jün13]. A major part of the information will consist of unstructured data, for example in the form of pictures, videos or presentations.

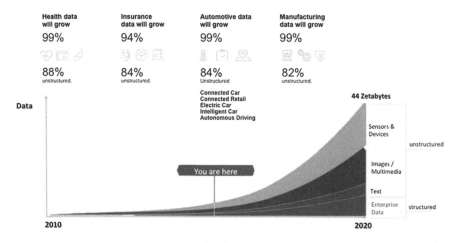

Fig. 4.3 Growth of available data by 2020 (IBM)

Apart from the health and insurance industries, growth will be found in all areas of industry, and certainly in the automotive industry as well. Some topics such as Intelligent Vehicles and Autonomous Driving lead to additional growth.

The technical literature usually distinguishes between the three "V's":

- Volume ... ever more expanding data pools
- Variety ... different formats, sources, types
- Velocity ... processing speed, real-time

As additional parameters there are often added Veracity, Trustworthiness, i.e. seriosity of the data source, and Value.

In order to handle these extremely high data volumes and different characteristics, a large number of technologies are available, which go far beyond the performance of classical data tools such as relational databases or spreadsheet-oriented evaluations. A good overview on the new tools is included in a guide to Big Data by Bitkom (Bitkom is the Digital Association Germany). Authors from different companies and technology providers have been involved, and the entire field of tools has been reviewed in a structured manner. The summary is shown in Fig. 4.4 [Web14].

The topic field is structured into six function clusters, to which available technology components have been assigned. This results in a modular system that can be used to configure the required solutions. For example, streaming tools could access data in so-called Hadoop stores, analyse these data with appropriate mining tools, and visualise the results in a dashboard. In the further process, encryption and the integration of adjacent systems are carried out. There are also offerings from different software providers for the technology elements, which are also listed in the Bitkom study. The technical details will not be discussed further. Operational and architectural aspects as well as new methods such as "Data Lake" are explained in Chap. 8.

Big Data projects offer significant potential to businesses, as the efficiency of today's software provides powerful tools for aggregating and evaluating data from

4.1 IT Solutions

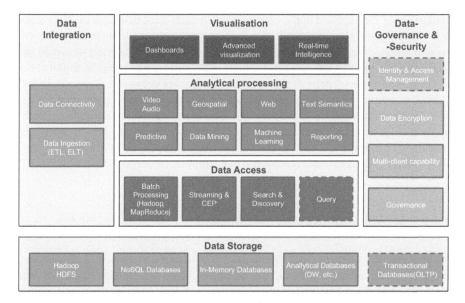

Fig. 4.4 Modern tools for data processing (Bitkom)

multiple sources, both within and outside the enterprise. These technological options go well beyond the tried and tested structured evaluations which have been used with Excel, Business Objects or Cognos until now.

Modern Big Data tools can process very performantly structured and unstructured data in a combined manner, independently extract patterns from data sets, and suggest new insights and options for action on very appealing graphical surfaces or prepared in apps on the smartphone. Typical applications for Big Data projects in the automotive industry are, for instance:

- Early detection of failures in body-in-white manufacturing
- Segmentation of customer interests and definition of the "next best action"
- Understanding of stock movements to reduce inventories
- Recognition of warranty patterns
- Increase the reuse of common parts
- Recognition of bundling potential in purchasing

These are examples of realised Big Data projects with very short payback times. Chapter 9 presents specific projects in this context.

Finally, the question arises here as well as to why the implementation of Big Data projects in the automotive industry is rather sluggish despite the fact that the technologies are available and the benefits are proven in many references. Similar to the Cloud subject in the previous section, an obstacle lies in the distributed responsibility for the data pools.

If, for example, the retrieval of spare parts is combined with the age of certain vehicles and their service and maintenance history and this information linked with the quality of the supplied parts, new patterns of early fault detection could be discovered, and preventive maintenance measures based on these failures could reduce vehicle breakdowns. Technologically, such an overarching data analysis would be relatively easy to carry out. Organisationally, such a Big Data project turns out to be problematic as various parts of the company have to act in concert. Expenses in the individual organisational units achieve benefits for the company which can not be offset against the expenses directly though. Due to this problem, unfortunately such projects often remain undone, and there is a lack of overarching motivation for change and improvement. It is this readiness to transform which the change in corporate culture needs to address – a topic in Chap. 7.

4.1.3 Mobile Applications and Apps

Another established technology, being available as the core element of digitisation projects, are mobile applications, the so-called Apps. Smartphone and Apps are mutually dependent; both have achieved massive growth rates. At the launch of the iPhone in 2007 and the Android smartphones in 2008, mobile applications were initially offered in the areas of games, news, weather and entertainment. Very soon, Apps were added as user interfaces to established Internet platforms such as eBay, Amazon and social networks. Due to their great success and the customer interest, apps from companies, initially from the areas of marketing and communication came to the market soon. On the two best-known store platforms of Apple and Google more than 2 million programmes are on offer, of which approximately 20% are fee-based. The following categories are on top, according to their proportionate number of apps: Education, Lifestyle, Entertainment, Business, Personalisation and Tools [App16], [Ipo6]. Many companies also offer Apps on the established platforms, and an increasing number of platforms from other providers are developed directly in companies.

The Apps are not developed by Apple or Google, rather by a global developer community. So-called store operators take care of the quality assurance and distribution of the apps through the store environment, and for this they receive a fee from the developers posting their Apps on the stores. A direct installation of Apps on iPhone or Android-based smartphones is possible only through unofficial or unsupported paths, so these are closed systems. The principle is similar with other providers.

The massive crowdsourcing for the development of Apps leads to an extremely high pace in the provision of new applications and adjustments, at low cost to the customers. Key to the success of the Apps is the easy installation and operation of the programmes, which do not require any special training. The Apps are available, specifically for the smartphone technology in use, in the respective stores to download by clicking. After downloading or installing the application, these are usually opened and tested immediately. Applications that convince the potentially interested user through functionality, stability and response times normally remain on the

4.1 IT Solutions 53

smartphones for later use. The entire App environment is characterised by very high dynamics, which is very much in line with the behavioural patterns of the digital natives as was covered in detail in Chap. 3.

The App environment is thus known to all users from daily use and is accepted despite all criticism for instance with regard to data security, terms of use and the filter of the store operators. Therefore, this type of IT usage is shaping the expectation of users in the company. Instead, many enterprise applications are still used by complex user menus on stationary workstations. With the progressive spread of the smartphone, more and more users in the companies want a simple and flexible App-oriented IT environment. Similarly, the customers of companies are looking for marketing or product information for instance or to ask questions via mobile applications.

The challenge to the companies in the context of these expectations is to connect the established IT structures with its proven applications and enormous data pools with the mobile App-oriented world. This is also referred to as integrating the so-called Systems of Record and Systems of Operations, i.e. the proven IT world, with the Systems of Engagement, the mobile, App-oriented world [Moo16]. Figure 4.5 illustrates this situation.

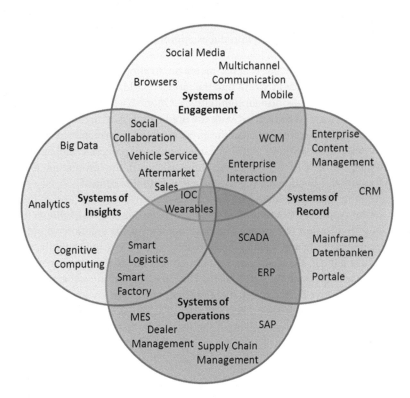

Fig. 4.5 The new IT structures in times of digitisation (Moore)

"Systems of Insights" are added here. They pursue the goal of gaining new insights from the huge data stocks within and outside companies [Whe15]. This is served by the Big Data technologies already explicated, as well as new solutions from the areas of Cognitive Computing and Collaboration, which will be discussed in the following.

New insights result from the combination of different data sources. For example as for the customer behaviour regarding a product, derived from company data of the CRM environment (Customer Relation Management), after-sales information and product discussion of user groups on Internet platforms. The findings are then available as up to date information to customer service representatives through a mobile sales application in the sales talk. This results in the integration of the established IT-world with the mobile capabilities and the Big Data and Cognitive Computing Technologies.

The direction and also the potential for a comprehensive integration and use of the new technologies are shown by this example. These opportunities must be developed quickly within the companies. For this purpose, the enterprise IT should provide a secure environment, which makes it easy to specialist departments and users to establish themselves in the mobile application world. This can also prevent an "uncontrolled growth" which is generated when each department establishes its own path to App development and to the hosting of solutions, as is quite common within the companies. Chapter 8 introduces appropriate approaches to an adequate IT environment.

It is important for the business departments not to simply transfer existing applications from the stationary IT world into the mobile world. Rather, the Apps should be used as bearers of transformation initiatives. On the basis of customer interests, existing processes are to be queried with the application systems supporting them. For which business purpose are specific processes and information required, and which customer value do they create? What is the overall process chain, and how do upstream and downstream organisational units work with the information and with which IT solutions? Section 8.4.4 provides answers to this context.

The questions raised are intended to give thought-provoking impluses to methodically use the Apps, for example in so-called design thinking workshops, to start overarching digitisation initiatives. The development of the solutions should then follow through an agile approach in the App development in which a first version of the App is available to the users very quickly, so that the user feedback can already be taken into account with the first update. This cross-organisational approach with overarching objectives, which can be implemented swiftly with new methods, should for successful digitisation become part of a new corporate culture. This is why Chap. 7 is devoted extensively to the subject of the transformation of corporate culture.

4.1 IT Solutions

4.1.4 Collaboration Tools

An important element for supporting cultural changes are collaboration tools. Telephone and e-mail are currently the established communication standard as the basis for cross-departmental and transnational cooperation in companies and also across corporate boundaries, for example, with partners. However, these technologies currently reach their limits, e-mail due to the pure flood of daily messages, and the telephone due to acceptance problems. Digital Natives prefer the group-wide exchange of information, for example, in social networks or even messaging services in interest groups, instead of communication in direct contact between communication partners. Thus, established technologies such as e-mail, calendar, video conferencing, document management and project management tools are increasingly being replaced or supplemented by:

- Social Networking
- Workflow Systems
- Wikis, blogs, bots
- e-Learning
- Messaging systems
- Whiteboarding, Desktopsharing, Teamrooms
- RSS feeds, tagging.

These technologies are by no means new yet rather tested in reference projects, and there are comprehensive offers available on the market. A good overview of application examples is provided, for instance, by a Bitkom study [Eng13]. This underlines the importance of these tools for communication and collaboration in teams. To understand the fundamentally different ways in which traditional communication and collaboration tools work, they are contrasted in Fig. 4.6 with the new WEB 2.0-oriented world.

The comparison shows that the new tools emphasise an open collaboration in the team, flexible fields of application and the integration of different tools and processes. The traditional communication applications were aimed at direct dialogue. Electronic calendars, project and document management as well as special software solutions for developers supported the collaboration. The new world is characterised by flexibility, openness and integration capability and thus transparency for entire business areas. Mobile Apps are used on mobile devices in the same way as the established private applications, so that no special training is required.

This results in a wide range of applications for collaboration tools, such as software development projects with distributed teams at different locations in Teamrooms, company-wide communication and opinion-forming on strategic initiatives via social networks, and documentation of work experiences with a new tool in Wikis. If necessary, examples of solutions and references can be found on the Internet. It is important to recognise the opportunities to shape new forms of communication and cooperation with IT tools in a timely and appealing manner in order to use them for digitisation projects specifically.

Consideration	Traditonal Collaboration Tools	Web 2.0 Collaboration Tools
Focus	Clear-cut Structure	Open-minded Structure
Governance	Command & Control via Direct Dialog	Social Collaboration
Core Elements	Electronic Calendars, Project & Document Management	Wikis, Social Networks, Unified Communication & Collaboration
Value	Single Source of Truth	Open Forum for Discovery & Dialog
Performance Standard	Stability, Controllability	Flexibility, Openness, Transparency
Content	Authored	Communal
Primary Record Type	Documents, Structured Data	Conversation (Text-based, Images, Audio, Video), Unstructured Data
Searchability	Easy	Hard
Usability	User gets trained on systems and has access to follow-on support	Intuitive handling due to resemblance of enterprise systems to social networks
Accessibility	Regulated, Contained, Workplace-restricted	Ad hoc, Open, Always On
Engagement	Top-down, Management-driven	Intrinsic, collegially-driven
Policy Focus	Closed System (Knowledge Retention)	Open System (Knowledge Spillover)

Fig. 4.6 Comparison of traditional and modern collaboration tools (Source: Author in reference to [Moo16])

4.1.5 Cognitive Computing and Machine Learning

A still young IT technology is the focus of many research projects, and increasingly in innovative industrial projects: the so-called Cognitive Computing, often equated with machine learning. Both topics relate to the field of Artificial Intelligence.

For machine learning, algorithms are programmed that recognise patterns and laws in large quantities of data and, on this basis, derive forecasts of events. These findings or these algorithms can then be transferred to new, comparable data and improve through further cases of operation. The software does not optimise itself independently, but appropriate rules must have been provided and programmed in advance. Fields of use of machine learning are, for instance, the assessment of user behaviour on internet platforms, the detection of credit card fraud, the optimisation of spam filters and handwriting recognition [Shw14].

In contrast to machine learning, the learning algorithms are not preselected in Cognitive Computing, yet open algorithms are used at a higher, more abstract level, similar to the functioning of the human brain. In simple terms, the systems form hypotheses on structures and statements from the recognised patterns in the data sets which are then validated with probabilities or hypotheses, similar to the human thinking process. These sequences are programmed on a meta-level.

Systems based on Cognitive Computing develop in the course of operation time, in dialogue with human experts, autonomously in the focussed topic area, which means they keep learning. Due to the immense performance of today's IT systems,

4.1 IT Solutions

which is growing exponentially, cognitive methods from increasingly large data sets with more complex algorithms can produce impressive results in an acceptable time.

A well-known example of the potential of Cognitive Computing is the Watson solution [Kel15] used by IBM for the first time in the US television show Jeopardy! in 2011. This was based on high-performance hardware. The system understood the human voice of a quizmaster in real-time, analysed the background, context and words of questions, and then, using the cognitive algorithms, sought in extensive data sets with factual knowledge, images and documents for solution hypotheses to answer the question. Ultimately, Watson was more successful in the competition than two previous champions.

Since the presentation of this system in 2011, the area of Cognitive Computing has developed rapidly. In general, they are now able to access, analyse, and process various problem situations independently, using extensive heterogeneous data sets. Conclusions and proposals are compared with expert knowledge from specialists. The system sustainably acquires the new knowledge and also refines it. The following aspects are common to all cognitive systems:

- Flexible, open algorithms – trainable and self-learning
- Updating of experiences – without human input
- Flexible fields of application and trainability in different subject areas
- Interaction with people – also with speech control; multilingualism
- Processing of large data volumes, both structured and unstructured

As Cognitive Computing is able to be used in a wide range of tasks and to continually improve in working on the subjects, flexible possibilities of use are created with immense potential. Namely the handling of administrative processes and the merging and updating of information can be completely automated with procedures from this environment and, after learning phases, replace jobs in these areas. The following examples are already implemented as fully automated operations:

- Processing of the granting of small loans
- Answering questions at the customer helpdesk
- Analysis of radiographs and patient records
- Invoicing
- Assignment and planning of logistics tasks
- Handling of procurement processes in web shops.

These are just some relatively simple reference projects. The technology, as well as the underlying IT systems, will become even more powerful, the algorithms more comprehensive, and especially the voice-controlled user interfaces more secure.

This gives the automotive industry a wide range of opportunities for using this technology in the areas of autonomous driving, voice-controlled user guidance, vehicle diagnostics and vehicle configuration. The personal assistant, both in the workplace and in the private sector, is also foreseeable, accomplishing some of the pending tasks automatically and supporting the remaining subjects. Overall, Cognitive Cumputing will play an important role in the development of a vision for the automotive industry and the development of a roadmap for digitisation, as well as in the case examples in the coming chapters.

4.2 Internet of Things

In addition to IT technology, sensor technology also continues to become less expensive and more powerful. As a result, more and more items of everyday life, such as clothing, kitchen appliances, heating systems and weather stations, have components for status detection and communication. This is all the more true of the properties and facilities of manufacturing companies, where the sensor system for control is often already available however is now becoming smarter and able to communicate.

The term "Internet of Things" (IoT) includes the integration of sensory data from different "things" via web-based applications. They aim to support users, improve processes, intervene to control, or gain new insights. Some examples of applications are the turning on of a heating system in the house when the homeowner has almost arrived, the automatic selection of the correct washing machine programme in the case of delicate clothes, or the follow-up delivery of groceries which the refrigerator has previously ordered after the user at the start of the journey home had been asked in the car what food was preferred.

The topic of the Internet of Things is seen as a megatopic because of the diverse fields of application in all industries as well as in the public and private sectors. Figure 4.7 gives an overview on the possible applications [Man15].

The overview highlights that the topic of IoT is present in many areas and acts as a driver of digitisation projects there. The economic potential is also considerable, which is due to, on the one hand, the transformation and thus the improvement of existing processes and, on the other hand, new business models. Different studies assume a potential in Germany of at least €20 billion, some of which anticipating much higher values [Wis15].

For the automotive industry, the topic of IoT is of central importance since this industry is involved in almost all the areas which are shown in Fig. 4.7. Examples are solutions for vehicles, in order to recognise emerging problems prematurely and to plan maintenance services to avoid failures. Furthermore, it is possible in cities for instance to connect the vehicle sensor system with traffic signals or parking space sensors in order to design the route according to the driver's wishes. In the industrial environment, information from logistics vehicles about delivery situations can be used to control the supply chain according to demand. Due to the breadth and importance in many business areas, IoT is certainly also becoming a driver of digitisation in the automotive industry.

Setting		Description	Examples
	Human	Devices attached to or inside the human body	Devices (wearables and ingestibles) to monitor and maintain human health and wellness; disease management, increased fitness, higher productivity
	Home	Buildings where people live	Home controllers and security systems
	Retail environments	Spaces where consumers engage in commerce	Stores, banks, restaurants, arenas—anywhere consumers consider and buy; self-checkout, in-store offers, inventory optimization
	Offices	Spaces where knowledge workers work	Energy management and security in office buildings; improved productivity, including for mobile employees
	Factories	Standardized production environments	Places with repetitive work routines, including hospitals and farms; operating efficiencies, optimizing equipment use and inventory
	Worksites	Custom production environments	Mining, oil and gas, construction; operating efficiencies, predictive maintenance, health and safety
	Vehicles	Systems inside moving vehicles	Vehicles including cars, trucks, ships, aircraft, and trains; condition-based maintenance, usage-based design, pre-sales analytics
	Cities	Urban environments	Public spaces and infrastructure in urban settings; adaptive traffic control, smart meters, environmental monitoring, resource management
	Outside	Between urban environments (and outside other settings)	Outside uses include railroad tracks, autonomous vehicles (outside urban locations), and flight navigation; real-time routing, connected navigation, shipment tracking

Fig. 4.7 Application areas of the Internet of Things (Manyka)

4.3 Industry 4.0

The term "Industrie 4.0" was created in Germany on the basis of a political initiative to further automate production in the sense of safeguarding workplaces and improving competitiveness. Compared to earlier concepts such as CIM or Lean Production, the new element is the continuous digitisation of product and production at maximum flexibility of the order structure and the supplier connection up to batch size 1. For this purpose, IT, sensor system and production technology must be linked in such a way that an integrated IoT solution is created.

"4.0" emphasizes the positioning of this phase of digitisation as the fourth industrial revolution, following the previous three phases of steam drive, assembly line and programmable controllers. In order to achieve the desired smooth interaction between technology, people and computers, a working group has developed recommendations for architectures, technical standards and norms [Kag13]. The work of the panel is being continued, and research projects and pilot projects are driving the implementation on a sustainable basis.

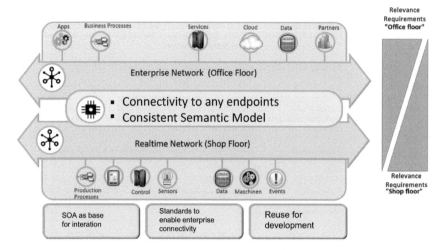

Fig. 4.8 Component model Industry 4.0 in office and shop floor layer (VDI/VDE)

The Industry 4.0 reference architecture is documented in a VDI/VDE status report [Ado15]. Figure 4.8 shows the component model with the distinction between Office and Shop Floor layers. While business processes in the office area are transaction-oriented, the shop floor area near the sensors and actuators is operated in real-time.

In the office or enterprise sector Cloud solutions establish themselves. Due to the required real-time capability, these can not simultaneously also provide the IT services in the shop floor area because the transmission speed is insufficient. For this reason, an architecture is establishing which provides for a separation of the two environments. For the production environment with sensors, actuators, control, field bus systems, etc., special Cloud solutions according to the principle of the so-called Edge or Fog Computing are applied [Rie15]. The connection with the Cloud environments is achieved via so-called edge gateways. These connect the special real-time Shop Floor protocols with the superimposed enterprise Cloud applications. Depending on the size of the IoT or Shop Floor environment, one or more gateway instances are required. Based on the idea of using several smaller edge servers on the shop floor, a load distribution is performed, which allows processing of real-time applications independent of the enterprise Cloud. As a result, Edge Computing also enables the implementation of machine-to-machine applications and the local preprocessing and handling of mass data. From Shop Floor, the concept of Edge Computing is also spreading into other application areas, such as connected services for cars, smart grid in the energy sector and in the field of Smarter Cities.

The topic of Industry 4.0 is in Germany in the many manufacturing companies in its implementation phase, and a large number of references are known. These often come from the areas of maintenance and services, where preventive measures help avoid the breakdown of production equipment by using Big Data concepts. A comprehensive implementation example for IoT/Industry 4.0 is shown in Fig. 4.9 [Man15].

4.3 Industry 4.0

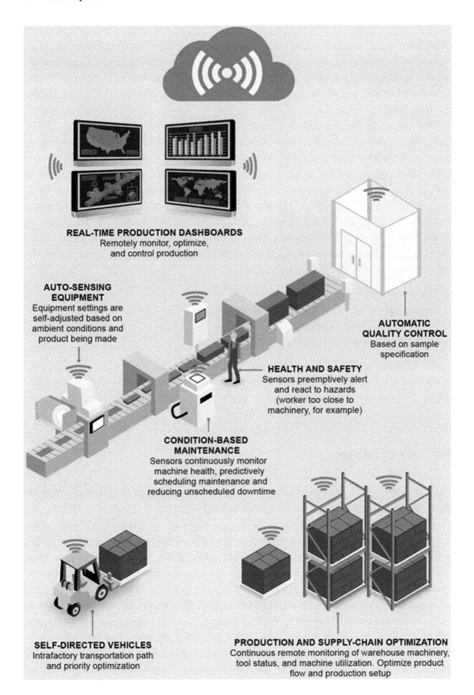

Fig. 4.9 Implementation of IoT/Industry 4.0 solutions (Manyka)

In a central control console, all information on current orders, machine statuses and logistics information for the supply of parts are combined on large displays. Preventive maintenance measures are derived from the entire data stock, and a continuous quality control takes place. Forklifts drive autonomously between warehouses and supply points. Sensors with corresponding application solutions ensure safety at work.

This scenario, allowedly, is a comprehensive and fictive, visionary example. There are, however, already many references from the area of flexible utilisation planning of plants that are integrated into the logistics concepts of the suppliers. This makes configuration flexibility of products possible right up to production start, as well as precise customer individualisation up to batch size 1. A good summary of reference projects, research projects and also providers can be found in a report of the German Federal Ministry of Education and Research [BMBF15].

By now, comparable initiatives such as Industrie 4.0, launched in Germany in 2011, are developing abroad as well. More detailed information on these initiatives can be found on the Internet, for example:

- USA: Industrial Internet Consortium (IIC)
- Japan: Industrial Value-Chain Initiative (IVI)
- South Korea: Smart Factories
- China: Five-Year-Plan initiative "Made in China".

All of these activities aim to increase the efficiency and process quality in production through digitisation projects, thus ensuring the competitiveness and sustainability of the national industries.

4.4 3D Printing

The 3D printing process can also be attributed to the subject of digitised production. By now it has outgrown the research and pilot project status and moved into industrial production. Gartner considers these processes in the Hype Cycle for future technologies as ready for production (see Fig. 4.1). Particularly in the automotive industry there are a wide range of possible applications, ranging from prototype construction and the printing of spare parts on demand, through to the production of special tools and the manufacturing of a customer-specific interior. Because of the anticipated further increase in the performance of the process and the improvement of the materials used, 3D printing with its entry into serial production has the potential of a "disruptive technology", i.e. a revolutionary change of previously established production processes. The growing importance and further development of this technology is underpinned by the appraisement of the related market volume, which includes the costs for printers, materials, software and services. In 2016 the total volume was around $7 billion, and at exponential growth a volume of approx. $21 billion is expected in the year 2020 [Ric16].

4.4 3D Printing

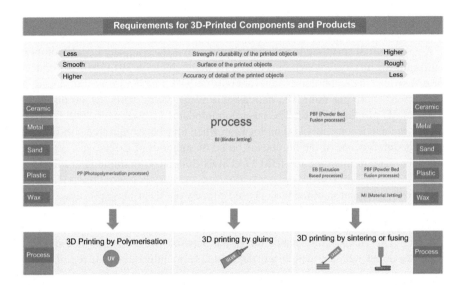

Fig. 4.10 Process models of additive production (Richter)

Under the generic term of 3D printing, different processes from the field of additive manufacturing are summarised. These methods have in common that in contrast to traditional abrasive methods, such as turning and milling for instance, in additive production, the target product is gradually built up from material layers. Three methods can be distinguished according to Fig. 4.10 [Ric16].

In the so-called PBF (Powder Bed Fusion) method, a thin material layer of plastic or metal is applied to a working surface and then melted or sintered with laser beams. After solidification, this process is repeated in layers until the target shape is achieved. With the EB or Extrusion-based processes, thermoplastic synthetic materials are placed down by means of a heated nozzle in strips or layers. The PP or Photopolymerisation process uses liquid materials, which are solidified in layers by UV radiation for example. In the BJ or Binder Jetting process, material powders and binders are alternately applied in layers, similar to inkjet printing. The component then consists of a multitude of powder and binder layers. It achieves high strength in this composite and is particularly suitable for large objects. Further methods are known, often also derivatives of the presented approaches, which are not dealt with in detail here and are instead referred to the specialist literature [Geb13], [VDI14].

The automotive industry has long since used additive manufacturing processes, and the degree of maturity is very advanced. Powder based technologies and the extrusion process are most frequently used. However, application fields are not large series with high batch sizes, which will continue to be dominated by traditional manufacturing processes, but rather products with many variants in smaller quantities. In addition to prototype construction, the production of special tools and spare parts are established fields of application, which take advantage of the high flexibility and short throughput time. This allows complicated component geometries to be manufactured economically [Hag15]. With the increase in performance of the technology, further fields of application are sure to emerge.

A pioneer of 3D printing in vehicle manufacture is the company Local Motors in California for instance, which introduced its first vehicle in 2014, being produced entirely using this technology. Subsequently in 2016, the company introduced an autonomously driving mini-bus, called Olli, all of whose parts are produced by means of 3D printing [Tri16]. The bus accommodates 12 passengers. Olli picks up its customers independently at their location when they have requested it through an App. The intelligence of the minibus is based on methods of Cognitive Computing (see Sect. 4.1), in this case the IBM WATSON Suite. The example is deepened in Sect. 9.2.

Since the production of the bus is based entirely on 3D printing processes, Local Motors has plans for interesting production concepts. Instead of mass production in central large factories, it is planned to have local 3D printing workshops near the customer. The vision is to establish a global network of these mini-factories in a market-oriented manner, which are flexibly able to integrate local requirements directly into the products.

In the light of this idea, a vision that was expressed some time ago could at least come true in facets. According to this, no more production lines would be used in the Wolfsburg Volkswagen factory by the year 2035, at most as museum pieces. Instead, there would be a tightly coordinated network of many small production plants with more than 10,000 3D high-speed printers, 100 design offices, 500 marketing companies and 300 assembly and test centres [Eck13].

There is of course still a long way to go to implement this vision. Traditional production processes are still superior to additive technologies, especially in the case of high quantities. However, in the fields mentioned above which require flexibility and speed at low number of pieces, 3D printing is already competitive today. With further performance expansion, the entry into series production is foreseeable, namely since the series will reduce in size due to the further increasing customer individuality and high segmentation. Especially in the area of the new electric drives, characterised by less and simpler components and smaller series, 3D printing will play an important role.

4.5 Virtual and Augmented Reality

In the following, a further technology area will be presented, which is to be classified as a core element of digitisation projects in production, but also in many other areas, such as sales and service, as well as development. "Virtual Reality" and "Augmented Reality" are often used as synonyms, but they are quite different. Virtual reality (VR) is an entirely computer-generated 3D representation of objects without a link to the real, physical world. The viewer is able to move in a virtual world, for example in a street or in a factory hall. Typical applications are 3D movies, computer games or even animated manuals. The user is limited to the role of a

4.5 Virtual and Augmented Reality

consumer, and the interaction takes place in the programmed environment via consoles or input devices.

Augmented Reality (AR) means the projection of a computer-generated scene into the real world. An interaction takes place in real-time and in high-resolution graphics, often in 3D representations. There are mixed forms between the two technologies in which, for example, real situations are reflected and virtualised by simulation, or alternative solutions are looked into. In this way, the virtual and the real world superimpose one another, possibly supplemented by relevant information. Augmented Reality is a technology which intelligently connects different data sources with heterogeneous data types and high data volumes with powerful output possibilities such as animations, text and speech. These solutions use techniques from Big Data, Analytics and Cognitive Computing and can thus be applied in many areas. Possible applications in the automotive industry are described in [Teg06].

Figure 4.11 illustrates the basic structure of a virtual reality application with the required system components by the example of a worker's place [Teg06].

The total solution consists of hardware and software components. In principle, the following must be distinguished: input systems for recording the real scene (in this case a camera), processing systems for tracking the situation, integration of virtual elements and further data for the preparation of the integrated overall scene (scene generator). Display devices show the user the overall scene (in this case with the help of a data glasses). The solution components may be configured in various technologies. The processing power for the execution of the AR task is available via Cloud solutions or special appliances, depending on the specific application or

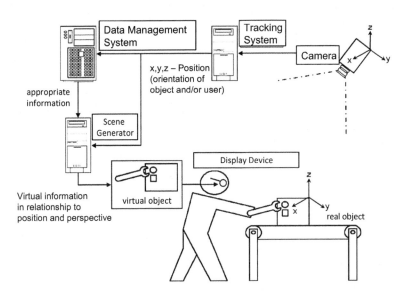

Fig. 4.11 Conceptual virtual reality application example

power demand on dedicated computers. The definition of appliances in this case means integrated devices, which for a particular application contain both the hardware and the required software, for example for the rapid analysis of large amounts of data or also for secure data transmission.

Besides the keyboard, touchscreen and mechanical devices, all types of sensors and cameras are also being used for the input. For the output or display, different devices are available, for instance:

- Monitors ... traditional screens, displays of smartphones
- Videoprojections .... Display on large monitors or projection screens
- Video glasses or Head Mounted Display (HMD) .... Glasses or a special data helmet
- Data glasses ... partial glasses, which allow to view the real scene in parallel
- Contact lenses with display .... Contact lenses with integrated displays
- Retinal data frames ... direct projection onto the user's retina.

There are numerous applications in the private sector and for companies, especially in the automotive industry. Examples for the AR deployment there are:

- Display of working instructions via data glasses during the execution of complicated assembly sequences
- Virtual test of assembly processes such as installation of the drive train
- Simulation of material flow concepts on virtual factory models
- Presentation of vehicle configurations as 3D models in a showroom in interaction with exhibits of different interiors
- Joint work of designers on the 3D model of a vehicle on a large screen to refine concepts
- Showing road surface information on the windscreen of a vehicle
- Voice-controlled interaction with an intelligent driver assistant
- Gesture control in the vehicle operation
- Interactive training of service staff during the execution of repair work.

For almost all examples, one can find on the Internet experience reports of the manufacturers who use Virtual and Augmented Reality. Due to the great benefits, the acceptance and further continuous improvements to the performance of the solutions, a massive increase in the spread can be expected. Thus, a large market growth is forecast for VR/AR solutions. Coming from approx. $2 billion in 2016 almost exclusively for VR solutions, a market volume of $150 billion is predicted by 2020, of which $120 billion for AR and $30 billion for VR solutions [Gor16]. The large proportion of VR is expected in films, computer games, marketing and distribution solutions, while the AR sector is growing significantly stronger by applications in the industrial environment. Other studies predict somewhat lower values of the future market volume, although on a considerable scale as well, so that these technologies are also to be considered as disruptive.

4.6 Wearables

The input systems are an important element of all AR solutions. Keyboards, sensors and camera systems are already established techniques. The following section briefly presents two relatively new technologies, which are receiving much attention as innovative and strongly growing solutions.

Wearable Devices, also being referred to by the short name of Wearables, are intelligent, small, body-worn devices such as data glasses, fitness belts, smart watches and sensors that are incorporated into clothes, as well as data gloves. Along with the growing interest and business potential of the Internet of Things and Augmented Reality, rapid growth is also forecast for the Wearables sector. Basically, Wearables are connected to the Internet, and there are two different versions. On the one hand, devices which assume a pure input/output function (I/O), such as clothing sensors or data glasses, and on the other hand devices which, in addition to the I/O capabilities, also have their own computing power in order to directly execute applications, such as smart watches.

Wearables are also becoming more popular in the automotive industry. Examples of using the data glasses were given in the previous section. Further options are to equip workers' clothes with sensors which measure their physical load in the work process. Using the data, workflows are then improved or even health-damaging processes are avoided by the use of suitable tools for the workers.

Further examples stem from the area of customer service for automobiles. The service employees call the next service request via their Smartwatch. Then, they are guided by "point to pick"-solutions to install the right spare parts and supported in the execution by instructions through data glasses. Further possible applications can be found during Sales activities. For example, many car dealers have established virtual sales areas. Interested customers can test selected vehicle configurations in a virtual environment. Powerful 3D glasses incl. loud speakers for the driving noise provide a comprehensive driving experience. The equipment can be changed in an interactive manner, and the bonnet and doors of the virtual vehicle can be opened by means of gesture control in order to inspect details.

Also namely in trade, a further technology allows to expand interaction with the customer and provide a personalised experience. So-called Beacons are used for this purpose. These are battery-operated transmitters with the size of a matchbox. These transmit signals at short intervals with their device-specific ID. Data transmission is done, for example, via the so-called Bluetooth low-energy technology, which operates very energy-efficient within ranges of up to 30 m [Stro15]. The signals are received using corresponding Apps. By analysing the signals from the Beacons in the showroom, the exact location of a potential customer can be recognised and used to provide information precisely about products in their field of view. In addition, the prospective buyer can also be guided through the showroom by relevant hints to offers that are relevant to his or her interests.

There are already many references in Apple Stores or at Starbucks for the mentioned application of Beacons in trade. The area of trade is certainly a suitable business field for this technology. In addition, many other areas open up interesting application opportunities as well, such as in logistics in the management of scheduling staff [Bvd16]. The use of Beacons in the automotive sector is just at the beginning.

4.7 Blockchain

The subject of Blockchain is positioned as a disruptive technology within the finance industry. As an idea published for the first time in 2008, more and more users and providers are turning to this approach with corresponding solutions. Many start-ups have been established around this topic, and a rapidly growing community is emerging. Starting from the financial sector, the basic approach is now being tested in many other process and business areas in the automotive industry in first pilot applications [Wil16].

The purpose of the Blockchain architecture is to enable direct and secure business relationships between two parties without intermediaries, for example, a money transfer from A to B without a bank being involved. Before and after a transaction, transfers have already been made, which may at least affect the parts of the transaction. The basic idea behind Blockchain is to transparently store a network of these transactions for all parties involved and to update these in a chronological way. In order to make the process tamper-proof, a multi-layered architecture was developed in which the changes of the transactions are filed in a block of the data set in encrypted form. The basic structure is shown in Fig. 4.12, taken from the public developer guidelines for Bitcoin, the Internet currency which is also protected by Blockchain methods [Bit16].

Throughout the process, verification mechanisms ensure that the payer is in fact the owner of the funds at the time of the transaction. Each new transaction is stored in a new block and appended to the previous blocks. This creates a chain of data blocks, which explains the name of the procedure.

Blockchain solutions suggest many advantages. Further to the security aspect, there are the cost advantages from avoiding the need for an intermediary organisation, and the transparency of the transactions. The disadvantages are in the effort and the processing time for the complete handling (storing, sending, updating) of the transactions. For this reason, the procedure is currently better suited for applications with individual business content and a small transaction volume, rather than for standardised mass processes. These are more cost-effective with specialised and optimised IT solutions. However, the application possibilities of Blockchain are of a universal nature and not restricted to the sphere of finance. The possibilities of use are characterised by any business relations between partners, which in turn are part of a chain, or the existence of predecessor and successor relations.

In the automotive field these are for example logistics processes in the supply chain, services in the field of warranty handling, vehicle control or handling of short-term loans in the area of mobility services, as well as payments in the area of

4.8 Robotics

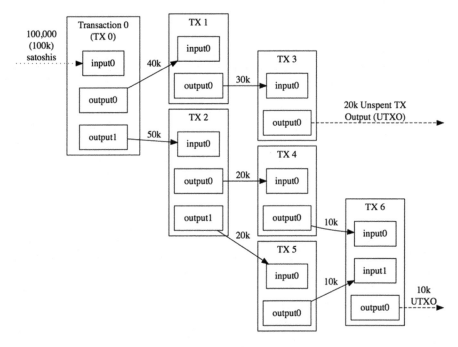

Fig. 4.12 Basic structure of the blockchain method (Bitcoin)

connected services. These examples demonstrate the flexible potential applications. A case example is explained in Sect. 9.3.

The spreading of Blockchains could accelerate if, with appropriate transaction volumes in the companies, general platforms were also established by third parties, which would offer the processing in a standardised and efficient way. On this basis, the importance will continue to grow. It remains to be seen whether a disruptive trend will develop from this, perhaps along with the digital Bitcoin currency. Potential cost, security and transparency benefits are the drivers.

4.8 Robotics

A long-established automation technology in the manufacturing industry are industrial robots. They have been used by the automotive industry in production for more than 50 years. Application fields are heavy and clocked activities which are physically exhausting to employees, such as in welding lines or assembly. Currently, the so-called robotics density (number of robots per 10,000 employees) in the automotive industry in 2014 averaged in the USA, Germany and South Korea at approx. 1100, while Japan has the highest density with 1400 and China is far behind at 300 [Har16]. Even if extensive experience is gained in the use of robots, this technology field nevertheless belongs to these fields of innovation which deserve a high level of attention as the possible uses for Robotics are currently greatly expanding.

Drivers are changing needs and application fields along with the extended capabilities of the devices. The high and further increasing computer performance, simple programming methods and the growing mechanical performance of the robots drive the spread of robotics into new business segments. The application is therefore also interesting with small batch sizes. Manufacturing of vehicles is more and more tailored to the customer and in small numbers. Hence it is foreseeable that in future fewer and fewer assembly lines will be used in production. Instead, autonomous production cells are created, in which the workers work flexibly without fix cycle times together with robots.

Robots are also increasingly used in areas still untapped by robotics so far, for example for the assembly of complex parts and also for work in the service sector. In addition to the established industrial robots, lighter collaborative systems and service robots will also be used in the future. The robots will have more sensor technology which is necessary to ensure safety in human-robot cooperation (HRC). With the help of sensors, the robots detect whether humans are dangerously close to their action radius and stop the movement immediately. Due to this sensitivity, robots will no longer be surrounded by safety fences in the future, called "Fenceless Production". The service robots also work without a safety fence, which means that robots are already in use in the care of the elderly people or in the cleaning of vehicles.

Furthermore, powerful innovative programming methods support the further spreading of robots. While in the past special programming methods and tools were required, the graphically oriented programming facilitates the use of robots. In the Teach-In process, experienced workers lead the robot directly and thus train it. In the future, gesture- and speech-controlled programming will be used to improve the possible applications and lower the costs for the introduction.

In addition to the technical feasibility, a comparison of the expected unit costs is an important decision criterion for robot use. Volkswagen has published a detailed analysis, summarised in Fig. 4.13 [Dol16]. Depending on the investment costs for the robot, cost per hour is below one and twelve Euros in the sample scenario. Comparable hourly costs for workers, as also stated in the study, are at full cost at around fifty Euros. Thus, the economic comparison also speaks in favour of the application of robots.

Against the backdrop of these advantages, Volkswagen plans to expand the use of robots in production, not only in established fields, but increasingly also in more complex areas and workplaces, for example, in the interaction with workers, i.e. human-robot cooperation. This strategy is also intended to counter the emerging shortage of skilled workers.

Due to similar drivers and advantages as in the industrial sector, the use of service robots continues to grow strongly. Completely new fields such as care services, concierge services, as well as cleaning work will be taken over by robots. Especially in these applications, Cognitive Computing (see Sect. 4.1) facilitates that robots become teachable, available for open dialogues and thus also fulfill social functions.

4.9 Drones

Assumptions Business Case	
Duration in operation	7 Years
Uptime	250 Working days/Year with 20 hrs/Working Day → 5000 hrs/Year * 7 Years = 35.000 hrs
Operating Costs	Electricity(1 - 13 KW) * 0,10 € * 35.000 hrs
Maintenance	5% of robot basic structure cost

Robot Investment Cost

30.000 €	75.000 €	112.000 €	182.000 €	217.000 €	250.000 €	400.000 €

= cost in euro per hour

0,90	2,10	3,20	5,20	6,20	7,10	11,40

Fig. 4.13 Overall cost comparison of different robot types (VW)

One example is the so-called humanoid robot "Pepper" by the company Softbank with IBM's cognitive base "Watson" [Wan16]. This autonomous robot is used in Japan as a concierge service in banks. After welcoming customers in the counter room and inquiring about the customer requirement, Pepper leads the customer to an advisor desk. Meanwhile, the analysis of public social media data and the existing customer data held in the bank are used to prepare the customer's profile and make it available to the consultant for the conversation. As a result, the conversation is conducted from the outset on a secure information basis. This example illustrates future possibilities in the field of Robotics, especially with the use of further digitisation technologies in the automotive industry (see also Chap. 9 with a practical example on "Pepper").

4.9 Drones

Drones are considered a further technology with high potential for innovation in many industries, and this is discussed below from the perspective of potential applications in the automotive industry. Drones are unmanned aerial vehicles which are operated by computer or remote control. They have long been seen in the military environment. As early as 1931, the British Royal Air Force used unmanned aerial objects as target drones [Pap17]. For some years now, drones have been spreading at a high rate of growth both in the private and the industrial environment. The devices come in many forms, ranging from a few centimeters in size to load carriers with spans of more than 30 m, such as the "Aquila" drone used for experiments by company Facebook. The flying apparatus which weighs over 400 kg, is powered by a solar engine and is able to remain in the air for 3 months. With the aid of this drone, people in remote regions shall be given access to the Internet. Google is working on a similar project [Han16].

This example illustrates the innovation potential which drones can have also for the automotive industry. There are many possible applications. In logistics, the delivery of parts could be carried out with such "flying robots", as well as the support for performing an inventory check on the stock in high-bay warehouses [Bmw16], and also traffic monitoring, which is already being tested in pilot applications. Further possibilities of use could be in service and also in the internal parts supply, so that this technology will play a role in the digitisation as well.

4.10 Nanotechnology

Nanotechnology is a topic area that deals with the research, production and application of structures which are smaller than 100 nanometers (1 tenthousandth of a millimetre) in at least one dimension. This technology has been rapidly developing since the 1980s and established as a cross-sectional topic in many industries. In addition to naturally occurring nanoscale materials, there are many artificially produced nanomaterials. One differentiates between carbon-based materials, metallic materials, dendrimers and composites [Wer16]. The effect of nanomaterials is essentially due to the fact that the ratio of surface area to volume is much greater compared to conventional materials, and thus the surface properties dominate the volume properties. Furthermore, the finely structured nanomaterials have fundamentally different chemical, physical and biological properties compared to coarser substances. The fields of application of nanotechnology can be roughly divided into the following fields:

- Nanoparticles ... "mini-materials" made of metals or composites
- Coating ... surface coatings such as paints or non-stick coatings
- Biology/Medicine ... special drugs and also implants
- Tools/Devices ... mini-sensors; mini-devices for medical or industrial areas.

In the automotive industry, nanoparticles are now an integral part of many vehicle components. For example, nanoparticles lead to better properties of car tyres and allow for more efficient catalysers and air filter systems. Even car wash facilities offer "nano treatments" in which special dispersions provide the paint surface with dirt-repellent properties. Further examples for use in the automotive industry are:

- Petrol additives to improve combustion
- Adhesives that replace welding joints
- High-strength materials as a substitute for traditional body parts
- Sensor systems in plants and also in the vehicle
- Coatings against corrosion
- Light-emitting diodes/LED for vehicle lighting

- Photochromic glass or mirrors
- Paints with a dirt-repellent surface or special colour properties.

The examples show the wide-ranging possibilities of nanotechnology. In terms of digitisation, future developments in the field of devices are of particular interest. In research projects, nano-devices are equipped with microelectronics, sensors and actuators. This creates quasi microscopically small "nano robots", called Foglets [Stor01]. Each of these foglets has its own computer power and can communicate with its neighbour or even with control devices.

For the foglets, many possible applications are conceivable. For example, in medicine, foglets could be injected into the bloodstream. They could remain there, monitor the state of health in situ and take preventive measures to maintain health, for instance by removing calcifications or even diseased cells. Similarly, "service foglets" could be used in cars while driving and, for example, stop wear problems in the oil circuit at an early stage.

Another example of use, which may be of interest also to the automotive industry, is that the foglets keep each other with their actuators, so that masses of foglets connect to solid material. This could lead to the development of programmable bodies or components, which is an application certainly still far ahead in the future. Nevertheless, in case of further progress in the relevant technologies, it can be assumed that this vision will also become a reality within the next 50 years.

4.11 Gamification

Since this chapter so far has described mainly engineering-oriented areas of innovation, the next section is devoted to new methods which increase the efficiency of processes and also the usability of new procedures and applications. One way of achieving this, is through so-called gamification. This means the approach to use gaming principles outside games, to transfer them to entrepreneurial interests and thereby to engage and motivate employees. The basis is an appealing game idea with the interests of the employees involved in mind. The game takes place in competing teams, whereby progress reports with ranking lists, special tasks in the game process with game-related bonuses, as well as transparency of the game advances and task fulfilment, are motivating elements.

Numerous project references for the use of gamification in companies namely are in human resources, in qualification, but also in change management. PwC Hungary, for example, uses the "Multiploy" game, which, with the help of Monopoly approaches and structures, helps new employees to get to know the company in a playful way, thus reducing their familiarisation phase. Walmart uses Gamification in security training of its employees, while company Qualcomm uses the method to improve its technical support [Mei16]. Gamification will also play a growing role in the automotive industry. Digital Natives in particular welcome the use of modern methods in companies. In addition to the direct benefits, the attractiveness of the work environment also increases.

The examples mentioned above show possible fields of application. A large number of other fields are foreseeable. Why not do the budget planning according to the principle of the show "Lion's Cave", or the qualification for a new application program as "Memory" in combination with "Activity" elements? Gamification certainly offers the potential to support the transformation of the corporate culture towards digitisation. Chapter 7 elaborates on the necessary cultural change.

References

[Ado15] Adolphs, P., Bedenbender, H., Dirzus, D.: Referenzarchitekturmodell Industrie 4.0 (Reference Architecture Model Industry 4.0) (RAMI 4.0) VDI/VDE Statusreport (2015). https://www.vdi.de/fileadmin/user_upload/VDI-GMA_Statusreport_Referenzarchitekturmodell-Industrie40.pdf. Drawn: 18.07.2016

[App16] NN: Number of available apps; AppBrain. http://www.appbrain.com/stats/. Drawn: 10.07.2016

[Bit16] NN: Bitcoin Developer Guide. https://bitcoin.org/en/developer-guide. Drawn: 28.07.2016

[BMBF15] BMBF (Hrsg.): Industrie 4.0 Innovation für die Produktion von morgen, Bundesministerium für Bildung und Forschung Bonn (2015). https://www.bmbf.de/pub/Industrie_4.0.pdf. Drawn: 18.07.2016

[Bmw16] BMWi. (ed.): Autonomik für Industrie 4.0; Bundesministerium fürWirtschaft und Energie, Berlin, September (2016). https://www.bmwi.de/Redaktion/DE/Publikationen/Digitale-Welt/autonomik-fuer-industrie-4-0.pdf?__blob=publicationFile&v=9. Drawn: 01.03.2017

[Bvd16] BVDW. (ed.): Proximity Solutions; Bundesverband Digitale Wirtschaft, Düsseldorf September (2016). http://www.bvdw.org/presseserver/Mobile/BVDW_Leitfaden_ProxiSolu_160909.pdf. Drawn: 01.03.2017

[Dol16] Doll, N.: Das Zeitalter der Maschinen-Kollegen bricht an. (The age of machine colleagues is dawning.) Die Welt, 04.02.2015. http://www.welt.de/wirtschaft/. Drawn: 28.07.2016

[Dum16] Dumslaff, U., Heimann, T.: Studie IT-Trends 2016 – Digitalisierung ohne Innovation? (Study of IT Trends 2016 – Digitisation without innovation?) Capgemini Studie. https://www.de.capgemini.com/resource-file-access/resource/. Drawn 07.07.2016

[Eck13] Eckl-Dorna, W.: Wie 3D–Drucker ganze Branchen verändern können. (How 3D printers can change entire industries.) Manager Magazin, 17.05.2013

[Eng13] Engel, W. (Hrsg.): Unternehmen 2.0: kollaborativ.innovativ.erfolgreich Leitfaden Bitkom (2013). https://www.bitkom.org/Bitkom/Publikationen/. Drawn: 12.07.2016

[Geb13] Gebhardt, A.: Additive Manufacturing und 3D Drucken für Prototyping – Tooling – Produktion. (Additive Manufacturing and 3D Printing for Prototyping – Tooling – Production.) Carl Hanser Verlag (2013)

[GfK16] NN: Understanding the driving force behind the connected customer. GfK: Tech Trends 2016. http://www.gfk.com/en/insights/report/download-tech-trends-2016/. Drawn: 07.07.2016

[Gor16] Goral, A.: DIGILITY – virtual and augmented reality as part of photokina. Schaffrathmedien, 21.06.2016. http://knows-magazin.de/. Drawn. 26.07.2016

[Hag15] Hagl, R., Hagl, R.: Das 3D–Druck-Kompendium: Leitfaden für Unternehmer, Berater und Innovationstreiber. (The 3D Print Compendium: Guide for entrepreneurs, consultants and innovation drivers.) 2. Aufl. Verlag Springer Gabler (2015)

References

[Han16] NN: Facebook-Drohne erfolgreich getestet (Facebook-drone successfully tested); Handelsblatt 22.07.2016. http://www.handelsblatt.com/technik/it-internet/solar-drohne-aquila-facebook-drohne-erfolgreich-getestet/13912824.html. Drawn: 01.03.2017

[Har16] Harhoff, D., Schnitzer, M., Backes-Gellner, U.: Robotik im Wandel. (Robotics in Change). Gutachten der Expertenkommission Forschung und Innovation (2016). http://www.e-fi.de/. Drawn: 28.07.2016

[Ipo6] How many apps are in the App Store about tech. http://ipod.about.com/. Drawn: 10.07.2016

[Jün13] Jüngling, T.: Datenvolumen verdoppelt sich alle zwei Jahre. (Data volume doubles every two years). Die Welt, 16.07.2013. http://www.welt.de/wirtschaft/webwelt/article118099520/. Drawn: 09.07.2016

[Kag13] Kagermann, H., Wahlster, W., Helbig, J. (eds.) Umsetzungsempfehlungen für das Industrieprojekt Industrie 4.0 (Implementation recommendations for the industrial project Industry 4.0); Final report Promotorengruppe Kommunikation der Forschungsunion Wirtschaft – Wissenschaft (editor) Frankfurt (2013). https://www.bmbf.de/files/Umsetzungsempfehlungen_Industrie4_0.pdf. Drawn: 18.07.2016

[Kav14] Kavis, M.J.: Architecting the cloud: design decisions for cloud computing service models (SaaS, PaaS, and IaaS), 1st edn. Wiley, New York (2014)

[Kel15] Kelly, J.E.: Computing, cognition and the future of knowing. White Paper, IBM (2015). http://www.research.ibm.com/software/IBMResearch/multimedia/Computing_Cognition_WhitePaper.pdf. Drawn: 12.07.2016

[Köh14] Köhler, T.R., Wollschläger, D.: Die digitale Transformation des Automobils (The digital transformation of the automobile). automotiveIT Verlag Media-Manufaktur GmbH (2014)

[Lev15] Levy, H.P.: What's new in Gartner HypeCycle for Emerging Technologies 2015. http://blogs.gartner.com/smarterwithgartner/

[Man15] Manyika, J., Chui, M., Bisson, P.: The internet of things: mapping the value beyond the hype. McKinsey Global Institute (2015). https://www.mckinsey.de/files/. Drawn: 07.07.2016

[Mei16] Meister, J.: Future of work: using gamification for human resources. Forbes/Leadership 2015. http://www.forbes.com/sites/. Drawn: 26.07.2016

[Moo16] Moore, G.: The future of enterprise IT. Report AIIM Organization. http://info.aiim.org/. Drawn: 10.07.2016

[Pap17] Papp, K.: Wissen – 8 Fakten über Drohnen (Knowledge – 8 facts about drones); INGENIEUR.DE 1.3.2017. http://www.ingenieur.de/Themen/Flugzeug/8-Fakten-ueber-Drohnen. Drawn: 01.03.2017

[Ric16] Richter, S., Wischmann, S.: Additive Fertigungsmethoden – Entwicklungsstand, Marktperspektiven für den industriellen Einsatz undIKT-spezifische Herausforderungen bei Forschung und Entwicklung (Additive production methods – development status, market perspectives for industrial use and IKT-specific challenges in research and development). Editor: Begleitforschung Autonomik für Industrie 4.0. Berlin (2016). https://www.vdivde-it.de/publikationen/studien/. Drawn: 21.07.2016

[Rie15] Rieger, S.: Die 5 wichtigsten Infrastrukturtrends 2016 (The 5 most important infrastructure trends 2016). Blog it-novum 17.12.2015. http://www.it-novum.com/blog/die-5-wichtigsten-infrastrukturtrends-2016/. Drawn: 18.07.2016

[Shw14] Shwartz, S., David, B.: Understanding machine learning – from theory to Algorithms. Cambridge University Press, New York (2014). http://www.cs.huji.ac.il/~shais/UnderstandingMachineLearning/understanding-machine-learning-theory-algorithms.pdf. Drawn: 20.08.2016

[Stor01] Storrs, H.J.: Utility fog: the stuff that dreams are made of Kurzweil essays. KurzweilAI.net July 5, 2001. http://www.kurzweilai.net/utility-fog-the-stuff-that-dreams-are-made-of. Drawn: 22.08.2016

[Stro15] Strobel, C.: Beacon-Technologie: Das große Ding der kleinenDinger (Beacon technology: The big thing of the little things). CyberForum (2015). http://www.techtag.de/. Drawn: 26.07.2016

[Teg06] Tegtmeier, A.: Augmented Reality als Anwendungstechnologie in der Automobilindustrie (Augmented reality as application technology in the automotive industry). Diss. TU Magdeburg (2006)

[Tri16] Trisko, A.: Olli: Selbstfahrender Bus mit Elektroantrieb aus dem 3D–Drucker (Olli: Autonomous-driving-bus with electric drive out of the 3D printer). Trends der Zukunft, 17.06.2016. http://www.trendsderzukunft.de/. Drawn: 21.07.2016

[VDI14] NN: VDI Richtlinie 3405: Additive Fertigungsverfahren Blatt 1–3 (VDI Directive 3405: Additive manufacturing processes Sheet 1 – 3); Düsseldorf 2014

[Wan16] Wang, B.: IBM putting Watson into Softbank Pepper robot. 10.04.2016. http://www.nextbigfuture.com/. Drawn: 28.07.2016

[Web14] Weber, M. (ed.): Big-Data-Technologien – Wissen für Entscheider (Big-Data Technologies – Knowledge for Decision-Makers) Bitkom-Arbeitskreis Big Data (2014). https://www.bitkom.org/noindex/Publikationen/2014/Leitfaden/Big-Data-Technologien-Wissen-fuer-Entscheider/140228-Big-Data-Technologien-Wissen-fuer-Entscheider.pdf. Drawn: 08.07.2016

[Wer16] Werner, M., Kohly, W., Simic, M.: Nanotechnologie in der Automobilbranche (Nanotechnology in the Automotive Industry). Hessisches Ministerium für Wirtschaft, Verkehr und Landesentwicklung. hessen-nanotech NEWS 1/2005. https://www.hessen-nanotech.de/. Drawn: 28.07.2016

[Whe15] Whei-Jen, Ch., Kamath, R., Kelly, A.: Systems of insights for digital transformation. IBM Redbook (2015). http://www.redbooks.ibm.com/abstracts/sg248293.html?Open. Drawn 10.07.2016

[Wil16] Wild, J., Arnold, M. Stafford, P.: Das Rennen um die Blockchain (The race around the Blockchain) Capital, 12.11.2015. http://www.capital.de/. Drawn: 28.07.2016

[Win14] Winkelhake, U.: IT Innovationen als Treiber von Wachstum im Automobil-Aftersales (IT innovations as drivers of growth in automotive aftersales). Contribution to the 5. Deutsche Fachkonferenz Aftersales Services (5. German Expert Conference on Aftersales Services). Süddeutscher Verlag Veranstaltungen GmbH, Mainz, 03./04.12.2014

[Wis15] Wischmann, S., Wangler, L., Botthof, A.: Industrie 4.0; Volks- und betriebswirtschaftliche Faktoren für den Standort Deutschland (Economic factors for Location Germany). BMWi (editor); Berlin (2015). https://www.bmwi.de/BMWi/Redaktion/. Drawn 18.07.2016

Chapter 5
Vision Digitised Automotive Industry 2030

In the previous chapters, the relevant drivers, determining factors and influencing criteria were described as the basis for the development of a vision and roadmap of digitisation in the automotive industry. The performance and the spread of IT are advancing with unreduced momentum and exponential growth. The basic statement of exponential performance remains as Moore's Law will continue to apply with new technologies such as neuromorphous chips which are moving towards product maturity. Digital Natives enter the labour market, and with their experiences and their own value system they bring new behavioural patterns to the corporate and customer world. There are a variety of "disruptive technologies", such as Cognitive Computing, 3D printing, Robotics with ever more flexible and efficient devices, or in the future the Nanotechnology with Foglets.

As a result, there are many technical possibilities to address digitisation in an expedited manner and to maintain and expand competitiveness. Traditionally, however, the automotive industry is comparatively slow in the implementation of comprehensive changes and transformations. This industry is accustomed to operate in the development periods of new vehicles with cycles of 4–6 years and is thus in no way compatible with the rhythm of, for example, the smartphone or App development, to which, as a rule, successor products are presented within 1 year. Parallel to the implementation of digitisation, the industry faces many other challenges. The market demands electric drive, connected services and autonomous-driving cars. The new car buyers are slowly approaching the life stage of "best agers", while the young people of today assign less importance to owning a vehicle. Many young persons do not even hold a driving license. Instead, mobility services and sharing models are in demand.

To date, the automotive industry has been in a secured position against new entrants due to the capital-intensive manufacturing facilities, the extensive marketing and sales structures and also the required after-sales services. However, this situation is changing fundamentally. New framework conditions and opportunities along the digitisation are now encouraging new competitors from other sectors as well to enter the automotive market. They put pressure on the established companies which must fundamentally transform themselves in order to maintain market positions. In addition to Tesla Motors with its focus on electric drive and direct distribution over the Internet, Apple, Google and Faraday Future are particularly noteworthy, which are pressing ahead a lot with autonomous driving. In China, corporations from outside the automotive industry, such as Baidu and Alibaba, signalise their upcoming market entry. These companies have comprehensive IT experience and bring this knowledge to the electrification, connected service and mobility services. These newcomers will surely use established suppliers, in analogy to Apple with their Chinese production partner Foxconn, to cost-effectively manufacture large portions of the vehicles. This allows to launch new vehicles faster. Large factories with a high manufactoring depth level are no longer in a distinguishing position in the market. Rather, IT-driven solutions, connected services and innovative drive technologies are increasingly in demand as a buying criterion.

In this situation, it is vital for the established automotive companies to tackle the necessary changes under a comprehensive digitisation strategy and roadmap. This is developed in the following. The starting point is an analysis of the future expectations of the market and the customers, as well as a brief assessment of the current strategies of selected manufacturers. This will be compared with a vision of how the automotive industry could develop and how it could look in 2030 with the implementation of digitisation initiatives. Understanding customer expectations is the key for recommendations to adapt the business strategy in line with an integrated digitisation strategy. For this purpose, a framework is developed in Chap. 6, which ensures that all necessary measures are addressed in an integrated program or on the basis of a holistic roadmap.

5.1 Development of the Automotive Market

The car market is currently developing differently, depending on the country. Despite all warnings, the market in China continues to grow, currently at a moderate single-digit annual growth rate. China is and remains the most important car market in the coming years with regard to sales volumes and also increasingly as an innovation driver. The USA, the second largest market in the world, is similar to Central Europe on a stable level following recovery. Challenges have to be addressed in Brazil and Russia. The market in these countries will continue to stagnate in the medium term. There is a wealth of opportunities in ASEAN, especially in Indonesia, and also increasingly in Africa. The global car market has a total volume of approx. 78 million units sold in 2016, and an annual growth of approx. 3.4%, thus overall a stable situation [Dud15]. This is underpinned in a study by McKinsey in collaboration with Stanford University, from which Fig. 5.1 is lent.

The study estimates on the basis of comprehensive expert interviews in Asia, Europe and the USA that the sales of the global automotive industry will go up from approx. $3.5 trillion in 2015 to about $6.7 trillion in the year 2030. This would correspond to an annual growth rate of 4.4%. Within this overall turnover, new mobility

5.1 Development of the Automotive Market

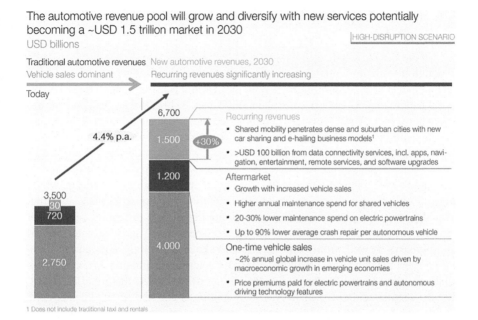

Fig. 5.1 Development of the automotive market by 2030 [Kaa16]

offers and revenues from Connected Services are growing at an annual rate of 30% and thus are the driver of the overall rise. The revenue share from traditional vehicle sales is moderately rising at 2%, similar to revenues from the after-sales business, which are rising from $720 to $1200 billion, despite the reduction due to lower service costs for vehicles with electric drive and less revenue from accident repairs due to falling accident numbers in the area of autonomous vehicles. These revenue reductions are more than compensated for by the increase in the number of vehicles and by additional services from the "shared vehicles" segment.

Even if growth in vehicle sales and revenue is projected, the automotive market is undergoing a massive upheaval, on the one hand due to the new market participants and technologies, on the other hand due to a fundamental change in customer behaviour. This is characterised by the following overarching trends:

- Urbanisation … with the consequences of traffic congestion and environmental pollution
- Cars generally losing their character of a status symbol … use instead of ownership
- Digital Natives behaviour pattern … focus on connectivity, sharing, mobile work
- Environmental awareness and ecological aspects of driving
- Flexible mobility services … mobility easy to retrieve, without brand reference
- Growing health and physical awareness … bicycle instead of car
- Lifestyle customisation / individualisation

These trends will have a major impact on future market development.

5.2 Future Customer Expectations in the Passenger Car Area

In addition to these trends, there are various consumer trends which reflect social developments and the spirit of time. With the "Digital Natives", one of these behavioural patterns has already been explained in Chap. 3. In addition, there are further consumer trends. These must also be analysed and reflected in future vehicle and mobility offers. A clear summary of other consumer trends is shown in Fig. 5.2. Some of these trends are briefly discussed below.

The consumer trend of "multigraphy" describes the social situation that today, compared to the times of our parents, there are more and shorter life stages, which are also each designed and equipped with mobility. For example, there are often several partnership phases and also several employers with different occupational focuses. The personal hobbies have become more demanding and are often assigned to life phases as well, from wind surfing and skiing to golfing. The requirements with respect to vehicles and mobility offers are determined situationally as per these relatively short-term life stages. For the automotive industry, this trend indicates, that the segmentation of vehicle types will continue, and high individualisation is also required of mobility offers. One example of this are the innovative Audi offers "Select" and "Shared Fleet" [AUDI16].

In the Select offer, the customer does not acquire a specific single vehicle yet the rights of use on up to three different types of vehicles from a pool of young attractively equipped used cars, which are then appropriately used one after the other, such as the convertible for the summer period, the estate with loading space for phases with renovation works, and the sedan for long business trips. Another example of flexible usage models is the "Shared Fleet" offer, which is aimed at companies with a vehicle fleet for employees. A car sharing solution enables via a booking portal the flexible use of individually selectable Audi models of employees, both for work and private trips, outside of working hours. Through this opportunity of pri-

Consumer Trends	Implications for (Vehicular) Mobility
Multigraphic	More fragmented life designs – Needs are becoming more situational. "Stage-of-Life-Products" are becoming more important than target group strategies (Age, social status, etc.)
Downaging	Consumers are feeling younger than their biological age, no Ghetto-Products, but experience products for a second awakening
Family 2.0	Network, Patchwork and Fragmented Families have a higher and highly differentiated need for mobility, which can be catered for solely by a family van, SUV or station wagon
Neo-Cities	Vehicular mobility which adjusts to the requirements of future green cities (zero-emission-cities)
Greenomics	Vehicular mobility which satisfies a need for a healthy and sensual lifestyle at the same time. Mobility solutions, which are ecological correct, but also sustainable for the consumer
New Luxury	Products, which increase one's quality of life. Nevertheless there is a trend away from prestige and status objects
Simplify	Simplification, time saving, simplicity, invisibility of technological processes
Deep Support	Support services which cater for the individual's need. Infrastructure of micro services which organise life between home and work
Cheap Chic	Affordable, "clever" products, which nevertheless satisfy wish for exclusivity, design and luxury

Fig. 5.2 Consumer trends in the automotive market [Win15]

vate use, companies can increase the utilisation of their fleet while offering their employees interesting mobility alternatives. Similarly, many offers may be devised to take up the multigraphy trend.

Another important consumer trend is the "downaging", especially relevant to the automotive industry, as the elderly are a very important group of solvent buyers. The life expectancy continues to increase, and people, even at the age of retirement, are of increasingly better physical fitness and more and more behave like younger people. This generation of "best agers" has developed into an active social class to which cars are an important part of their lifestyle. More than a third of all new car buyers are older than 60 years [Wit15]. In addition to an overall concept that emphasises the sporting lifestyle, one purchasing criteria is also comfort which meets the personal needs. A higher, more comfortable entrance into the vehicle, seat- and steering wheel adjustment as well as electronic assistance systems are frequently selected equipment features. These are interesting for the margins of automobile manufacturers. Previously often ridiculed staid sedans with the crocheted paper roll cover on the parcel shelf are definitely no longer in demand. As the downaging trend will grow in the future along with the corresponding economic power, the automotive industry should continue to focus on this customer group with vehicle and mobility offerings, in particular in terms of user-friendliness.

The "neo-cities" trend reflects, on the one hand, the trend of increasing urbanisation and the growth of the population in the big cities and, on the other hand, the efforts of many cities to become ecologically cleaner or "greener", up to "zero-emission" targets. Copenhagen was certainly a pioneer with its dedicated promotion of bicycle traffic within inner-city areas, but nowadays many other cities are taking initiatives in this field. One example of this is the eco-programme "Future London – Footprints of a Generation" [Wen12]. The aim of the project is to promote the green lifestyle in the metropolis. In addition to advice on sustainable behaviour, there are tips for Londoners on stores with organic food and fair trade products.

The offerings of the automotive manufacturers must fit in with such environment. To embrace this trend through green mobility offers could even lead to additional market opportunities. London is just one example. In many other cities, too, efforts are made to reduce the individual car traffic through different restrictions. For example, in Sao Paulo, vehicle usage is only permitted on every second day, being administered by the even/odd vehicle identification number. In Beijing, the total number of vehicle licences is severely limited, and in Singapore only vehicles with at least three occupants are allowed at peak times. This, incidentally, brings up quite new odd jobs: getting on the car as a "third man", so the required number of passengers is reached. These examples reveal that this trend also has a major impact on the vehicle industry.

The same applies to the trend summarised under "Greenomics", which not only affects mobility yet also affects other industries as well as cities and regions. A healthy and sustained lifestyle is establishing in all population sections. In nutrition, sports and travel, these aspects are more and more included in the planning and purchase decision. This is also increasingly true for vehicle acquisition or rather the fundamental decision whether a car is purchased at all, or mobility offers are used instead. A purchase decision is grossly influenced by fuel consumption and climate-

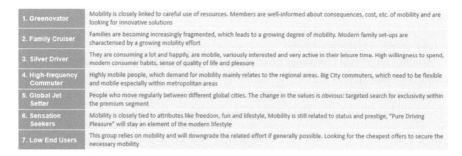

Fig. 5.3 Mobility types in the "mature markets" [Win15]

relevant values. In the case of mobility offers, simplicity of use and sustainability are the most important criteria, and the price/performance ratio is rather of minor importance, whereas in contrast to earlier days, engine output and maximum speed almost do not matter anymore. This fundamental change in values has captured many people, especially in the established countries, and will develop even more in the face of climate change and resource shortages. This is also an important trend for the automotive industry.

The consumer trends shown in Fig. 5.2, such as "New Luxury", with the attitude turning away from prestige towards immaterial values such as quality of life, "Simplify", with the strong trend towards the focus on the essentials, "Deep Support" with the desire for complexity reduction and simple support services, or "Cheap Chic" with its focus on quality and premium at reasonable prices, are also very relevant to the automotive industry and provide hints with regard to the future orientation of products and range of offerings. Also, there are certainly further trends, such as the "DINKs" (double income, no kids), which are an interesting customer segment just because of their purchasing power. Furthermore, individual consumer trends and the megatrends described are often overlapping.

For this reason, it is important to carry out an appropriate target customer segmentation, depending on the market. The relevance of the outlined trends varies according to the "maturity" of the market and the specific market position of the manufacturer. For example, for some Chinese or Indian customers in some customer segments, prestige and engine performance are buying criteria, while the German "Greenomics" focus on sustainability and CO_2 emissions. For the Internet and sports-minded Brazilians with tight budgets, smaller SUVs with easy-to-use Connected Services are important. On the basis of the trend patterns, mobility types can be defined, which, in turn, represent possible customer target segments, which are to be addressed. In the "mature markets" such as the countries of central Europe, Japan and the USA, between the following mobility types shown in Fig. 5.3 can be distinguished in future, refining the insights from Fig. 5.2 and the cited study.

One of the most relevant mobility types is the "Greenovators" (an artificial word derived from Green and Innovator), which will account for more than 30% of the population in North America, Western Europe and Japan by 2020 [Kaa16]. As the term expresses, this group is focusing on sustainability and innovation. Quality of life, conservation of resources and environmental compatibility, in line with the interest in new drive technologies, are important criteria in decisions to purchase a vehicle. This does not necessarily mean ownership, but also mobility- or sharing models. Important are coherent ecological concepts. "Silver Driver", "Global Jet Setters" and "Sensation Seekers" certainly are typical consumer groups of established markets. In contrast, the emerging markets are not so finely segmented. These are more about attracting the basic consumers, yet in parallel the prestige-oriented luxury buyers as well.

However, further deepening of this aspect is not expedient at this point, so please refer to available studies on the Internet and in the literature, e.g. [Kaa16]., [Wen12], [Sta15]. This abstract's objective was to highlight the importance of this topic. From the outset, the customer and market understanding must be taken into account in the business and digitisation strategy and roadmap to be derived from it. The change is massive and challenging in particular for established manufacturers who have to carry out this analysis on an in-depth basis, since the proven customer segmentation of the past is certainly not the target orientation of the new market players. Against this background, the question arises as to how well automobile manufacturers are currently positioned when it comes to customer orientation and especially digitisation.

5.3 Digitisation Situation in the Automotive Industry

To this end, a brief study on the digitisation circumstances of some established automobile manufacturers was carried out. A large number of information sources, such as annual reports, investor relationship publications of companies and specialist articles, were analysed and evaluated on the basis of uniform criteria (as of 08/2016). For this purpose, the following parameters have been defined, since these allow, from the author's point of view, the assessment of the respective digitisation strategy:

- Orientation towards customer expectations
- Transformation Sales / After-Sales
- Development of company culture towards digitisation
- Digitisation of business processes; Industry 4.0 up to Business 4.0
- Connected Services
- Mobility Services
- Autonomous Driving
- Collaboration with incubators
- Transformation of IT

Especially the changing customer expectations should be the basis of the business strategy and the derived digitisation strategy. Proper products, offered through preferred sales channels and flanked with desired after-sales services are success requirements. To this extent, it was analysed as to whether the manufacturers define a clear market and customer segmentation and are already active in projects for transformation of the sales channel. Today, the Internet already plays an important role at least for product analysis and comparison in the run-up to product acquisitions. The importance of the Internet as a sales channel will continue to rise significantly, thereby heavily changing the role of car trading.

An important success criterion for digitisation is the transformation of the whole enterprise culture towards a willingness and also "an appetite" to tackle this topic with openness and joy. This is not about individual organisations or just "the IT". The topic is a broad project and concerns everyone in the enterprise. Especially Germans like to remain in established procedures and tend to see changes as a threat rather than an opportunity. The transformation measures must however begin with changing this attitude, so that company-wide digitisation can be achieved, thus enhancing competitiveness, or at least maintaining it vis-à-vis the new competitors. Thus, in this field, the analysis evaluates the extent to which transformation initiatives are already being implemented in the companies.

Another realm of research is the digitisation of processes, both in production with Industry 4.0 initiatives as well as in all other business areas. From this perspective, it can be assumed that nearly all business processes can already be automated with today's IT solutions. However, economic considerations are frequently in the way since necessary solutions are in many cases still more expensive than human labor. With the increasing efficiency of the applications, the degree of automation of the business processes will increase significantly in all domains. In this respect, the research evaluates the extent to which the manufacturers are already in the subject of process digitisation.

Many customers expect at least the Apps that are used by the smartphone to be available in the vehicle too. The handling and the flexibility of Connected Services provided by manufacturers should be just as easy – including updates of the solutions to eliminate problems or to add additional functionalities. Many manufacturers or solutions currently offered are often still far from this. Nevertheless, this vision is the benchmark, and the analysis rates the extent to which Connected Services are available and what the response of the customers is.

Furthermore, it was examined whether the manufacturers already offer mobility services or at least have established closer partnerships with corresponding service providers. The entry into autonomous driving and cooperation with incubators to speed up digitisation activities was analysed as well as transformation initiatives to dynamise IT and position it as a core component of a reoriented company. It is important that the traditional IT work more closely with the IT in the car development in order to create mutual synergies and ensure compatibility between the "embedded IT" in the vehicles and the traditional IT.

The difficulty with the evaluation of the publicly accessible sources for the analysis is that the sustainability of the projects and initiatives reported can not be assessed due to a lack of internal detailed knowledge, and the unclear probability of realization of announcements, which roll over each other in the "hype topic" of

5.3 Digitisation Situation in the Automotive Industry

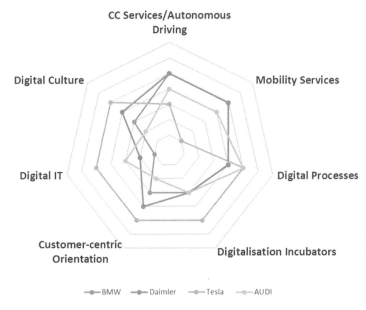

Fig. 5.4 Digitisation depth premium manufacturers (basis: Internet research author 08/2016; details in Annex A2)

digitisation. The following presentation and the benchmark are to be looked at against this background. Nevertheless, since this applies to all analysed manufacturers, the overall trend should be meaningful. A summary of the research results is given in Figs. 5.4, 5.5 and 5.6. Details and background of the research can be found in Annex A2.

Figure 5.4 shows that the premium manufacturers are more advanced than the volume producers in the digitisation activities. In particular, there are more reports on initiatives to transform the company culture, and also the cooperation with incubators is established. Tesla Motors is the benchmark amongst the premium manufacturers. As a new company in the "born on the web" industry, it is precisely its customer centricity, the digital company culture and the automated processes based on a highly efficient IT Cloud which the established manufacturers cannot cope with. Sales exclusively via the Internet, just supported by a small number of showrooms mainly in large shopping centers, as well as the update of the vehicle software "over the air", are benchmark as well. It is interesting however that the established manufacturers apparently are ahead with offers on mobility services, or Tesla does not yet see a focus here but can catch up relatively quickly.

For the volume producers, the result is a mixed impression (Fig. 5.5). Volkswagen seems to be well on track in cooperation with incubators, autonomous-driving plans, and initiatives to transform corporate culture, while the company lags behing with respect to mobility service offerings, business process digitisation, customer orientation, and transformation of the IT. In the field of process digitisation, Toyota

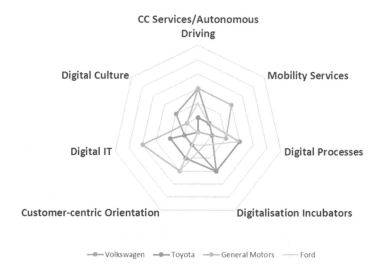

Fig. 5.5 Digitisation depth volume manufacturers (basis: internet research author 08/2016, details in Annex A2)

reports broad initiatives. Ford and General Motors are leading the field in customer-oriented sales/after-sales.

The study summarises the current digitisation state of affairs in the automotive industry in Fig. 5.6. The subject was launched by all companies with high priority. In summary, the following backlogs and uncertainties arise from the study:

- Definition of target customers or markets
- Adaptation of business strategy and focus
- Definition of adjusted sales structure incl. target figures
- Integral digitisation strategy – vision; roadmap
- Roadmap for the transformation of corporate culture

These deficits are to be addressed through an integrated business and digitisation strategy. Before this is developed as a proposal, the following is a description of how the automotive industry could look by the year 2030.

5.4 Vision Digitised Automotive Industry

The author draws now a vision of the automotive industry in 2030. This outlook is based on many years of experience and practical insights in the IT sector in various business areas of various manufacturers in Germany and abroad. These experiences

5.4 Vision Digitised Automotive Industry

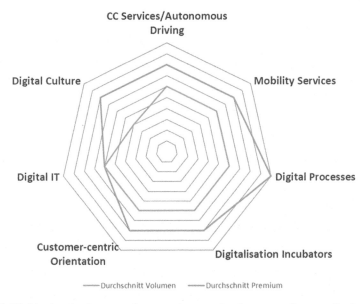

Fig. 5.6 Digitisation depth mean values premimum and volume manufacturers (basis: Internet research author 08/2016, details in Annex A2

are supplemented by cooperation in expert committees, and the implementation and evaluation of IBM Automotives studies [Sta15]. Furthermore, the drivers and influencing variables of the digitisation and the changing market expectations and situations explained in the preceding chapters are taken into account. Against this background the following hypotheses will characterise the digitised car industry by the year of 2030:

- In the industrialised countries, especially in the big cities, the ownership of cars will fade into the background, and the market will be characterised by mobility services. In today's emerging economies there is still a buying market, particularly in the basic and luxury segments, whereas in the megacities vehicle ownership is also retreating into the background.
- Mobility services are mostly accessed via brand-independent platforms. These include intermodal connections, i.e. the integration of different types of traffic and also the integration of inner city offers.
- There will be new forms of mobility services. For example subscription or flat rate concepts, similar to today's mobile phone contracts. Also price stagings depending on vehicle type or the willingness to drive together in vehicles, similar to a shared taxi will come up.
- In order to optimise capacity utilisation and reduce congestions and waiting times, mobility providers are using a higher-level of traffic control, comparable to today's airspace monitoring of air traffic.
- Autonomously driving vehicles, in particular buses, taxis and also the cars of new mobility providers, make up at least 30% of the vehicles in the large cities,

covering more than 50% of the mobility needs in the cities, taking into account the increased levels of utilisation.
- Electric drive and extensive Connected Services, also from vehicle to vehicle, are integrated in all new vehicles and support new mobility concepts. Updates of the embedded software are supplied in short cycles "over the air".
- Connectes Services of the vehicles and Apps on the smartphone are fully synchronised so that the same solution environments are available in both worlds in the same functionality and on the same data status.
- The "embedded vehicle IT" is based on central servers, additionally with backup servers for security and as an accident recorder, quasi a "black box", similar as in today's aircrafts.
- Vehicles are no longer designed on component platforms, but the core element becomes the central computer, similar to the approach of smartphones. Many equipment elements are connected by software as required, similar to today's server equipment.
- Augmented Reality is established as an essential part of the car development process, so that the number of prototypes required is halved, and test drives take place to a considerable extent in virtual environments.
- The sale of vehicles takes place at least 50% via brand-independent portals directly over the Internet. Virtual Reality solutions support the vehicle configuration. The presentation is done in the "home megaplex centre", quasi a temporary virtual private sales room.
- The number of traders in the industrialised countries is massively reduced. Successful traders are involved in the provision of mobility services.
- The production structure of the manufacturers is adapted to the markets: in the emerging markets, the focus is on mass or rather assembly line production; in the "mature countries" for ever more customised vehicles, manufacturing islands with a high level of robotics in close collaboration with workers characterise the manufacturing environment.
- 3D printing of components and spare parts is part of new production and logistics concepts.
- There is an surplus of production capacity. New providers use this in open production networks.
- IT services are at 80% based on Cloud environments. These are not operated by the manufacturers yet rather by special providers from their mega data centres. Desktop systems are completely replaced by mobile terminals, which are operated by voice and gesture control.
- Assistance systems are established in many business areas and also directly in the vehicles. These support the users proactively and constantly learn to increasingly better satisfy the customer requirements individually.
- At least 50% of the business processes of the automotive enterprises are automated.

The main hypotheses are explained and substantiated in the following.

5.4.1 Mobility Services Instead of Vehicle Ownership

The urbanisation keeps progressing massively. Today more than half of the world's population already live in so-called megacities of more than ten million inhabitants. This trend continues, and in 2030 more than two-thirds of the world population will be living in such cities, with likely some giant cities with more than fifty million inhabitants, as well as a large number of new megacities which as per today one would not necessarily expect to grow to this level [Gri15], [Dob15]. As many of us may have personally experienced during long hours in traffic jams, today's traffic situation is no longer acceptable at peak times in the big cities, such as Sao Paulo, Beijing, Mexico City, Paris and Moscow for instance. Environmental impacts and also time losses due to the viscous "flow" of traffic are forcing new technologies and mobility concepts. Also the other consumer trends towards sustainability and acceptance of sharing rather than ownership especially appeal to younger customers, at least in the cities,.

That is why in 2030 the private car traffic in the cities will be reduced significantly, and the street scene will be characterised by autonomous-driving cars used in open mobility concepts. The charging infrastructure will be established, along with greatly improved ranges, especially due to improved battery technologies [Thi15]. In the cities, the image of car ownership will be similarly stigmatised as is smoking in "mature countries". The situation in the "new emerging countries", such as Indonesia, Namibia, Colombia, China and India, and also inevitably in rural areas, such as the "vastness" of the USA for example, will be different. In these regions there continues to be a buying market, whereby the vehicles are bought on the one hand with smaller, highly efficient engines and appropriate equipment in the low, highly competitive price segment, and on the other hand the luxury segment with corresponding prestige character will be in demand.

The mobility services are easily accessible via internet platforms. In addition to vehicle mobility, other services are also offered, for example, "traffic-type-overarching" or intermodal bookings, incorporating the complementary transport systems, such as ferries or suburban railways, which are necessary, or just as an alternative, to get to the destination. Optionally, further offers such as ticket systems for the theatre, hotel bookings or Smart Home functions can also be requested, for example, for the timely start of house heating.

The use of the mobility service platform is voice-controlled. The solution recognises multiple users and supports them individually during the booking process, taking into account habits and preferences from previous bookings. Different service levels for mobility are offered, ranging from luxury class cars from specific manufacturers with chauffeur for individual journeys, to shared services with several passengers, using available ride capacities, regardless of any specific model, with some halts along the way. The price structure will be based on the respective service level, with subscription and flat rate models being available as well. Today's providers such as Uber or Lyft already offer creative commercial concepts and also very flexible options to require a service, and it can be assumed that this trend leads to further new business models.

Similar to today's air traffic, there will be a higher-level traffic control, also comparable to the digital shadow in the subject area of Industry 4.0 (see Sect. 5.4.9). In this higher-level traffic management system, all vehicles are recognised precisley by their current location and usage, not only autonomously driving cars, yet private vehicles with a driver can arrange to get recognised as well and thus receive information on traffic management and also become a potential provider of rideshare opportunities. The traffic management system will handle mobility service requests according to the required service level, and also organise the routing of traffic with minimum congestion. This information can also be used in new service models.

The use of mobility services via the platform and with the supported technologies is very convenient for the customer, and especially in the cities, more and more users will accept this offer instead of car ownership. Enterprises with company car fleets will establish alternative models and offer their employees the necessary mobility services instead of investments in individual vehicles. Due to the high levels of utilisation, prices for the services will be on the decrease and lead to growing acceptance, in the sense of the chicken/egg effect. From the author's point of view, brand-independent platform operators will gain an edge over the manufacturers as operators, since the breadth of the offering, similar to today's overnight accommodation platforms, is rather accepted by customers in the "neutral" offer. Furthermore, customers request mobility services independent of the brand, i.e. brand loyalty becomes of minor relevance.

5.4.2 Connected Services

Another important business area for the automotive industry is Connected Services. Today approx. 30% of all new vehicles are equipped with this technology. In 2030, however, all new vehicles will be "connected". The market for these services is establishing itself and is expected to grow from approx. €100 billion in 2020 to a volume of €500 billion by 2025. Over the service life of an interconnected vehicle, it is possible to generate additional revenues in this field of approx. €5000 [Gis15]. These figures impressively underline the importance of this topic. Connected Services not only are an essential prerequisite for the efficient design and comfortable use of mobility services but also provide the basis for new business models. Figure 5.7 examplarily shows an overview on possible services and functions [Wee15].

In addition to the mobility solutions presented in the centre of the display, solutions are also presented which are of interest to drivers, such as navigation, weather and office services, as well as services to interact with urban infrastructure, such as toll and road conditions, and also services in after-sales, such as remote diagnostics and maintenance. Moreover, examples from the infrastructure, such as car parks, as well as from the insurance sector for personalised policies taking into account the individual driving behaviour, are shown. This diversity of topics underscores the

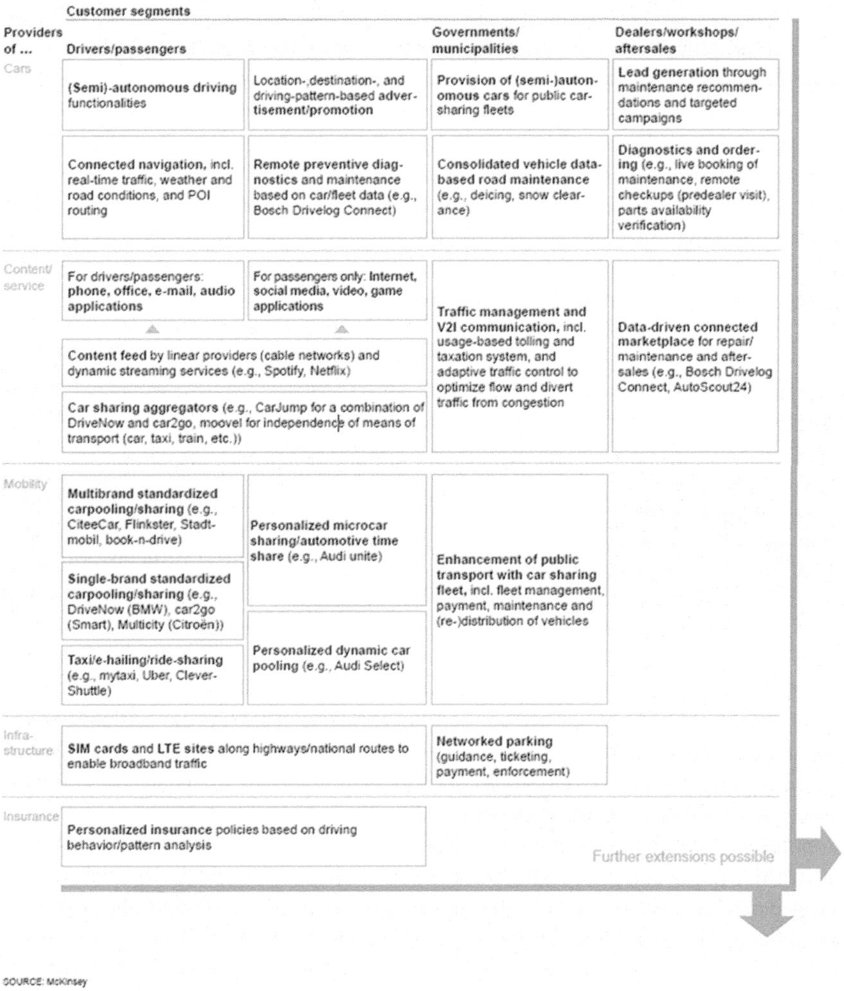

5.4 Vision Digitised Automotive Industry

Providers of...	Customer segments			
	Drivers/passengers		Governments/municipalities	Dealers/workshops/aftersales
Cars	(Semi)-autonomous driving functionalities	Location-, destination-, and driving-pattern-based advertisement/promotion	Provision of (semi-)autonomous cars for public car-sharing fleets	Lead generation through maintenance recommendations and targeted campaigns
	Connected navigation, incl. real-time traffic, weather and road conditions, and POI routing	Remote preventive diagnostics and maintenance based on car/fleet data (e.g., Bosch Drivelog Connect)	Consolidated vehicle data-based road maintenance (e.g., deicing, snow clearance)	Diagnostics and ordering (e.g., live booking of maintenance, remote checkups (predealer visit), parts availability verification)
Content/service	For drivers/passengers: phone, office, e-mail, audio applications	For passengers only: Internet, social media, video, game applications	Traffic management and V2I communication, incl. usage-based tolling and taxation system, and adaptive traffic control to optimize flow and divert traffic from congestion	Data-driven connected marketplace for repair/maintenance and aftersales (e.g., Bosch Drivelog Connect, AutoScout24)
	Content feed by linear providers (cable networks) and dynamic streaming services (e.g., Spotify, Netflix)			
	Car sharing aggregators (e.g., CarJump for a combination of DriveNow and car2go, moovel for independence of means of transport (car, taxi, train, etc.))			
Mobility	Multibrand standardized carpooling/sharing (e.g., CiteeCar, Flinkster, Stadtmobil, book-n-drive)	Personalized microcar sharing/automotive time share (e.g., Audi unite)	Enhancement of public transport with car sharing fleet, incl. fleet management, payment, maintenance and (re-)distribution of vehicles	
	Single-brand standardized carpooling/sharing (e.g., DriveNow (BMW), car2go (Smart), Multicity (Citroën))			
	Taxi/e-hailing/ride-sharing (e.g., mytaxi, Uber, Clever-Shuttle)	Personalized dynamic car pooling (e.g., Audi Select)		
Infrastructure	SIM cards and LTE sites along highways/national routes to enable broadband traffic		Networked parking (guidance, ticketing, payment, enforcement)	
Insurance	Personalized insurance policies based on driving behavior/pattern analysis			

SOURCE: McKinsey

Fig. 5.7 Services and functions in Connected Services [Wee15]

fact that Connected Services are desired by almost all customers and seen as an important differentiating feature in choosing a car.

In addition to the examples shown, further possibilities arise through innovative use of the immense amount of data available via the Connected Services. A large volume of data is generated by the signal transmitters integrated in the mechatronic components for process monitoring and control. The growing number of sensors and cameras required to run systems for driving assistance up to autonomous driving provide a high volume of data as well. This information can be connected and evaluated with data from the vehicle environment and the Connected Services. This results in valuable new insights, such as patterns of driving behaviour, details for the generation of high-precision maps, or precise information on the background of wear and failure. This information can, in turn, form part of new business models.

These examples underpin the potential of Connected Services to open up new business areas. It is therefore to be assumed that by 2030 new providers are established in this sphere and in the business use of "Big Data". In addition to the purely vehicle-oriented offers, new business opportunities will be created through the integration of Smarter Cities and accompanying services such as in insurance and marketing. This naturally raises the question for the manufacturers of which positioning in this new business environment can be taken. During the Automotive News World Congress 2016 in Detroit, Audi America's Chairman, Scott Keogh, summed it up by saying that through the competition and the decision on the leadership in Connected Services and its relating platforms it is being decided of who will in the future be the one who operates the "profitable bar in the hotel".

5.4.3 Autonomous Driving

Similar to Connected Services is the topic of Autonomous driving as an objective and vision with every manufacturer and also with new competitors and suppliers on the agenda of many initiatives and projects. After first research projects and pilot activities were published in the 1990s, the topic received special attention when Google announced such vehicle project in 2010 and subsequently introduced a fully autonomously-driving car in 2014 without pedals and steering wheel [Cac15]. Meanwhile the Google test fleet operating in California around Mountain View, where the Google HQ is located, is a common sight in the everyday street scenery. Other manufacturers are also in road tests with pilot vehicles; for example, Delphi and Audi with tests in the USA, Volvo in Oslo, Daimler-Benz in Germany, also with autonomously-driving trucks, and first driverless taxis in Singapore [Gro16]. Autonomous driving therefore constantly develops into production maturity. On the way to this, more and more automation and assistance functions are offered in today's production vehicles. For this technology field, standardised classification steps are established in gradation of the evolutionary transition of the control function from humans to automats. In addition to definitions of the German Bundesanstalt für Straßenwesen (Federal Institute for Highways), the classification of the SAE (Society of Automotive Engineers) is established, as shown in Fig. 5.8 [SAE14].

In the first three stages, starting from "technology-free driving" through assistance up to partial automation, the supremacy over the vehicle while driving is in all situations in the hands of the driver, such as acceleration and steering, or the observation of the traffic situation, with all the necessary responses. Only general functions, such as keeping lane and distance, are taken over automatically by a system in level 2. In Level 3 to Level 5, three stages of automation are classified, from highly automated through to complete automation in which the vehicles without steering wheel also move without a driver. The development towards autonomous driving takes place with the established manufacturers in evolutionary steps. In series vehicles, especially in the upper segments, systems of Levels 2 and 3 and thus the introduction to automation are established. Market demand and customer acceptance are

5.4 Vision Digitised Automotive Industry

Level	Name	Narrative Definition	Execution of Steering and Acceleration/ Deceleration	Monitoring of Driving Environment	Fallback Performance of Dynamic Driving Task	System Capability (Driving Modes)
Human driver monitors the driving environment						
0	No Automation	the full-time performance by the *human driver* of all aspects of the *dynamic driving task*, even when enhanced by warning or intervention systems	Human driver	Human driver	Human driver	n/a
1	Driver Assistance	the *driving mode*-specific execution by a driver assistance system of either steering or acceleration/deceleration using information about the driving environment and with the expectation that the *human driver* perform all remaining aspects of the *dynamic driving task*	Human driver and system	Human driver	Human driver	Some driving modes
2	Partial Automation	the *driving mode*-specific execution by one or more driver assistance systems of both steering and acceleration/deceleration using information about the driving environment and with the expectation that the *human driver* perform all remaining aspects of the *dynamic driving task*	System	Human driver	Human driver	Some driving modes
Automated driving system ("*system*") monitors the driving environment						
3	Conditional Automation	the *driving mode*-specific performance by an *automated driving system* of all aspects of the dynamic driving task with the expectation that the *human driver* will respond appropriately to a request to intervene	System	System	Human driver	Some driving modes
4	High Automation	the *driving mode*-specific performance by an automated driving system of all aspects of the *dynamic driving task*, even if a *human driver* does not respond appropriately to a *request to intervene*	System	System	System	Some driving modes
5	Full Automation	the full-time performance by an *automated driving* system of all aspects of the *dynamic driving task* under all roadway and environmental conditions that can be managed by a *human driver*	System	System	System	All driving modes

Fig. 5.8 Definition of automation levels for system-assisted driving [SAE14]

		Technological Maturity	Potential for Innovation
Technology within Vehicle	Sensory Monitoring of Driving Behavior	High	Low
	Sensory Monitoring of Environment	Medium	High
	Control Unit and Vehicle Software	Medium	High
	Human-Machine Interaction	Medium	High
	Actuator Technology	High	Medium
	Trip Data Storage	High	Low
Technology outside of Vehicle	Location and Map Material	Medium	High
	Car2X Communication	Medium	High
	Telecommunication Infrastructure	Medium	Low

Fig. 5.9 Maturity and innovation potential of automated driving solutions [Cac15]

constantly increasing in line with the offered comfort and attractiveness of prices. The required technology components for achieving further automation levels are available, but these are in different degrees of maturity, as shown in the overview in Fig. 5.9 [Cac15].

The figure shows the technologies, both inside and outside the vehicles which are required to establish solutions for automated driving. The respective technical maturity as well as the innovation potential of the components are assessed. The degree of maturity is, for example, in sensor technology, in the communication capabilities and in the map material in the upper middle range or "near-series state", and the innovation potential is mostly high. From the experts' point of view, there is thus no

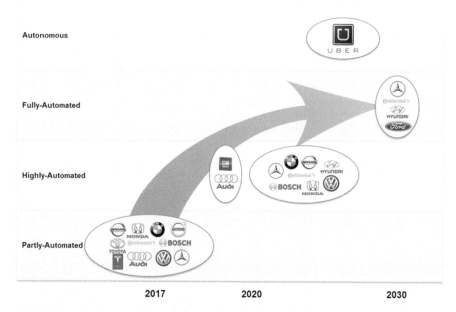

Fig. 5.10 Roadmaps for automated driving [following Cac15]

obstacle to a market introduction of highly automated driving in 2020. Consequently, many manufacturers and also suppliers are envisaging fully-automated vehicles by 2030, evaluated in the IAO study, and shown in Fig. 5.10.

The graph shows the announcements made by manufacturers and suppliers as to when production maturity of vehicles in different degrees of automation will be achieved. It can be assumed that due to the competitive pressure and the large number of announcements, the target of fully automated vehicles will certainly be reached by 2030, despite minor setbacks. In addition to the evolutionary steps of the established manufacturers, new suppliers, such as Uber and Tesla, and certainly further suppliers as well in the future, such as Baidu and Alibaba for instance, will "attack" directly at the level of Autonomous Driving. By 2030, the currently widely discussed legal framework will have been created precisely with regard to liability issues. The operational performance of the communication infrastructure will also be established to ensure the necessary vehicle/vehicle and vehicle/manufacturer backend dialogues, as well as communication from the vehicle to any new business model partners.

This development towards autonomous driving is clearly marked and developing continuously, so that this technology can not be qualified as disruptive in character as there is no surprise moment. Autonomous Driving is however bringing outright new possibilities to redesign the mobility services offerings. In this field, new business models are to be expected, which will put established offers into question entirely. Some ideas are these:

- Autonomously-driving shared taxis
- Urban district mobility … residential areas intelligently sharing autonomous vehicle fleet

5.4 Vision Digitised Automotive Industry

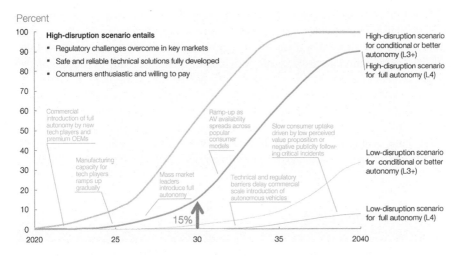

Fig. 5.11 Share of autonomous vehicles in the new car business over time [Kaa16]

- Company platforms … mobility instead of company cars
- Logistics cloud … e.g. autonomous delivery services for spare parts
- Supply chain platforms … autonomous delivery of production lines
- Medical service by health insurers … autonomous vehicles for elderly persons to transport to the doctor
- Autonomous delivery services of purchases
- …

The common characteristic of these ideas is that services and autonomous mobility are bundled into new, comfortable offers and then become easily accessible via internet-based platforms. Operators of the platforms are often IT-affine new competitors, which then due to the scalability of the Internet solution and the fast market maturity of the new service ideas quickly take significant market shares from often established providers. The manufacturers must (re-)position themselves here as well and prepare for this type of competition if they are to succeed.

With the constantly improving general conditions and also the growing possibilities on the basis of this technology to open up new business opportunities, the share of autonomous vehicles in the new car business is continuously increasing. Figure 5.11 shows a forecast.

The picture depicts the development of the share of autonomous cars in the new car business by the year 2040. Four different scenarios are analysed, which each differ in the vehemence with which the new technologies and business models are absorbed by the market. Furthermore, it is assumed that today's barriers in the areas of law as well as remaining technological hurdles are resolved. Within this framework, a market share of 15% is predicted in the middle scenario by 2030, between 4% in the conservative scenario and 50% in the progressive scenario. Beyond the year 2030, another significant share of business is seen, partly with exponential growth.

It can be assumed that most autonomous vehicles are not used by private individuals, but are used within new mobility and service models, especially in large cities. From this, the author deduces his hypothesis that in 2030 at least 30% of the vehicles drive autonomously in the Megacities. This is taking into account the higher levels of utilisation covering more than 50% of the mobility needs in the cities, so that through the autonomy and the new mobility services offered by platforms a "disruptive" upheaval in the automotive industry is imminent. As a result of the increased use of vehicles in mobility services, the new car market will at least be significantly reduced in the megacities, certainly for the time being still being compensated by the growing demand in the emerging countries.

5.4.4 Electromobility

The currently increasing traffic density, namely in the big cities, leads to massive environmental pollution, for example by carbon dioxide, particles and noise, too. This development is not acceptable anymore and many efforts are underway to change this situation. In addition to offering mobility services, the automotive industry also aims to reduce consumption and pollution levels by using lighter materials, design measures and the use of smaller engines. These technological methods are not to be detailed within the framework of this book. However, electromobility is to be discussed here as another area for solutions to environmental issues, since this technology also plays an important role in the topic of digitisation. In addition to the environmental aspects, a driver of electric mobility also is the fact that the supply of fossil fuels is limited.

At the beginning of the car introduction, electric drives were the preferred drive technology around the turn of the century, and around the year 1900 most cars were driven electrically in the USA. In competition, however, the combustion engine quickly began to dominate. The arguments were the higher range and a fast growing petrol station infrastructure and thus the very parameters, which are the focus of the current discussion about the spread and acceptance of the electric drive. Furthermore, in this discussion the environmental balance of electric drives in comparison with combustion engines is now questioned frequently. In this context, not only the energy consumption for the driving but the entire life cycle of the vehicle type should be compared, for example, with respect to production, service and also scrapping. Furthermore, it is decisive for the assessment of the environmental balance of how the required electricity is generated. Regenerative processes, which increasingly contribute to the energy mix in Germany, are particularly environmentally friendly. With the proportion of this "green power generation" currently being achieved in Germany and observing the entire life cycle, electric mobility achieves significantly better environmental values than internal combustion engines [VDI15].

Combustion engines using natural gas or biofuels are also considered clean. Advantages of the electric motor are in its high energy effectiveness with an efficiency of more than 90%, while combustion engines perform at around 35% [VDI15].

In addition, the assembly of electric vehicles is considerably simpler, since many parts such as gearbox, fuel system and exhaust system are eliminated, making them considerably easier to manufacture and also significantly cheaper in terms of maintenance. The special feature of the "engine brake", i.e. he fact that the vehicle can be decelerated without using brakes, adds to the high efficieny and also to low brake wear; the brake energy can be fed into the battery, while in the conventional brake system it just fizzles out as heat.

The electrical energy required for the operation is currently provided by batteries as energy storage for almost all vehicles. The energy density of today's batteries is considerably lower than that of diesel or petrol. This low density leads to very large-volume, heavy batteries and to a relatively short range of electric vehicles. For this reason, many research projects around the world aim to improve battery technology, thereby improving the range at acceptable weight, volume and price. An alternative to the battery, and a parallel development focus, is to produce the energy on board the vehicles via hydrogen fuel cells. The maturity of this technology clearly lags behind that of the battery.

As a result of the massive research efforts undertaken by all manufacturers as well as the clear consensus among policy makers and customers that electricity is the preferred climate-friendly and sustainable solution, it is to be expected that by 2030 a considerable proportion of the new cars will have a pure electric drive. Available serial cars of some manufacturers underpin this hypothesis. Especially Tesla Motors, currently benchmark in the range of its standard vehicles, with its investment in charging infrastructure and the construction of a gigantic battery factory together with Panasonic in Nevada, is the driver of the development [Lan16]. Many manufacturers offer hybrid drives, i.e. the combination of combustion and electric drive, to bridge the range problem. From the author's point of view, these will by 2030, with the solution of energy supply, be less in demand and play just a minor role.

In Germany, according to the ideas of the Federal Government, one million electric vehicles will be on German roads by 2020, with the aim of reducing the carbon dioxide emissions by 2020 by 40% below the 1990 level. China plans to launch five million electric vehicles by 2020 and initiated support programmes and infrastructure projects. Also the UK, France, Norway and Japan have ambitious targets, and purely battery-driven electric vehicles gain market significance. There is extensive literature and numerous studies on electromobility, which also provide prognoses for development [VDI15], [Kor12], [Wie13], [Bra16]. For example, Fig. 5.12 shows the situation of the most important markets.

The graph shows the sales figures in major automotive markets in a year-on-year comparison of 2014/2015 as sums for purely electrically-powered vehicles versus hybrid models. By far the leading market for electromobility by volume and also growth is China, thus underpinning the strategy of technology leadership. The considerable state support shows effective there, with the goal of achieving a strategic position through technology leadership. The US is behind certainly also as a result of the currently low petrol prices, while in Europe the UK, France and Norway are characterised by high growth rates and Germany by growth at a low level. The

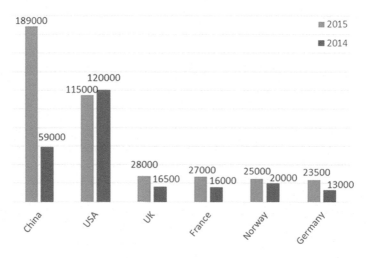

Fig. 5.12 Electric vehicle sales situation (incl. hybrid vehicles) on a year-to-year basis 2014/2015 [Bra16]

above studies assume that the market will grow heavily, and the share of new car business is predicted to be at 30% by 2030.

The trend is clearly confirmed this way and very interesting and important from the point of view of digitisation, as the considerably simplified structure of electric vehicles also results in new business models in production, assembly and logistics, as well as in sales and after-sales. The software ratio in the vehicles is also increasing massivelly, for example for electronic controllers. This results in options in the vehicle configuration, for example, to change parameters for the driving behaviour via the setting of software parameters. In 2030, necessary updates of the software will, in analogy to the update of the software of smartphones, only be "over the air" without much effort, flexible and demand-orientated, without a visit to the garage which nowadays is often still required. Because of this affinity for digitisation subjects, manufacturers should align their plans for electromobility and digitisation, so as to exploit synergies and strengthen their competitiveness. This is particularly important as electromobility allows the possibility of revolutionary approaches and thus lowers market entry barriers. The opportunity will be taken by new competitors, namely from China.

5.4.5 Centralised Embedded IT Architecture

Electromobility is another significant step of the car towards the "Driving IT System". This statement is supported by a survey of the so-called "embedded IT" which is established in today's serial vehicles. The development of the software is shown in Fig. 5.13.

5.4 Vision Digitised Automotive Industry

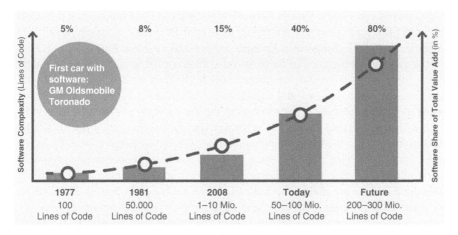

Fig. 5.13 Volume of software in cars [Sei15]

The engine and chassis control system is becoming ever more complex. Also, more and more automation, assistance and comfort functions such as, for example, automatic wiper control, ABS, or air conditioning systems with individual climate zones are added to the vehicles. This leads, as the figure shows, over time to an increase in the volume of software used in vehicles, measured in "Lines of Code". At the same time, the complexity and the software costs as a portion of the vehicle costs increased exponentially over time from 5% share in 1978 to 40% in 2015. This trend will continue, and in the coming years the software ratio will grow up to 80% for hybrid and electric vehicles. To illustrate this immense amount of software: In 2015, the Ford GT included more Lines of Code than the Boeing Jet Airliner [Ede15]. Similarly, the number of control units installed in the vehicles, quasi individual computer systems, for controlling vehicle system units further develops. Now more than a hundred control units are used in today's premium vehicles.

It is therefore clear that IT will become the formative technology in the automotive industry, and the majority of manufacturers in 2030 will be IT companies with "attached vehicle production".

This inevitable development is meanwhile recognised, but is often not taken into account adequately. Adequate measures and transformation initiatives are missing in many manufacturers. Today, ECUs (Electronic Control Units) are mostly seen as part of traditional development projects and are typically assigned island-wise as "blackbox" to provide required functions as a supply part, for example, an integrated hardware/software solution for climate control. Independently of one another, further locally optimised "control device islands" are created, for example, to enable comfortable automatic driver-specific seat settings or to control engine optimisation or driving behaviour. In this way, today's embedded IT architecture, which has evolved over decades, is based on the vehicle structure and often oriented on the organisational structure of the vehicle manufacturers. In addition the control units are supplied by different suppliers. The required wiring between the control units and connected actuators, signal transmitters and control elements has resulted in kilometres of cable trees. Various network topologies and communication protocols are used. The manufacturers are trying as a total integrator to ensure a fail-safe interaction of all IT components through comprehensive integration tests.

The grown architectures of the embedded IT have led to a massive and barely manageable complexity, and it requires a very high effort for development, integration test, operation, updates and adjustments. Extensions within this architecture, not to speak of fundamental adjustments in the life cycle of a vehicle, are only possible with considerable effort. The error rate and failure frequency due to IT and electronic errors is high. The implementation of new functions, especially in the area of "drive by wire", i.e. the transmission of mechanical control to electronics, such as braking or steering, are delayed by the performance deficiencies of today's embedded IT. This unsatisfactory overall situation has been recognised, and attempts are made to achieve harmonisation and simplification, through standardisation for instance. In particular, the Automotive Open System Architecture (AUTOSAR) initiative is to be mentioned here, in which numerous manufacturers have formed a consortium [Sch12].

From the author's point of view, even these standardisation efforts will not suffice for the evolutionary improvements of the established IT architecture to continue successfully into the year 2030, since their sustainability for future requirements, e.g. for electric drive and autonomous driving, no longer exists, at least not in economic terms. Therefore, it can be assumed that by then completely new "disruptive" architectures have established themselves [Wei16]. At the very least, new market participants will follow this path and thereby overtake established manufacturers who for (too) long stick with the "status quo", thus adapting themselves more sustainably to the future requirements of the automotive industry. Characteristics of the architecture which will be established in 2030 are under technological and functional aspects:

- Integration of embedded IT to manufacturer backend via open platform
- Hardware: central computer, plus a black box
- Middleware for the integration of sensor system and applications
- Software/Applications: groups / domains of Functions ... multitier architecture
- Embedded broadband communication: Ethernet as the main technology
- Car-to-Backend, Car-to-Car and Car-to-Infrastructure communication near real-time
- High proportion of equipment/functional elements can be connected by software
- IT updates as well as on-demand control of functions "over the air"
- Openness to the connection of neighbouring IT – for example smartphones, mobility service providers, charging stations, transport infrastructure
- Vehicle concepts built on a central computer

Based on future customer expectations and new business models, the implementation of these aspects happens by taking into account the tried and tested concepts and experiences of traditional IT, such as decoupling, virtualisation, separation of data storage and logic. Established Open Source solutions are part of the technology

5.4 Vision Digitised Automotive Industry

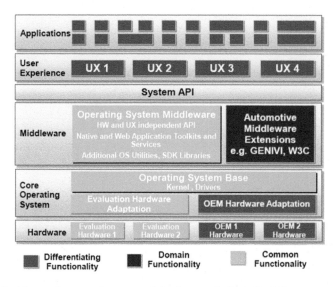

Fig. 5.14 Architectural concept for an open infotainment platform [Bre15]

platform in order to exploit the innovative power of the interested "crowd" and also to achieve cost advantages. The implementation also takes account of the concerns for a smooth interaction of mechanics, electronics and software. The new approach also focuses on harmonisation and standardisation, thus reducing technological diversity. The simple overall approach and the low complexity that can be achieved in this way lead to comparatively high operational reliability, low costs for development and operation, and easy expandability in order to be able to map future requirements as well as new business models.

In light of these objectives, initiatives by providers and manufacturers are underway to design appropriate architectures. Figure 5.14 shows an example of a Linux-based concept for an open infotainment platform.

Based on the operating system, the middleware level provides standardised services as a central element, which integrate the applications via the standardised interface (API) – also in different user-specific representations, called User Experiences (UX). Similarly, the various control units, actuators and signal transmitters are flexibly integrated via the hardware layer. Within the shown architecture, the latest technologies such as WiFi, Bluetooth, multimedia and so-called "location based services" (LBSs) are also supported [Bre15].

Many manufacturers pick up these new integrated architectural concepts and try to drive the evolutionary further developments of existing approaches in this direction, e.g [Ain13, Ber16]. However, from the author's point of view, a new approach is more promising. New vehicle series, but especially electric vehicles, should be used to form a new architectural approach in a "green field". An integral approach by Audi AG is shown in Fig. 5.15.

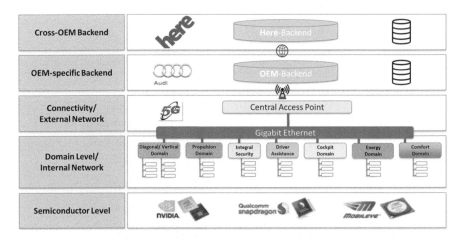

Fig. 5.15 Structure of an End-2-End architecture [Hud16]

The IT architecture in the vehicle, shown as an overview in the centre of the figure, is based on a domain structure, for example for services in the area of the cockpit, the drive and also for driver assistance. The in-vehicle communication is based on fast Ethernet technology. Different control units and signal generators are combined in the semiconductor level and are integrated via a flexible, non-displayed, adapter layer. Outside the vehicle the integration into the so-called backend systems of the manufacturer (OEM) takes place via fast 5G communication standards. In the OEM backend are applications that work together with the Connected Services of the vehicle, for example, for a concurrent diagnosis to prophylactically detected maintenance requirements, or to provide the driver with individual operating instructions. The OEM backend holds protected data, for example on the vehicles, on customers or also on motion profiles. Also provided is an integration from the vehicle or manufacturer environment, and the integration of the map service "here" is shown exemplarily at this point.

These integration options are also used, for example, to integrate vehicles into mobility services or to display information to the driver from the surroundings, for instance to parking facilities or even restaurant offers, in the vehicle display. On the basis of the architecture, also the communication from vehicle to vehicle and from vehicle to infrastructure is implemented and basis of new services.

The future software and application structure requires high computer and communication performance. For this reason, a changed IT hardware structure will be established in the vehicles by 2030. Instead of numerous distributed control units, often with a single function reference, a few central high-performance computers, secured by backup systems, will provide the required performance. This consolidation will simplify networking. Software development and testing are carried out to a high degree automatically.

These topics, briefly presented as a perspective, shall not be discussed in more detail here. Instead, please refer to the relevant specialist literature, for example a roadmap study with further sources [All15]. For the digitisation focus of this book, it is important that manufacturers take measures to establish a future-proof architecture for embedded IT. A new approach seems to offer time and cost advantages over an evolutionary method. With the architecture, a comprehensive integration and communication capability must be achieved in order to enable, for example, mobility services, autonomous driving and the integration into new services and thus new business.

5.4.6 Prototype-Free Process-Based Development

The development of new vehicles from the idea through to serial production is generally described in the so-called Product Development Processes in accordance with the German VDA Standard 4.3 [VDA11]. The development area is divided into components and vehicle subsystems, for example, there are organisational units for the chassis, powertrain and interior. The embedded IT is often a separate organisational unit.

Vehicle development has been taking place for years with the help of IT solutions. For example, CAD applications with different versions are used for manufacturing drawings incl. calculations and simulations. They often work with workflow solutions as well as Bill of Material Systems. The basis for the development is the mechanical vehicle structure and a component-oriented approach. The embedded IT is designed along with the respective components, often as an isolated solution and without an integrated IT overall architecture. This approach does not adequately reflect the increasing penetration of vehicles with electronics and software. The result are current development times of several years for new vehicle projects, massive efforts for adjustments, changes and inadequate use of the IT possibilities.

As a way out, functional views are becoming increasingly established in addition to the traditional component-oriented approach. This means, assemblies of a vehicle are viewed integrally under functional aspects, thus making mutual influences and effects transparent, for example, in the determination of consumption or also of driving behaviour and for checking the interaction of mechatronic components. This concept of a function-orientated approach will be established universally by the year 2030. The current IT applications are being expanded to support this methodology, and new vehicles are automatically developed using cognitive engineering solutions based on functional structures. Augmented Reality solutions quickly show the first presentations of the new vehicle design and, in connection with virtual environments, the construction of prototypes and test runs largely become obsolete [Run16]. The development period of new vehicles will drastically shorten, according to the author's opinion to below 1 year. Customer-specific adaptations of existing vehicles can be entered daily.

In addition to tool support and automation, a further prerequisite for this is to give up in development projects the method of the so-called "waterfall" procedure and to rather achieve rapid results with agile methods close to the requirements with interdisciplinary teams and creative approaches.

This also applies to the safe elimination of vehicle problems, which are identified in the so-called fault elimination process. To this end, in 2030 company data from various sources, information from after-sales and service, as well as publicly accessible data, are automatically and continuously assessed by means of extensive "analysis agents" in order to detect problems and faults of serial vehicles in the market at an early stage. These findings are continually considered in the development, where they are prompting rectifications. These are bundled with further adaptations directly into production and are transmitted, if necessary, as preventative measures in after-sales and service as "Update over the Air" into the respective vehicles.

This close fault elimination process, the functional orientation and also the comprehensive use of Virtual Reality technologies towards prototypless development as well as the automation of the development processes by "IT-machines" are focussed on in the industry by some manufacturers, and corresponding transformation projects are being implemented.

A major challenge remains that all these initiatives are still based on the traditional component-oriented vehicle approach. The embedded IT is managed as a separate module. From the author's point of view, this approach does not adequately take account of the massive entry of IT into the vehicles. More and more parts are determined by IT components. Future overarching IT architectures with common service elements, which are then used in different components, must then be maintained by means of cross-references or, again, special IT lists. This leads to considerable complexity and a great deal of effort, especially with respect to Bill of Matarial explosion for material requirements planning and also with regard to the updating in after-sales for maintenance.

In the view of the author a disruptive approach can therefore be foreseen here as well. It is to be assumed that in 2030, at least with electric vehicles having 80% IT-related value added share, the central processor of the embedded IT will be the basis of vehicle architectures and thus of the Billl of Materials Systems. Similar to the development of smartphones, the processor will become the dominant component of these vehicles and will replace the mechanical basic structure. To support this hypothesis, Fig. 5.16 shows the hardware architecture of a smartphone.

The picture shows that the central processor dominates the sub-assemblies of the device. Both sensor and camera systems are controlled as well as communication, energy management and the user display. All these functions already play an important role in vehicles today. The typical driving topics in close connection with assistance systems and the control of the electric drives come in here as well. There are only few mechanical parts left that are not integrated with IT.

In this respect, in the view of the author, the trend towards processor-oriented development is also to be foreseen as the basis for BOM processing, material requirements planning and further follow-up processes. It is up to the established manufacturers to decide to embrace this disruptive approach of a processor-oriented development. The development of new electric vehicles could provide a suitable entry. It can be assumed that new manufacturers are pursuing this path and thus create further pressure to drive the transformation of the development process.

5.4 Vision Digitised Automotive Industry

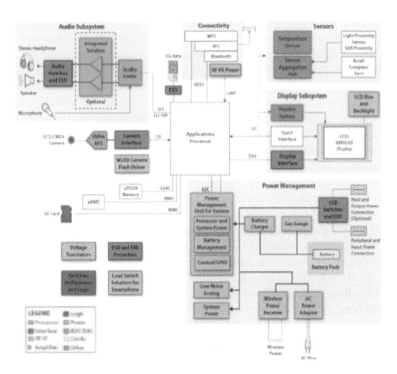

Fig. 5.16 Hardware architecture of a smartphone [Ram13]

	Marketing	Sales	After Sales /Service	Connected Services	Mobility Services	Intermodal Mobility	Third-Party-Business
Importer		Focus Manufacturer					
Dealer					New Sales Sectors		
Vehicle Customer		Focus Dealer					
Mobility Customer							

Fig. 5.17 Transformation of automobile sales (Source: Author)

5.4.7 Internet Based Multichannel Distribution

Another sector which has to undergo comprehensive transformation in the context of digitisation is the sales department. The pressure for change is driven by new technologies around the Web 2.0, smartphones and social media, through changed customer expectations and also changed buying behaviour. This situation was explained in detail in Chap. 2 and also in Sect. 5.2. Furthermore, the transformation is absolutely necessary because the established structure no longer meets the new market requirements and the changing supply. This situation is illustrated in Fig. 5.17.

In the overview, the long-established main processes of sales are set against the acting parties. In today's distribution structure, shown in the upper left part of the matrix, the manufacturers do not have a direct end customer link. Marketing, sales, after-sales and service support are provided to importers by the manufacturers in the countries and markets. For example, when launching a new vehicle, manufacturers arrange advertising and television spots and provide marketing material for sales support. The manufacturers also provide vehicle configurators and call centers for customer support. These services go to the dealers in the countries through the importers, who often make their own market-specific additions. Dealers, in turn, are often organised in commercial chains and are thus large enough to carry out their own sales campaigns or to use own local application solutions, for example so-called dealer management systems (DMS). The dealers are currently the end-user of the sales channel of the manufacturers and in direct contact with the vehicle customer. In addition to vehicle sales, the focus of the dealer is in particular the after-sales service business, including the lucrative spare parts business.

In the future, this sales structure will change drastically as new offers and business segments will arise alongside the vehicle and service business. If manufacturers wish to offer Connected Cervices, mobility services, the provision of intermodal transport, i.e. the use of different means of transport during a trip, as well as third-party business, such as the booking of hotel accommodation with the help of the Connected Services of the vehicle, appropriate sales channels have to be established. These must then be specifically tailored to the customer. In the future, these will be both vehicle buyers and new customers for whom the new offerings are interesting, irrespective of the specific vehicle. This situation with the new distribution fields is also shown in Fig. 5.17. The larger area of the new distribution sectors in the simplified presentation illustrates the considerable need for change in sales.

A very heterogeneous ownership structure is complicating the necessary adjustments to the current sales structure. Different ownership structures are established among manufacturers, often in mixed forms, which differ even further depending on the markets. In many cases, manufacturers own importers in strategically important markets and also hold selected trading companies in some markets, often also established as "flagship stores" in large cities. In some cases, exclusive contractual relationships with independent trading partners are established, or the distribution is exclusively via free companies. This heterogeneous ownership structure and the indirect ways of customer access and mixed customer management as well are complicating the necessary transformation. The following unambiguous trends for the orientation towards the year 2030 have to be considered in the transformation of sales:

- The sale of vehicles and spare parts via the Internet will rise massively.
- Big Data and Analytics through the evaluation of different data pools (for example, social media, manufacturer data, dealer information) create detailed findings about potential new customers so that these can be "developed" until purchase with personalised offers in the sense of "next best action".
- Virtual Reality will be of great importance in the sales process, also used for example in the configuration test and for virtual test drives.

5.4 Vision Digitised Automotive Industry

- The number of dealers will decrease significantly. Dealers are successful if they are part of larger organisations or commercial chains and are directly involved in the sale of offers from the new business segments.
- Customers expect to get the same experience on all sales channels, and that the level of information is synchronised, for example, to interests or inquiries.
- The Internet-based sales are carried out to a large extent via multi-brand platforms. Complementary products such as financing, insurance, and service are also handled via these platforms.
- The financing organisations and Banks are, just as the main business part, an integral part of sales platforms.
- Vehicle ownership will shrink in favour of mobility services, particularly in large cities.
- The traditional vehicle business with the focus on ownership takes place in "emerging markets".
- Mobility services are provided to a high proportion by autonomously driving vehicles. Electric vehicles make up a high percentage of fleets.
- Mobility services are managed to a significant extent via non-brand platforms. Also complementary services such as intermodal transport, booking of tours or chauffeur service as a premium option for special events.
- Brands play a minor role in mobility services. Main consideration will be a competitive price/performance ratio at the desired service level.
- The loyalty to mobility platforms is achieved through commercial models and customer programmes.
- Manufacturers will develop new business fields. For example, integrated Connected Services can be used to refer vehicle users at a "handling fee" as customers to hotels or restaurants.
- Manufacturers sell insights gained from vehicle and motion data to insurance companies or component manufacturers for instance.

These trends are clearly visible, they will intensify in the future and thus drive the transformation of sales. It is absolutely essential for the established manufacturers to swiftly and successfully push the necessary restructuring since the new business segments will in the future make a considerable proportion of the sales and profit share of the industry, and these topics are particularly competitive and being addressed by new entrants as well. This situation is underlined by some current studies and books, e.g. [Bra15], [Lau16], [Köh14]. The expected business development and division is impressively illustrated in Fig. 5.18, taken from one of these studies.

It is predicted that the revenue share from pure vehicle sales, including related Connected Services, will decrease continuously and will be 50% by 2035. By contrast, business with mobility services and new sales spheres, such as data trading and intermediary channels, will grow to as much as 50%. The study also examines profit shifts. In this respect the prognosis is that the margins of today's profit drivers, after-sales including spare parts and financial services, will decrease considerably, and profits shift to the new business segments. These statements confirm the substantial pressure on manufacturers to position themselves clearly in the changing business environment and to organise sales in an adequate manner.

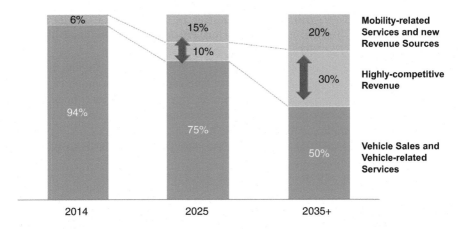

Fig. 5.18 Development of global industry turnover passenger cars by 2035 [Bra15]

From the author's point of view, the battle with competitors like Google, Apple, Uber, Alibaba and Baidu will take place with the focus on mobility services in sales. Undoubtedly, the products and offerings must be developed and made available by the manufacturers. It is however important to establish these at an early stage close to the customer in order to secure "air sovereignty". Other fields are also contested. For example, established online retailers such as Ebay and Amazon compete for market shares in the spare parts business and so-called "Fintechs", quasi platform banks such as Auxmonex, Kreditch or also LendingClub, provide fierce competition to the financial services of the manufacturers. In addition, it can be assumed that other start-ups and also established companies from other sectors, such as electricity suppliers, retailers or railways, try to enter the profitable new area of the automotive market.

In order to stay successful, the established manufacturers must transform the sales structure into a "multichannel" structure with many customer accesses, thus attracting customers directly via online channels as well as through traditional indirect approaches. The customer should at all times through a variety of different ways be addressed in a uniform approach, based on a consistent information base, and be in dialogue with the manufacturer. To achieve this, manufacturers are required to dissolve established organisational islands and to create a new integrated structure, involving the importers and dealers, the after-sales organisation, the in-house financial organisation and new web-based services. In the context of digitisation it is necessary to create a cross-departmental integration of processes, applications and data. In addition, a culture must be created for the provision of new products, such as integrated mobility services or the trading of data, to cooperate with strategic partners and incubators in order to quickly develop competitive offerings for the traditional car business. Dealers and importers must also be actively involved in this transformation. While first beacon projects have started at the manufacturers, dealers and importers have so far shown little readiness to transform [Lau16].

5.4.8 Digitised Automotive Banks

As already outlined in the description of the sales trends, the automotive banks are also undergoing a complete transformation. The wave of digitisation has already captured the general banking sector as one of the first industries years ago. The majority of all banking transactions are processed online today, and as a result, the traditional counter areas for private customers have been reduced to a minimum. A similarly complex restructuring is now underway at the banks of the automobile manufacturers. On average, 75% of all car sales are financed in Germany by a loan, of which 46% are accounted for by automotive banks [AKA16]. The overall objective of the automotive banks is the promotion of sales through attractive financing solutions. In addition, they finance investments by "their" manufacturers as well as by importers and selected dealers. The used-car business is becoming increasingly important. More and more private banking and complementary mobility services are being offered as well. In many cases is, in the author's opinion, namely in these new fields of business, the overall strategy and orientation remains unclear in comparison with the manufacturer's activities.

The traditional distribution path of the automotive banks leads to the customer via the car dealer. Interested car buyers in this day and age get information online about equipment features of the vehicle, financing and service options as well as prices. The purchase is carried out on the basis of the information obtained beforehand, in the majority of cases still after a test drive at the dealer. The product information and prices offered on the web by the atomotive banks are often of a generic nature and not specifically tailored to the current customer situation. This continues in many ways, and financing packages available at the car purchase are completed in tandem, partly with added insurance or service packages.

Direct, online-based interactions between the automotive banks and end users are currently taking place in complementary business segments, for example in the savings sector, but without any connection to the other business relationship between the customer and the manufacturer. There is no integrated customer profile about his private vehicle ownership, including vehicles of different brands, service history and other business connections. An extended view on the customer in the run-up to the car purchase on the basis of comprehensive network information is not available either at the dealer or at the automotive bank for the potential creation of a customer-specific leasing offer. This situation opens large improvement potentials for process adaptations and digitisation measures. Figure 5.19 gives a good overview of how the customer-oriented processes will be digitally supported in the future.

The entire leasing process, shown on the left in the picture, from the inquiry, detailed coordination and provision of financing documents, as well as the conclusion of the contract, can be processed online via smartphones, if necessary in dialogue with a consultant via live chat. The same technology in the interface with the same "look and feel" is used for the processing of payments, questions for clarification and also the re-marketing of used vehicles. In the background, a central support organisation is available around the clock, across all processes. A parallel continuous analytical system at any time provides an integrated view on the customer for

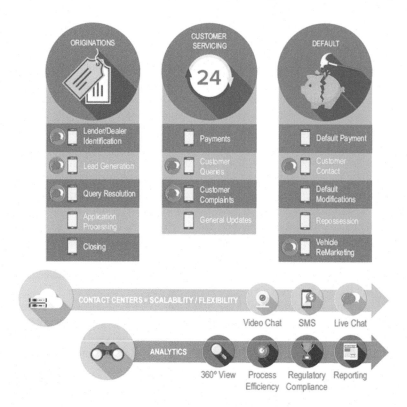

Fig. 5.19 Digitisation of the customer processes in automotive banks [Pan14]

the support as well as for the sales and service organisation. With this solution, technical problems on the ground but also customer dissatisfaction can be detected at an early stage in order to then implement preventive measures to improve the situation. This full digitisation of the customer processes significantly increases efficiency and quality in process handling. The same also applies to the digitisation of the other business processes of automotive banks. More details on the topic of process automation can be found in Sect. 5.4.10.

5.4.9 Flexible Production Structures/Open Networks/Industry 4.0

Sales reliably works together with Production in the programme planning and the feasibility testing and fine trimming of customer orders. This interface will also be more closely integrated in the future due to the increasing individualisation of the vehicles and will be supported by Big Data methods. This is just one facet of the digitisation of production. This is driven by the Initiative "Industrie 4.0" and the accompanying Internet of Things approaches already with the implementation of

5.4 Vision Digitised Automotive Industry

projects (see Sects. 4.2, 4.3 and 4.4). It needs to be investigated whether these initiatives cover the entire field or the whole digitisation potential. This assessment is based on the following hypotheses as to how the production may look like in 2030, and which parameters are relevant. The following constitutes the author's appraisal:

- In the high-wage countries, assembly-line production will be replaced by highly flexible production islands in which robots and workers produce customer-specifically configured vehicles in close cooperation.
- Assembly line production will mostly be in "emerging markets". High-volumes of vehicles are produced there for the local mass market and also for the fleets of mobility services providers for use in the established markets.
- Production make-to-stock is massively declining, and production to individual customer requirements predominates, especially in the premium segment.
- Industry 4.0 technologies are installed widely and enable the cost-effective production of customer-specific vehicles with lot size 1.
- Suppliers are already closely involved in the planning, so that the customer-specific production is continued in the supply chain.
- "Digital shadows" of production and logistics chains are established for proactive control. By using cognitive solutions, analyses and preliminary planning are possible on this basis, which create the required reaction times in the supply chain.
- A company-wide production monitoring is established for control and surveillance. From the overall view, manufacturers can zoom from individual plants down to individual tools or components. Self-learning application systems provide the control team with action plans which are implemented step by step, and automatically.
- New suppliers will be established, especially for electric vehicles, which do not have their own production capacities and rather solely use contract manufacturers as supply partners.
- There is an surplus of production capacity which makes it easy for new suppliers to find manufacturing partners.
- In addition to robots, 3D printers determine production. These produce just in sequence also hand-in-hand with workers in the production area.
- A high proportion (50 + %) of spare parts are produced on demand locally in the large markets based on 3D printing and flexible production cells.
- In Production, Augmented Reality technology is used to a high degree. Simulation of plant layouts, feasibility testing, management of workers and also training are typical fields of application.
- Recycling of end-of-life vehicles will increase massively. Ecologically degradable substances, such as those based on hemp, are used.
- In the further development of the RFID technology, the vehicles get a "lifetime chip" which is already used in production where the vehicle autonomously moves between the manufacturing cells. This chip remains in the vehicle and is then in active dialogue in after-sales and also for individualisation in Connected Services.
- Materials will contain communicable elements which can then be incorporated into the control system, for preventive maintenance works for instance. Foglets (see Sect. 4.9) are tested in pilot areas.
- Application software as well as robots are programmed in human language.

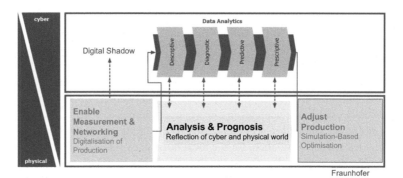

Fig. 5.20 Measures for production adjustment [Bau16]

Many research projects and pilot projects are under the heading "digital shadow" tackling one of the main topics of the digitisation of manufacturing [Dom16], [Sch16]. This shadow means a virtual image of the production with all the data relevant to the value creation of the lines and also of adjacent systems. From this data, a time-based virtual model is created, often also called a digital twin, which is used for evaluations and analyses. The approach is illustrated in Fig. 5.20.

In the lower part of the picture the physical level of production is shown. This provides the basic data for the generation of the digital shadow and of the virtual image plus continuously further current data, for example, regarding utilization, machine and logistics data. This information is analysed in the virtual model, and operating instructions are given, but also preventive measures for production improvement are derived, supported by simulations. In the development of the measures, the systems take into account production-relevant requirements, for example with respect to the inventory level or the set-up sequence. This integrated control and optimisation based on a digital shadow is still in the research stage, and there are first pilot testings. By 2030 though, this approach will be widely established.

Today's initiatives and projects among the established manufacturers have often been launched in the context of evolutionary improvements, for example for the continuous raising of the degree of automation, and are being addressed even more focused within the framework of the Industry 4.0 initiative. However, this is often a question of just individual projects, and there is a lack of a paramount objective, which is proposed in the implementation recommendation for Industrie 4.0 [BMB13]. The aim is in the horizontal integration to address the entire value chain across all company organisations and beyond company borders. The vertical integration seeks to facilitate a dialogue between the company's management level and individual machines. The challenge thereby is to establish digital consistency and continuous integration of the engineering throughout the entire product life cycle and the manufacturing system.

In this holistic sense, the digitisation progress in the production of the established manufacturers today is too slow from the author's point of view, and there is a risk that new entrants will right from the outset begin at a significantly higher level

5.4 Vision Digitised Automotive Industry

in their new production facilities. For example, Tesla Motors' factory in which it will produce its Model 3 is designed to operate without direct worker participation, since people along the line slow down the speed to that of humans [Pri16].

In order to achieve this goal, "disruptive" concepts are required which are addressing a challenging overall objective. Established manufacturers find this particularly hard to do. One reason is that the highly automated lines are trimmed to efficiency and are closely integrated in their processes, for example in the communication structure and in the logistics networks. Far-reaching changes to these structures often mean operating risks which one wishes to avoid, so one is content with smaller optimisations.

In addition to the technological challenges addressed in Chap. 6, another important reason for the slow progress is the heterogeneous organisation and the different objectives of the project participants. It starts with the IT. In many cases, the plant IT is part of the production organisation and is responsible for IT solutions around the plant and plant-oriented objectives, whereas the central IT as a staff function often reports to the financial sector and has more general objectives. The application solutions of the central IT are often developed separately and use different technologies than the plant solutions. The communication technology in the areas is also often different. It is an exciting point in the projects when IT people and the production managers are to communicate project objectives using different technical terms. This example illustrates the need for an overarching governance model that brings together all parties involved in a project with a common understanding and the same prioritisation. It is important at this point to recognise that the evolutionary approach of the established manufacturers is too short-sighted from the author's point of view and is also too slow to adequately counter the attack by the industry newcomers and to secure competitiveness on a sustained basis. Rather, holistic concepts are required, which are implemented together in a straightforward manner.

5.4.10 Automated Business Processes

Today the production processes are being pushed forward with the focus on the Industry 4.0 initiative for digitisation. In addition, there is high potential in all other company divisions as well to increase efficiency and quality through digitisation and thus to secure competitiveness. Overall, according to the author, the following trends are characteristic for this topic, so that the vision for the year 2030 is as follows:

- In the administrative areas, for example in the financial, personnel and administrative areas, 80% of the business processes are executed automatically without manual intervention.
- In business areas where higher interaction and coordination are required, such as in engineering, quality assurance, order management or marketing, many business processes are automatically processed, but the degree of automation will be less, around 50%.

- The "process automata" are based on cognitive software technologies. These are integrated into the respective existing systems of the process areas and further develop by "self-learning".
- The blockchain principle is also part of the company-internal platforms and enables a simplification or shortening of business processes by the elimination of testing tasks.
- The machines are integrated into platforms. The service call is performed via voice control based on intelligent mobile devices.
- In line with the concepts of the platform economy of today's Web 2.0 economy, internal company-wide central business platforms will be established for handling administrative tasks. These service platforms support all brands of larger manufacturers. Brand-specific organisations that provide these services today, will cease.
- The platforms will provide open interfaces so that the platform's functionality can be easily expanded by third-party components.
- Today's rollout projects of large software programmes will completely disappear and be replaced by "roll-in" projects, i.e. the extension of functions by third party components can easily be performed.
- Employee workplaces are equipped with assistance systems. These proactively support the user while, for example, prioritising of work and preparing information search, as well as handling many tasks automatically, such as the booking of trips or the invitation and coordination of all parties involved in a meeting.
- The number of employees in the administrative and in the indirect sectors will decrease significantly.

The high degree of automation of the business processes and the establishment of business-internal business platforms, which are used as a "shared service" by all corporate brands, represent a significant increase in efficiency and quality compared to today's handling of business processes with a relatively high proportion of manual work. Similarly, assistant systems at the workplace greatly enhance productivity in day-to-day work. The first approaches on this path are now being evaluated in initiatives and tested in first pilots with small work contents. The approach of in-house platforms for the finance sector is shown as an example in Fig. 5.21.

The core functionality of the in-house platform includes, for example, access management (single sign-on), security, data integration as well as analysis, business process control- and automation functions. Through an integration layer, existing financial applications are flexibly linked, so that existing know-how and investments made are protected. The App and mobility layer provides access to the financial solutions via mobile devices such as smartphones or cameras for the recording of gesture control. This allows users to work with the existing "legacy system environment" through modern access methods. New functionalities are integrated from the outset as Apps via this level. The overall concept is based on a "roll-in" approach. The companies can connect themselves to the platform via defined interfaces and gradually start using the platform functionalities and step by step reduce or deactivate the legacy systems. The platform concept thus enables an evolutionary transformation from a heterogeneous legacy system environment to a global, harmonised

5.4 Vision Digitised Automotive Industry

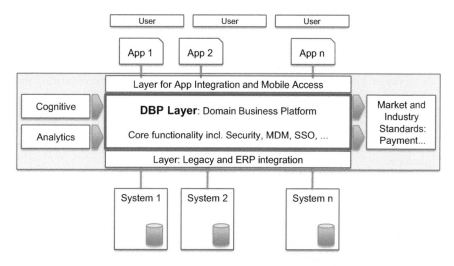

Fig. 5.21 Concept of an internal company financial platform (Source: Author)

solution platform, which can be expanded flexibly and appealingly to users through Apps. This platform also enables step-by-step automation of business processes.

The concept explained examplarily can be applied to many company sectors such as purchasing, human recources and quality management. There are many companies that face the challenge of entering new solutions from an existing, historically grown application environment. A proposal for the implementation of the approach is given in Chap. 6. In addition to the in-house solutions, cross-industry platforms will also be established as a marketplace between manufacturers and suppliers, for example in the handling of logistics services. The goal here is to increase the efficiency of the transactions and the process transparency.

5.4.11 Cloud-Based IT Services

In the implementation of business platforms as well as in all other digitisation activities, efficient IT services are a basic prerequisite. The provision of these services will in the future be based on fundamentally different structures and on the basis of new methods and concepts compared to the current situation. Typical for the IT in 2030 are, according to the author, the following aspects:

- The manufacturers obtain 80% of the required computer and storage capacity from Cloud environments "out of the socket" on the basis of flexibly agreed upon service levels.
- Cloud environments are run by special providers from mega computing centres. These are flexibly linked to the manufacturer's IT systems in so-called Hybrid Cloud concepts.

- Unlimited data storage will be available almost free of charge on the Internet. Data storage is carried out via software layers (software defined storage).
- Mobile devices such as smartphones, wearables and smartscreens are integrated into furniture, clothing and machines. Computers controlled via voice command or gestures have replaced desktop systems entirely. The performance of the devices equals that of today's supercomputers.
- IT projects with a long runtime as well as comprehensive rollout projects are a matter of the past. Instead, App-like solution components are created within days, and roll-in concepts are implemented on the basis of integration platforms.
- Agile project methods replace the "waterfall method" [Zwe16].
- Programming or software development is through voice control based on microservices architectures.
- Applications for the evaluation of the massive data stocks with recommendations up to the automatic implementation of reactions will make up a high proportion of the software offerings.
- The creation and operation of platforms for both in-house services and marketplaces will account for a large proportion of IT services.
- The monitoring of IT infrastructures is carried out by means of so-called agent systems which are active in the application systems. These detect emerging problems early and automatically take corrective action.
- One of the essential IT cost drivers are the necessary security systems for the defense of cyber attacks. Quantum computers are first used primarily in the field of security, for example in cryptographic procedures for company protection.
- Public internet is available free of charge with sufficient bandwidth. Speeds of over 500 megabits per second are established.
- A new Internet structure based on IPv6 addressing will be established. A distinction is made in this regard between a private and a commercial area, which facilitates different services.
- The "Internet of Everything", a comprehensive linkage of all kinds of things, is established [Tho15].

Figure 5.22 shows in summary the challenges facing IT to establish a solid platform or basis for the transformation and digitisation.

The IT is represented in the figure as platforms, evolutionarily beginning from a centrally organised structure in stage 1, over distributed client/server architectures currently prevalent in stage 2, up to the future "3rd Platform" which must be able to process "Millions of Apps, Billions of Users, Trillions of Things" in the companies. This can, as discussed in the trends, only be achieved with very flexible hybrid architectures which can then scale in terms of both the computing power and the storage capacity by means of Cloud environments. Big Data technologies with deep analytical abilities are to be employed, based on which, for example, solutions from the field of artificial intelligence can be used for automation and the support of users. The mobile workplace with the always-on mentality is supported by the IT environments as well as new forms of collaboration based on "social business" solutions.

5.5 General Electric – An Example of Sustainable Digitisation

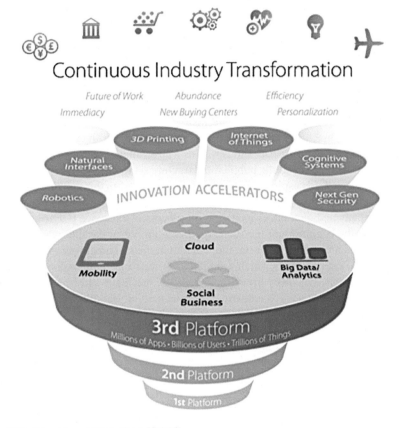

Fig. 5.22 Cloud-based IT platform [Sch15]

A highly efficient, agile IT is the basis of digitisation initiatives, which are often based on the relevant technologies already described in Chap. 4, such as 3D Printing, Robotics or even Cognitive Solutions as an innovation driver. This is not about one-time projects, but the key is in the continuous transformation both in the company and in the entire value-added network, taking into account the transformation among the customers and the suppliers. This holistic approach is explained in the following by means of a case study.

5.5 General Electric – An Example of Sustainable Digitisation

To explain the topic of digitisation exemplarily with well-known "Valley based" companies such as Apple, Google or even startups, would actually be obvious. Uber, with its affinity to the industry targeted in this book, also is a valid candidate however will later be taken up in Chap. 6. As a comprehensive example of a profound and sustainable transformation based on digitisation, rather the company General Electric (GE) is acknowledged here. This company, with an age of almost 130 years, is a founding member of the electrical industry and the only company that has been listed in the Dow Jones Index from its beginnings in 1896. With its origins in the lighting sector based on Edison's light bulb which was patented in 1879, the company developed into a globally active conglomerate with more than 300,000 employees and acting in more than 100 countries. The product portfolio includes plants, machines and components, for example, for the healthcare sector, the basic material and aerospace industries, as well as integrated solutions for industrial plant construction and the public infrastructure sector, as well as related maintenance services. GE's aspiration is technological leadership through innovation. In this context, continuous development and transformation is a fundamental feature of the company and guarantees its long-term successful existence [GE16].

At about the beginning of the second millennium, when the Internet had established itself as a business platform, more and more new competitors were entering this market of capital goods which supplemented their machines and systems with software solutions. For this purpose, data from the machines and from the application environment was used to improve utilisation possibilities and reduce downtime on the basis of innovative algorithms.

This trend was perceived by GE as a risk and at the same time an opportunity and led to the decision that a drastic change in strategy and direction was needed. In order to prevent the risk of "commoditisation", CEO Jeff Immelt clearly explains the GE strategy in 2011: "The goal is to create a global network of connected machines that GE uses to provide customers with significant operational improvements." In 2013, the CEO added: "We know there will be a partnership between the industrial and the Internet world. We just can not accept the fact that the data collected in our world are used by other companies. We must be part of this development." [Buv15].

These findings invigorate the motivation to what can be described as a total reversal in the direction of a company which is characterised by industrial goods towards a data-based service company. In order to speed up and secure this transformation, GE has initiated significant actions and bold steps, which are surely of exemplary relevance; here is a summary [GE16], [Buv15], [GE16], [Pow15].

- Clear definition of a vision underpins clear objectives. A top-down approach to communicate and cascade the vision and goals.
- Establishment of a transformation programme for the development of corporate culture around innovation initiatives and the "FastWorks" programme (tools and methodology) to establish a start-up and entrepreneurship behaviour across all levels.
- Founding of a software house for the development of innovative software solutions in the fields of Big Data, Analytics, Cognitive, Mobile, App and Industrial Internet with a basic funding of $1 billion.
- An investment of $1.5 billion to acquire smaller software companies from the Analytics and Big Data sphere in order to expand the company's internal capabilities.

5.5 General Electric – An Example of Sustainable Digitisation

- Hiring senior executives with appropriate digitisation experience, for example, as head of the new GE software company, as CTO and as sales manager for the new offerings and as the leader of the transformation. Parallel hiring of software architects, programmers and project managers to develop internal skills.
- Opening up of proven software solutions for interested software houses, so that they develop their own solutions to create a GE Ecospace, for example around the IoT platform "Predix". To quote GE Executive Dave Barlett: "We want Predix to become the Android or iOS of the machine world".
- Cooperation with technology companies such as Amazon Web Services, CISCO and Intel, as well as incubators such as LemnonsLab, RockHealth or Breakout Labs, as well as crowdsourcing partners, to discover new ideas and to temporarily bring specific expertise into projects.
- Creation of a venture capital unit (GE Ventures) and establishment of a start-up network in relevant fields of innovation and thus early adoption of ideas and trends.

The overview gives a good idea of the multitude of actions and measures that drive GE's transition towards a data-based service company. It is important to assign clear performance figures to the respective initiatives and to communicate them. For example, Fig. 5.23 shows the goals for the Internet of Things platform from the GE Annual Report 2016.

The picture shows some measurable specifications for moving forward with the digitisation in the field of the Internet of Things. In 2016, at least 200,000 machines should be interconnected, the Ecospace around the Predix platform be expanded to 20,000 developers and 50 partners, and more than 100 Apps available in the industry store. With these clear guidelines, GE underlines the strategy of providing sus-

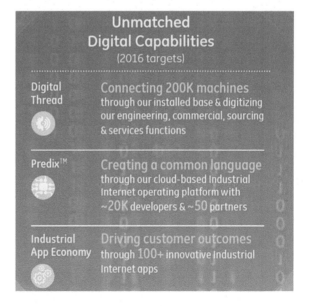

Fig. 5.23 Goals for stabilising the IoT platform at GE [GE16]

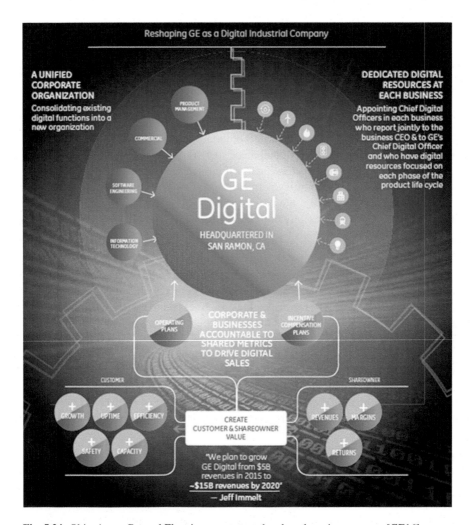

Fig. 5.24 Objectives – General Electric en route to a data-based service company [GE16]

tainable benefits to its customers in the area of IoT services. In the annual report, it is also said that the platform towards the Digital Shadow or Twin will be enhanced in order to create a basis for forward-looking, intelligent solutions (see Sect. 5.4.9).

The strategy for comprehensive digitisation is not limited to the IoT example yet also applies throughout the whole company. Figure 5.24 shows a central picture of the annual report 2015.

The picture shows a comprehensive overview of the strategic initiatives in the transformation programme. Next, it is important to establish digital knowledge in each organisational unit. For this purpose, digitisation leaders with staffing shall be appointed in each unit. At the same time, the currently decentralised organisations are to be consolidated in one unit. The distribution responsibility for digital offers is assigned to all organisations in a matrix responsibility. According to CEO Jeff Immelt, GE aims to triple the direct revenue of the digital division from $5 billion in 2015 to $15 billion by 2020.

All these measures are bound into an integrally structured approach which is driven top-down [Buv15]. Although GE has already made a significant step forward in digitisation, it is important to understand that the transformation is being pushed forward continually, and that the environment must still be monitored very carefully in order to detect emerging disruptive trends, to address them, and to adjust structures and focuses in order to create new business potential.

Annex A2

For a more detailed assessment of the digitisation situation of some established automobile manufacturers, the author conducted a study on the basis of a variety of information sources such as annual reports, investor relationship publications of companies and specialist articles. The analysis was based on the following criteria (as of 08/2016):

- Connected services/autonomous driving
- Mobility services
- Digital processes
- Digitisation incubators
- Customer-focused orientation
- Digital IT
- Digital culture

Manufacturers were graded according to the evaluation criteria and rated in the groups of volume and premium manufacturers in a range from 1 (bad) to 7 (very good). A result of this are the spider diagrams shown in "Figs. 5.4, 5.5, and 5.6: Digitisation depth of vehicle manufacturers: premium and volume manufacturers".

The following table summarizes the individual valuations.

	BMW		Daimler		Audi		Tesla		Volkswagen		Toyota		General Motors		Ford	
Connected services/ autonomous driving	IFTTT ConnectedDrive Intel&MobileEye Cooperation	5	MB me connect Future Truck 2025 F015 Luxury in Motion	5	Audi connect Speed Up! 2025 Audi RS7 piloted driving concept	5	EVE Connect Tesla Summon Tesla Motors App	4	LG Cooperation Together 2025 Car-Net	3	KDDI Cooperation SmartDeviceLink Automotive Grade Linux	4	Cruise Autmation Company Lyft	4	FordSync MyFord SmartDeviceLink	3
Mobility services	DriveNow moovit	5	Car2Go moovel myTaxi	5	Audi select Audi shared fleet Audi on demand	5	AirBnB Cooperation Super-& Destionation Chargers	4	GETT	1	Uber TMF Pilot	2	OnStar Maven CarUnity	4	Sync FordPass Wink Cooperation	2
Digital processes	Smart Logistics Smart Watches Plant Digitalisation NUMBER ONE > NEXT	4	Digital Design Digital Prototyping Smart Factory	4	Digitalisation Production Process Digital Economy Award IoT Platform Speed Up! 2025	5	Connected Production Multi-Task Robots In-House Component Production	5	VR in Production und Sales Smart Factory 3D Printing Together 2025	5	TOAD Big Data Control Tower	2	Digitalised Competitor Benchmarking Digital Test Tracks	3	3D Printing Digital connected production Virtual Testing	4
Digitalisation incubators	BMW iVentures BMW Start-Up Garage BMW Future Lab	3	Daimler Mobility Services Center of Competence Digital Vehicle and Mobility	3	Audi Electronics Venture Audi Urban Future Initiative Audi Business Innovation GmbH	3	Born Digital	3	Digital Lab Group Future Lab Data Lab	5	Smarter Mobility Society Toyota Connected Inc. Toyota Research Institute	4	-	4	Ford Smart Mobility LLC	1
Customer-centric focus	BMW Retail Online BMW Brand Stores Product Genius	3	MB me finance & MB me inspire Best Customer Experience (2020) Mercedes me Store	3	Audi City myAudi	4	Exklusive Online & Direct Sales over-the-air-Updates WeChat Payment (China)	5	Online Booking Testdrive After Sales Digital Reception (UK)	2	iOS App Customer Configurator Direct Acceptance App	2	myOpel	3	FordPass myFord FordHubs	4
Digital IT	Postgraduate Program IT for Engineers	1	Project 100/100	1	-	2	Born Digital	3	200 Mio€ for Digitalisation Projects	5	Advanced IT for Manufacturing (NA) Single Point of Truth iOS App (Toyota Motor Europe)	2	Dissolution HP Outsourcing Consolidation DCs Central Datawarehouse	3	-	1
Digital culture	SVP Digital Strategy Bündelung aller Digitalisierungs-projekte	3	VP Digital Vehicle and Mobility DigitalLife@ Daimler Digital Life Day	3	Chief Digital Officer Audi Speed Up! 2025	2	Born Digital	5	Chief Digital Officer Department Digitalisation Strategies Organisation 4.0	5	-	3	VP Urban Mobility Programs	1	Digital Worker Program	2

References

[Ain13] Ainhauser, C., Bulwahn, L., Hildisch, A., et al.: Autonomous driving needs, ROS BMW Car IT GmbH (2013). https://www.bmwcarit.com/downloads/presentations/AutonomousDrivingNeedsROSScript.pdf. Drawn: 04.09.2016

[AKA16] AKA: Automobilbanken 2016 AKA – Arbeitskreis der Banken und Leasinggesellschaften der Automobilwirtschaft (Working Group of Banks and Leasing Companies of the Automotive Industry), White Paper (2016). http://www.autobanken.de/download/102788/hash/673149728cc7085ef6b5677d62398d00. Drawn: 11.09.2016

[All15] Allmann, C., Broy, M., Conrad, M. et al.: Eingebettete Systeme in der Automobilindustrie Roadmap 2015–2030 (Embedded Systems in the Automotive Industry Roadmap 2015–2030), Gesellschaft für Informatik e.V. (Society for Informatics), SafeTRANS e.V.; Verband der Automobilindustrie e.V. (German Association of the Automotive Industry); (2015). http://www.safetrans-de.org/documents/Automotive_Roadmap_ES.pdf. Drawn: 04.09.2016

[AUDI16] Audi: Audi Business Innovation GmbH, overview: Audi select; Audi shared fleet. https://www.audibusinessinnovation.com/de/service/de_audibusinessinnovation/Audi-mobility innovations.html. Drawn: 07.08.2016

[Bau16] Bauerhansl, T., Krüger, J., Reinhart, G.: WGP-Standpunkt Industrie 4.0, 2016 (WGP Position Industry 4.0, 2016), Wissenschaftliche Gesellschaft für Produktionstechnik (Scientific Society for Production Engineering). http://www.ipa.fraunhofer.de/fileadmin/user_upload/Presse_und_Medien/Pressinformationen/2016/Juni/WGP_Standpunkt_Industrie_4.0.pdf. Drawn: 11.09.2016

[Ber16] Bernard, M., Buckl, C., Doors, V., et al.: Mehr Software (im) Wagen: Informations- und Kommunikationstechnik (IKT) als Motor der Elektromobilität der Zukunft, Abschlussbericht BMWi Verbundvorhabens „eCar-IKT-Systemarchitektur für Elektromobilität" (More software [in the] car: information and communication technology as a driver of the electromobility of the future, final report of the BMWi project "eCar-ICT-Systemarchitektur für Elektromobilität"), Editor: ForTISS GmbH (2016). http://www.fortiss.org/forschung/projekte/mehr_software_im_wagen/. Drawn: 01.09.2016

[BMB13] Umsetzungsempfehlungen für das Industrieprojekt, Industrie 4.0 (Implementation recommendations for the industrial project, industry 4.0; Abschlussbericht Promotorengruppe Kommunikation der Forschungs-Union Wirtschaft – Wissenschaft (Final report promotor group Communication of the research union Economy – Science), (Editor), Frankfurt (2013). https://www.bmbf.de/files/Umsetzungsempfehlungen_Industrie4_0.pdf. Drawn: 18.07.2016

[Bra16] Bratzel, S.: Elektromobilität im internationalen Vergleich. (Electromobility in an international comparison). Bilanz 2015 und Prognose, Center of Automotive Management CAM (2016). https://kommunalwirtschaft.eu/images/presse/pdf/87ee9d8567ba0c8a9bf0f868c40ba56b-Pressemitteilung-Elektro-Jan-2016-v01.pdf. Drawn 01.09.2016

[Bra15] Brand, F., Greven, K.: Systemprofit 2035: Autohersteller müssen den Vertrieb neu erfinden (Car manufacturers have to reinvent sales); Oliver Wyman study (2015). http://www.oliverwyman.de/who-we-are/press-releases/2015/oliver-wyman-studie-zum-automobilvertrieb-der-zukunft-systemprof.html. Drawn: 11.09.2016

[Bre15] Brendon, L.: How open-source collaboration is transforming IVI and the automotive industry, Embedded Computing Design (2015). http://embedded-computing.com/articles/how-open-source-collaboration-is-transforming-ivi-and-the-auto-industry/. Drawn: 04.09.2016

[Buv15] Buvat, J., KVJ, S., Bisht, A.: Going Digital: General Electric and its Digital transformation, Capgemini Consulting (2015). https://www.capgemini-consulting.com/resource-file-access/resource/pdf/ge_case_study_28_5_2015_v4_1.pdf. Drawn: 24.09.2016

[Cac15] Cacilo, A., Schmidt, S., Wittlinger, P. et al.: Hochautomatisiertes fahren auf Autobahnen – Industriepolitische Schlussfolgerungen (Highly Automated Driving on Highways – Industrial Policy Conclusions), Fraunhofer-Institut für Arbeitswirtschaft und Organisation IAO; Study for BMWi (2015). https://www.bmwi.de/BMWi/Redaktion/PDF/H/hochautomatisiertes-fahren-auf-autobahnen,property=pdf,bereich=bmwi2012,sprache=de,rwb=true.pdf. Drawn: 24.08.2016

[Dob15]   Dobbs, R.; Manyika, J.; Woetzel, J.: No ordinary disruption: the four global forces, breaking all the trends, Public Affairs, 2015

[Dom16]   Dombrowski, U., Bauerhansl, T.: Welchen Einfluss wird Industrie 4.0 auf unsere Fabriken und Fabrikplanung haben? (What impact will Industry 4.0 have on our factories and factory planning?) 13. Deutscher Fachkongress Fabrikplanung (German Technical Congress Factory Planning), 20./21.04 2016 Ludwigsburg

[Dud15]   Dudenhöffer, F.: Weltautomarkt 2015: Niedriges Wachstum lässt sinkende Branchengewinne erwarten (World Automotive Market 2015: Low growth anticipates declining sector profits, report GAK (2015). https://www.uni-due.de/~hk0378/publikationen/2015/20150119_GAK.pdf. Drawn: 07.08.2016

[Ede15]   Edelstein, S.: Ford's new GT has more line of code than a Boeing jet airliner, Digital Trends, 21.05.2015. http://www.digitaltrends.com/cars/the-ford-gt-uses-more-lines-of-code-than-a-boeing-787/#/2. Drawn: 01.09.2016

[GE16]   General Electric: GE Fact Sheet (2016). http://www.ge.com/aboutus/fact-sheet. Drawn 24.09.2016

[GE16]   General Electric: Digital Industrial – GE Annual Report (2015). http://www.ge.com/ar2015/assets/pdf/GE_AR15.pdf. Drawn: 24.09.2016

[Gis15]   Gissler, A.: Connected vehicle – succeeding with a disruptive technology, Accenture Strategy (2015). https://www.accenture.com/t20160504T060431__w__/us-en/_acnmedia/Accenture/Conversion-Assets/DotCom/Documents/Global/PDF/Dualpub_21/Accenture-Connected-Vehicle-Transcript.pdf. Drawn: 22.08.2016

[Gri15]   Grimm, M., Tulloch, J.: Allianz Risk Pulse; Leben in der Megastadt: Wie die größten Städte der Welt unsere Zukunft prägen (Living in the Megacity: How the largest cities in the world shape our future), Allianz SE (2015). https://www.allianz.com/v_1448643925000/media/press/document/Allianz_Risk_Pulse_Megacitys_20151130-DE.pdf.   Drawn: 22.08.2016

[Gro16]   Grosch, W.: In Singapur fahren die ersten Taxis ohne Fahrer (In Singapore, the first taxis are driving without a driver), VDI Verlag GmbH; ingenieur.de; 26.08.2016; Risiken, Trends, Herausforderungen. (The automotive industry in [radical] upheaval, chances, risks, trends, challenges. Lecture Automobil Elektronik Kongress, 14./15.6.2016 Ludwigsburg

[Hud16]   Hudi, R.: Die Automobilindustrie im (radikalen) Umbruch Chancen, Risiken, Trends, Herausforderungen, Vortrag Automobil Elektronik Kongress, 14./15.6.2016 Ludwigsburg

[Kaa16]   Kaas, H., Mohr, D., Gao, P. et al.: Automotive revolution – perspective towards 2030, McKinsey & Company (2016). https://www.mckinsey.de/files/automotive_.revolution_perspective_to wards_2030.pdf. Drawn: 07.08.2016

[Kor12]   Korthauer, R., Fischer, H., Funke, C. et al.: Elektromobilität – Eine Positionsbestimmung (Electric mobility – a positioning), editor: ZVEI – Zentralverband Elektrotechnik- und Elektronikindustrie e. V. (Central Association of the Electrical Engineering and Electronics Industries); (2012). http://www.zvei.org/Publikationen/ZVEI_Elektromobilit%C3%A4t_ES_25.10.12.pdf. Drawn: 01.09.2016

[Köh14]   Köhler, T., Wollschlager, D.: Die digitale Transformation des Automobils (The digital transformation of the automobile), automotiveIT. Verlag Media-Manufaktur GmbH, Pattensen (2014)

[Lan16]   Lange, C.: Tesla eröffnet Gigafactory: Batteriefabrik startet Produktion in Nevada (Tesla opens Gigafactory: Battery factory starts production in Nevada), Auto-Service.de; 01.08.2016;http://www.auto-service.de/news/tesla/84202-tesla-gigafactory-batteriefabrikproduktion-nevada.html. Drawn: 01.09.2016

[Lau16]   Lauenstein, C.: Digital customer experience: Wer baut hier die meisten Leuchttürme? (Who builds the most lighthouses here?) Post Capgemini Consulting (2016). https://www.de.capgemini-consulting.com/blog/digital-transformation-blog/2016/digital-customer-experience. Drawn: 11.09.2016

References

[Pan14] Pandey, I., Jagsukh, C.: Digitizing Automotive Financing: The road ahead Cognizant 20–20 Insights, White Paper (2014). https://www.cognizant.com/InsightsWhitepapers/Digitizing-Automotive-Financing-The-Road-Ahead-codex949.pdf. Drawn: 11.09.2016

[Pow15] Power, B.: Building a Software Start-Up Inside GE, Harvard Business Review, 01/2015. https://hbr.org/2015/01/building-a-software-start-up-inside-ge. Drawn: 24.09.2016

[Pri16] Prigg, M.: Tesla's Model 3 production line will be an 'alien dreadnought': Elon Musk reveals humans will be banned as they will slow progress to 'people speed', dailymail, 05.08.2016.   http://www.dailymail.co.uk/sciencetech/article3726179/Tesla-s-Model-3-production-line-alien-dreadnought-Elon-Musk-reveals-humans-banned-slow-progress-people-speed.html. Drawn: 11.09.2016

[Ram13] Ramesh, P.: Signal Processing in Smartphones – 4G Perspective, SlideShare (2013). http://de.slideshare.net/ramesh130/signal-processing-in-smartphones-4g-perspective. Drawn: 07.09.2016

[Run16] Runde, C.: Whitepaper Virtuelle Techniken im Automobilbau, Technologien – Einsatzfelder – Trends (Whitepaper Virtual Techniques in Automotive Engineering, Technologies – Application Fields – Trends), Virtual Dimension Center VDC (2016). http://www.vdc-fellbach.de/files/Whitepaper/2016-VDC-Whitepaper-Virtuelle-Techniken-im-Automobilbau.pdf. Drawn: 07.09.2016

[SAE14] SAE: SAE: Society of Automotive Engineers, Standard J3016: Taxonomy and Definitions for Terms Related to On-Road Motor Vehicle, Automated Driving Systems (2014). http://www.sae.org/misc/pdfs/automated_driving.pdf. Drawn: 24.08.2016

[Sch12] Schmerler, S., Fürst, S., Lupp, S., et al.: AUTOSAR – Shaping the Future of a Global Standard.   http://www.autosar.org/fileadmin/files/papers/AUTOSAR-BB-Spezial-2012.pdf. Drawn: 04.09.2016

[Sch15] Schulte, M.: FutureBusinessWorld 2025 – Wie die Digitalisierung unsere Arbeitswelt verändert (How digitisation changes our world of work), Whitepaper 2015, IDC Central Europe   GmbH.   http://www.triumphader.de/C125712200447418/vwLookupDownloads/B4DAB93CACA6F4F4C1257E2100305F61/$File/IDC_White_Paper_Future_Business World_2025_TA.pdf. Drawn: 21.09.2016

[Sch16] Schürmann, H.: Smartfactory – Auf dem Weg in die Zukunft (Smartfactory – On the way to the future), VDI Nachrichten, April 8, 2016; No. 14. http://www.iqm.de/fileadmin/user_upload/Medien/Zeitungen/VDI_Nachrichten/Downloads/Exklusiv_Zukunft_der_Fertigung.pdf. Drawn: 11.09.2016

[Sei15] Seibert, G.: Wie verändern digitale Plattformen die Automobilwirtschaft How do digital platforms change the automotive industry); Accenture (2015). http://plattform-maerkte.de/wp-content/uploads/2015/10/Gabriel-Seiberth-Accenture.pdf. Drawn: 01.09.2016

[Sta15] Stanley, B., Gymesi, K.: A new relationship – people and cars, IBM study (2015). IBM Institute for Business Value. http://www935.ibm.com/services/us/gbs/thoughtleadership/autoconsumer/. Drawn: 16.08.2016

[Thi15] Thielmann, A., Sauer, A., Wietschel, M.: Produktroadmap Energiespeicher für die Elektromobilität 2030 (Product roadmap energy storage for electromobility 2030); Fraunhofer Institut für System-und Innovationsforschung ISI; Karlsruhe (2015). http://www.isi.fraunhofer.de/isi-wAssets/docs/t/de/ publikationen/PRM-ESEM.pdf. Drawn: 22.08.2016

[Tho15] Thompson, C.: 21 technology tipping points we will reach by 2030, TECHinsider, 12.11.2015.   http://www.techinsider.io/21-technology-tipping-points-we-will-reach-by-2030-2015-11. Drawn: 21.09.2016

[VDA11] VDA: Das gemeinsame Qualitätsmanagement in der Lieferkette (Joint Quality Management in the Supply Chain), Verband der Automobilindustrie eV (2011). http://vda-qmc.de/fileadmin/redakteur/Publikationen/Download/Risikominimierung_in_der_Lieferkette.pdf. Drawn: 07.09.2016

[VDI15] VDI: Elektromobilität – das Auto neu denken, (Electro mobility – rethinking the car), Bundesministerium für Bildung und Forschung (Federal Ministry of Education and Research),

Redaktion VDI Technologiezentrum GmbH (2015). https://www.bmbf.de/pub/elektromobiltaet_das_auto_neu_denken.pdf. Drawn: 30.08.2016

[Wee15] Wee, D., Kässer, M., Bertoncello, M., et al.: Wettlauf um den vernetzten Kunden – Überblick zu den Chanchen aus Fahrzeugvernetzung und Automatisierung (Race for the interconnected customer – overview on the chances from vehicle interlinking and automation); McKinsey & Company (2015). https://www.mckinsey.de/files/mckinsey-connected-customer_deutsch.pdf. Drawn: 22.08.2016

[Wei16] Weiß, G.: Zukünftige Softwarearchitekturen für Fahrzeuge (Future Software Architectures for Vehicles), Fraunhofer-Institut für Eingebettete Systeme und Kommunikationstechnik ESK (Fraunhofer Institute for Embedded Systems and Communication Technology ESK) (2016). http://www.esk.fraunhofer.de/content/dam/esk/dokumente/PDB_adaptives_Bordnetz_dt_web.pdf. Drawn: 01.09.2016

[Wen12] Wenzel, E., Kirig, A., Rausch, C.: Greenomics – Wie der grüne Lifestyle Märkte und Konsumenten verändert. (Greenomics – Like the Green Lifestyle changes markets and consumers), REDLINE Verlag, 2. edition, 2012

[Wie13] Wietschel, M., Plötz, P., Kühn, A., et al.: Markthochlaufszenarien für Elektrofahrzeuge, (Market ramp up scenarios for electric vehicles), Fraunhofer-Institut für System- und Innovationsforschung ISI (Fraunhofer Institute for System and Innovation Research ISI) (2013). http://www.isi.fraunhofer.de/isi-wAssets/docs/e/de/publikationen/Fraunhofer-ISI-Markthochlaufszenarien-Elektrofahrzeuge-Zusammenfassung.pdf. Drawn 01.09.2016

[Win15] Winterhoff, M., Kahner, C., Ulrich, C., et al.: Zukunft der Mobilität 2020 (Future of mobility 2020) Die Automobilindustrie im Umbruch, Studie Arthur D Little (2015). http://www.adlittle.de/uploads/tx_extthoughtleadership/ADL_Zukunft_der_Mobilitaet_2020_Langfassung.pdf. Drawn: 07.08.2016

[Wit15] Wittich, H.: Neuwagenkäufer 2015: Das sind Deutschlands Rentner Marken (New car buyers 2015: These are Germany's retiree brands), auto motor and sport; 28.12.2015. http://www.auto-motor-und-sport.de/news/deutschlands-rentner-marken-neuwagen-2015-10343615.html. Drawn: 07.08.2016

[Zwe16] Zweck, A., Holtmannspötter, D., Brown, M., et al.: Forschungs- und Technologieperspektive 2030 (Research and Technology Perspective 2030), BMBF-Foresight Zyklus II, VDI Technologiezentrum (editor), Commissioned by the Federal Ministry of Education and Research.http://www.vditz.de/fileadmin/media/VDI_Band_101_C1.pdf. Drawn: 21.09.2016

Chapter 6
Roadmap for Sustainable Digitisation

The first chapters of this book have presented drivers, influencing variables and technologies of digitisation. In addition to IT as a digitisation driver, the innovative technologies and solutions relevant to the automotive industry were described, and namely the influence of digital natives as future employees and customers was discussed. Based on this, Chap. 5 examined the changes in customer expectations and purchasing behaviour, analysed the current degree of digitisation in major manufacturers and described in detail a vision of the digitised automotive industry by the year 2030.

Between the achievements so far and the upcoming developments and possibilities to 2030 is a considerable distance, which is to be designed purposefully on the basis of a comprehensive roadmap. In the following, detailed proposals will be developed based on the author's studies and projects over many years.

6.1 Digitisation Roadmap as Part of Company Planning

Digitisation initiatives are a cross-cutting issue in the business model of a company and influence all major business processes. To this extent, a digitisation roadmap covering all fields is not to be handled in isolation however as an integral part of a long-term company-wide strategic planning process. Figure 6.1 structures a suitable approach.

In the first step, the understanding of the market situation and customer requirements form the basis for the following strategic decision of which markets or customers will be addressed with which products and solutions. The business structure and the associated processes must then be aligned in a streamlined manner as efficiently as possible to implement the strategy. Building on this optimised structure, a vision for the direction and goal of digitisation has to be developed. Digitisation fields with procedures and roadmaps are to be defined for the implementation. IT and corporate culture are not part of individual planning steps yet overarching thematic fields whose inclusion is a prerequisite for the successful transformation and implementation of the digitisation roadmap.

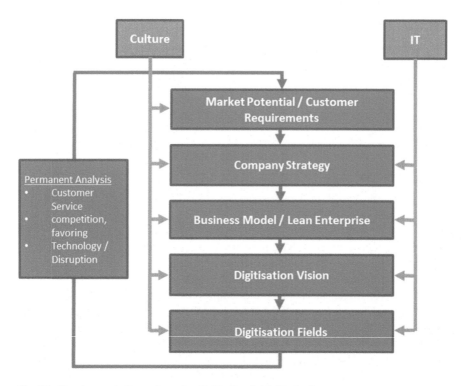

Fig. 6.1 Development of a roadmap for digitisation fields (Author)

However, the planning process is not to be carried out just once with results being implemented as one-off measures. Rather, it should be established as a control loop and normally be cycled every year. Due to the specific dynamics of digitisation, a half-year cycle is recommended at least in the "run-up" phase of implementation. Parallel, it is necessary to analyse continuously which movements are occurring in the markets and in customer behaviour, and also whether there are any disruptive trends on the basis of new technologies or business models. The findings lead to continuous strategy and planning adjustments.

This briefly outlined sequence gives the structural framework for the following explanations. The sections below detail the individual steps, while the subjects of corporate culture and IT are discussed in Chaps. 7 and 8.

6.1.1 Assessment of Market Potential and Customer Requirements

Detailed customer and market understanding is the essential basis for formulating a company strategy. This insight is not new and surely taken into account by the automotive manufacturers today, and there is extensive information on this in the companies so that the discussion of the topic can be kept brief and exemplary here.

6.1 Digitisation Roadmap as Part of Company Planning

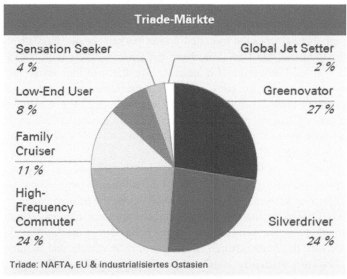

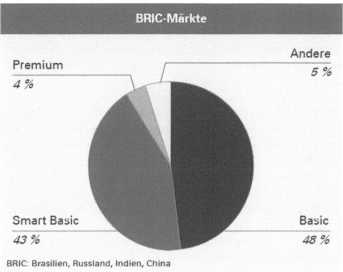

Fig. 6.2 Distribution of mobility consumption by type of mobility in 2020 [Win15]

Section 5.2 already described as basis of the "Auto-Vision 2030" the relevant consumer trends and the changing customer behaviour as well as the buyer types deriving from these. It is important to translate this information into an estimation of the market potential by regions and target customer segments. Traditionally, manufacturers are focusing on the potential vehicle sales, broken down, for instance, according to vehicle types, drive technologies and equipment preferences. However, due to the changes in the automotive value-added areas, further new business potentials are also to be evaluated. These include autonomous driving, mobility services and new digital business segments, which are evolving through the sale of data or through commissions for drivers as customers of restaurants, hotels or trade.

Following the customer segmentation in Chap. 5, Fig. 6.2 illustrate the breakdown of automotive markets by type of mobility in the main markets in 2020.

The graph distinguishes the so-called Triade markets, i.e. the NAFTA region, the EU and the industrialised East Asia, from the BRIC markets with Brazil, Russia, India and China. The established markets are divided into different mobility types with respect to mobility consumption. For the definition of these types and the differentiation in buying behaviour, please see Sect. 5.2. The Greenovators, the High-Frequency-Commuters and the Silverdrivers already cover 75% of the market volume between them. If the Family Cruisers and Low-End Users are included, it even is as high as 95% of the potential.

The segmentation of the BRIC markets is less heterogeneous, and the two segments of Basic and SmartBasic cover 91% of the volume. In these emerging markets, there are still many first-time buyers who usually enter the Basic segment and subsequently develop into the slightly higher Smart Basic segment as a second buyer.

These market segments and the purchasing behaviour of the mobility types must be addressed by the manufacturers with appropriate vehicle and mobility offers. Greenovators are particularly interested in buying smaller vehicles with the latest low-emission drive technologies and electric drives. However, they are also open to alternative mobility services instead of vehicle ownership. High-Frequency Commuters, the daily commuters, focus on security, efficiency and reliability in a purchase decision, while in the BRIC markets in the Basic segments, the price-performance ratio and also the image of the vehicle, the "prestige potential", are crucial factors in the purchase decision. To this end, namely in the emerging countries a distinction through attractive Connected Services is important.

The appraisal of the market potential must, in addition to the traditional studies on possible vehicle sales in the individual segments, also include an analysis of the other business opportunities of the automotive industry. In addition to financial and after-sales services, these include the Connected Services, mobility- and other car-related services as well as the new digital business around the trading of data. The need for this expanded analysis is underpinned by the worldwide development of sales and earnings distribution in the automotive industry by 2030, as estimated by PricewaterhouseCoopers, shown in Fig. 6.3.

In addition to the sales performance (revenue), a forecast of the profit appropriation (profits) in the overall market is also shown. Important for the established manufacturers is the realisation that the sales of vehicles will reduce significantly, by turnover from 49 to 44%, and by profit contribution from 41 to 29% in 2030. After-sales services are also declining slightly in sales to a 13% share; profits will fall even more to 10%. Financial services are seen as relatively stable in business performance, while the insurance business is down by 2–3% in both sales and earnings. On the other hand, mobility services show the strongest growth, with a 10% turnover share in 2030 and a high contribution of 20% to earnings. Digital services in the software sector are also growing significantly, while the digital hardware business is declining. Interestingly, the study assumes that today's manufacturers will be able to address only about 70% of the total market volume of $7.8 trillion in 2030, while more than 45% of the total volume could be targeted by new entrants [Vie16].

In summary, it is important to note that in the future the vehicle-related business will fade into the background, and the importance of mobility services and digital

6.1 Digitisation Roadmap as Part of Company Planning 131

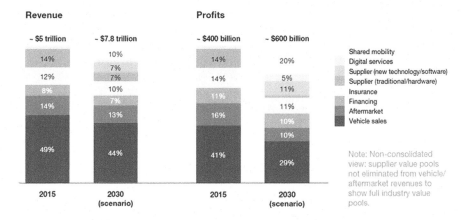

Fig. 6.3 Estimation of the sales and earnings distribution of the automotive business sectors [Vie16]

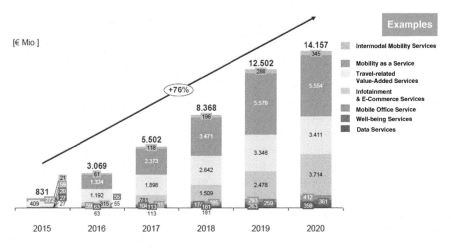

Fig. 6.4 Estimation of the global business performance of complementary services of the automotive industry [Bra14]

services will have a massive impact on growth and earnings. Complementarily, Fig. 6.4 shows a more detailed forecast of the development of the new service business segments.

According to this, the complementary services of the automotive industry will continue to grow in business volume up to €14.1 billion Euro by 2020. The largest proportion of €5.5 billion will be provided by mobility services, followed by additional services such as car park reservations or current map and traffic information with a potential of €3.4 billion. The potential for infotainment and e-commerce solutions is seen on a similar scale, while intermodal transport, mobile office solutions as well as health care and data services by the year 2020 are each less than 3% market share in the complementary services.

These statements also suggest, as numerous other studies show as well, that the traditional business of the automotive industry will decrease in the future and that, in return, there will be significant growth in the areas of mobility services, connected services and third-party business in cooperation with new partners, see [Röm16], [Thi15] for instance. It is therefore imperative for the manufacturers to define a reorientation of the company strategy and to swiftly begin the implementation along with the digital transformation.

6.1.2 Adaptation of the Company Strategy

As explained in detail, in the emerging markets vehicle ownership is still the main buying motif, however, in saturated industrialised markets, with an increasing urbanisation, a clear trend towards the use of mobility services is emerging, particularly among younger customers. In order to meet the mobility requirements, easy-to-demand rideshare opportunities are used in different sharing models via smartphone applications and are the basis for the success of providers such as Uber, Lyft, DriveNow or Car2Go. Furthermore, the cars are developing to driving IP addresses and are bringing more and more powerful applications into the vehicle via their interlinkage. In addition to functions for simplifying vehicle operation, as well as monitoring with proactive maintenance options, applications from third-party providers are increasingly integrated, for example, for the continuous medical monitoring of a driver's diabetes data, to fully voice-controlled office functions for postal and calendar organisation.

The vehicle data as well as information about the driving behaviour are of interest to insurance companies, marketing agencies and also service chains for instance and can thus be marketed. The business environment to be deriving from the described market conditions is summarised in Fig. 6.5.

Around the customer, the possible products and services offered by the mobility business are shown in the inner circle. These consist, apart from the traditional business components of vehicles, maintenance services and spare parts, but also financing, mobility services and digital services with various content and marketing information.

In the creation of these products and services, various companies are involved, represented in the outer ring. In addition to manufacturers with their suppliers, there are trade organisations and increasingly also internet-based platforms with different trading focus, such as mobility services, spare parts and financing. Furthermore, content providers offer contents such as map material, entertainment or traffic management as digital services. After all, insurance companies, marketing agencies, fuel and electricity suppliers as well as commercial companies and restaurants are part of the ecosystem named mobility.

The large number of players in this heterogeneous business environment offers wide-ranging entry-level options and illustrates the risk arising from new players

Fig. 6.5 Ecosystem of the automotive industry with focus on the customer (Source: Author)

and new business models. This is what the manufacturers have to adjust to and therefore should work on answers to the following questions for example:

- Which business fields in the ecosystem mobility should be addressed? Which core business areas will be established? Currently recognisable fields are:
  - Vehicle Development
  - Manufacturing, Trade
  - Mobility Services
  - Connected Services for in-car applications through to separate Appstore
  - Data Trading, e.g. with marketing agencies and insurance companies
  - Third-Party Business, e.g. commission fees, trade with Apps
  - Financing including complementary fields such as insurance
  - After-Sales Services, spare parts business
- When will which share be achieved with which business segment and which sales and earningsshare will be achieved?
- How is the respective product distributed? What is marketed to the customers through dealers, what directly via Internet shops or commercial platforms?

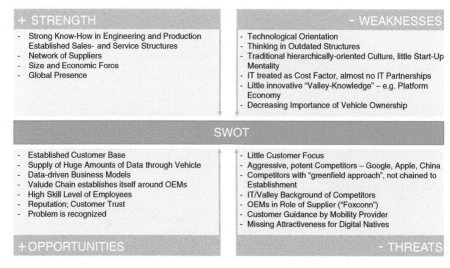

Fig. 6.6 SWOT analysis on the digitisation of the automotive industry (Source: Author)

- How much is the added value in the respective business area? Examples of this, with affinity for digitisation, are:
  - Establishment and operation of a platform for mobility services incl. Fleet of vehicles
  - Establishment and operation of Connected Services and Digital Services
  - Construction and operation of commercial platforms for vehicles, used cars, spare parts, financing
- What are core components of the company in the long term? Will enterprise parts which are not future-proof be discontinued and in return companies with strategically important business content be acquired?
- Collaborate with partners and in which areas? What forms of partnership will be established?
- Does the company structure need to be adapted, and will the new business areas have to be outsourced to new companies?
- What are the main measures for changing the company culture?

The examples show that the established manufacturers need extensive adjustments to the company strategy, and based on this a comprehensive transformation, in order to successfully open up the new business segments. Further to the market and customer information, the decisions on reorientation require a detailed analysis of the company's established own strengths, the current weaknesses, a realistic assessment of the opportunities and the upcoming threats. The well-known SWOT analysis is an important decision-making base for this. SWOT stands for Strengths (S), Weaknesses (W), Opportunities (O) and Threats (T). As an example, Fig. 6.6 summarises an analysis based on the author's assessment.

6.1 Digitisation Roadmap as Part of Company Planning

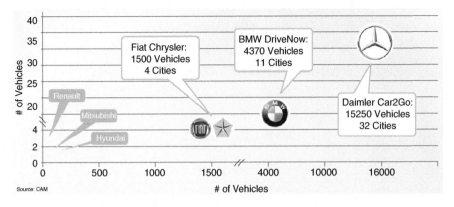

Fig. 6.7 Car-sharing activities of the automotive industry [Bra16]

The strengths of the automotive industry are rooted in its many years of experience in the development and manufacture of vehicles, the established structures and in the vehicle-related innovative power. This also gives rise to opportunities to secure customer base and customer access in the future digital business too. Weaknesses are the fact that the business model established over decades has to be adapted to the new business segments and that the current culture is traditionally hierarchical and not start-up like.

Successful companies find it particularly difficult to assess the new threats, including the taking up of disruptive trends for the targeted business and also the identification of future major competitors [Chr16]. Almost all established manufacturers have developed new strategies that emphasise an orientation towards electric drive, autonomous driving, mobility services, connected services and the digitisation of internal processes (see Sect. 5.2).

With the announced strategies, the focus on customer expectations and future business structure is, however, scarcely recognisable, and usually there are no clear plans as to which portion of sales is to be achieved in which business area by when and how. At least partial targets are quantified. For example, Daimler-Benz is planning to increase its car-sharing business Car2Go from 150 million Euros in 2016 to one billion Euros in 2020 [Eis17]. Daimler is thus leading the car-sharing initiatives of the established manufacturers, summarised in Fig. 6.7.

In 2017, Car2Go was represented in 32 cities with a fleet of more than 15,000 vehicles. BMW and also Fiat are active with their own projects, while Renault, Mitsubishi and Honda are only engaged at a small scale with pilot testing. Volkswagen is starting a catch-up with Greenwheels and its share in the Israeli mobility service provider Gett. In presenting its "Together-Strategy 2025", Volkswagen also announced the founding of a new division named "Moia" for mobility solutions. The focus is on online solutions, car-sharing offers and robot taxis, i.e. fully autonomously driving taxis. In the new field, "turnover in billions"

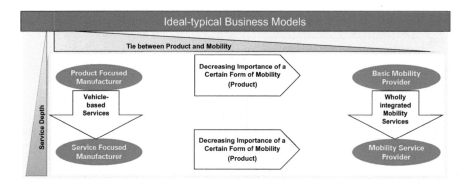

Fig. 6.8 Possible future business models of the automotive industry [Win15]

by 2025 is being strived for. By this time Volkswagen also wants to sell two to three million electric cars, which equates to a 25% share of group sales [Mor16].

The activities of the established manufacturers are relativised though, if put into perspective with the current Uber figures. This company was active in 425 cities in 2017 and supposed to generate sales of $4 billion in 2016, a doubling compared to the preceding year [ECO16a].

Against this background, the question arises as to whether the established manufacturers are concentrated and nimble enough to successfully compete against the newcomers for the new business segments. From the author's point of view, more aggressive strategies should be established, which address the new business segments in the selected markets with the focus on the target customers by means of measurable sales targets in the individual areas. Due to the gradually accelerating demand for vehicles with electric drive, autonomous driving and mobility services, higher sales volumes are recommended in the new business segments within shorter periods than in the announced manufacturer strategies.

6.1.3 Business Model and Lean Enterprise

Sales planning as per the company strategy is based on assumptions regarding market development and the implementation of the strategy. For the implementation then again an appropriate corporate structure is to be established which supports the selected business model. With regard to the business models, the manufacturers have to decide on their basic orientation. The options are illustrated in Fig. 6.8.

It is about whether a more vehicle-oriented or rather a more mobility-service-oriented business model is the goal. To service-oriented manufacturers, both are on a par. In contrast to this, for companies with a clear focus on mobility services, the vehicle is of lesser importance, as vehicles in their own fleet tend to be subordinated or positioned just as an accompanying marketing instrument, as Car2Go or

6.1 Digitisation Roadmap as Part of Company Planning

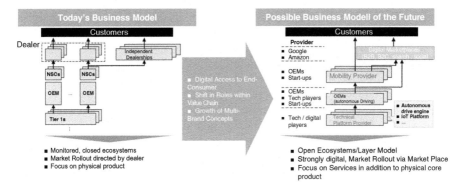

Fig. 6.9 Change of business models in the automobile trade [Kos16]

DriveNow are handling it. Or other, third-party vehicles are used in the sharing anyway, as with Uber or Lyft.

Many manufacturers have announced a "service-focused manufacturer" strategy. The understandable objective is to drive the established vehicle-related business as well as the new business segments. In implementing this strategy, conflicting objectives are to be balanced. Electric drive and traditional drives, as well as mobility services and vehicle sales, for example, are each other's opposite. The friction losses resulting from these target conflicts during the implementation of the strategy should be minimised and circumvented by organisationally separate business areas. Especially for mobility services and digital services the establishment of new business units is recommended.

It is remarkable from the author's point of view that none of the established manufacturers have announced so far a "product-focused manufacturer" strategy. It may however be a sensible option to go this route. Especially the long-term production experience, established supplier networks and also production facilities in competitive regions offer the right conditions to successfully implement this strategy.

However, it should also be borne in mind that the price pressure on the vehicles will increase and therefore low costs will be the decisive success factor. One reason for the pricing pressure are the apparent overcapacities that are further growing with the transition to electric drives which are considerably easier to produce. A second reason is the shift in vehicle sales from individual customers to fleet customers, namely to providers of mobility services that demand corresponding price advantages through the bundling of volumes and acquire the vehicles with little brand loyalty yet rather cost-benefit-oriented. Despite these challenges, it should be possible for established manufacturers to successfully pursue this strategy with a clear focus.

Aside from this fundamental decision to align the business model and the company structure, massive changes in sales are also ahead. In addition to traditional vehicle-related sales, structures for connected services and digital services are to be created, if these are part of the strategic orientation. The adaptation requirements of the sales structure, also explained in the Vision 2030 in Sect. 5.3.7, are illustrated in Fig. 6.9.

The current structure in motor vehicle trade in comparison to the predicted structure from 2025 is shown. In the current model, manufacturers (Original Equipment Manufacturers, OEMs) supply the cars manufactured with their component suppliers to the importers (National Sales Companies, NSCs), which pass on the vehicles to affiliated or free dealers for sale to customers.

In the medium term, it is foreseeable that the importer and dealer structures will change fundamentally, and that both the vehicles and also mobility services will in large part be distributed via Internet platforms or marketplaces which have several manufacturer brands on offer. The integration of the suppliers also takes place via trading platforms. The manufacturers also supply their products directly to end customers via digital channels, and a so-called multichannel distribution scenario develops. In this environment, additional companies such as platform operators, service providers for autonomous vehicles, free mobility providers and also new manufacturers are positioning themselves. Manufacturers, and especially traders and importers, need to reinvent themselves in order to continue to play a significant role in the future value-added system. Since the structure has to be very much oriented on internet-based platforms, digitisation plays an important role for this transformation.

Regardless of the digitisation of other business areas, business modelling, the design of efficient business processes and the subject of Lean Enterprise continue to be of great significance. Before initiating digitisation initiatives, it is urgently recommended to optimise existing structures and processes first rather than just digitising the status quo, even if productivity gains can be achieved by automating existing processes. In fact, the methods for Lean Management presented in many specialist contributions as well as in VDI guidelines should be used in order to align processes as efficiently as possible. The VDI 2870-2 for instance provides a good overview on the designing of integrated production systems. The following states the design principles together with examples of relevant methods [VDI13]:

- Avoiding wastage... Low cost automation, assessment of wastage
- Continuous improvement process... Benchmarking, ideas management, audit
- Standardisation 5S, process standardisation
- Zero-fault culture... 5xWhy, Six Sigma, Poka Yoke, Ishikawa diagram
- Flow principle... Value stream planning, one piece flow, first in-first out
- Pull principle Just in time, just in sequence, Kanban, Milkrun, Supermarket
- Employee orientation and goal-oriented leadership... Hancho, goal management
- Visual management... Andon, Shopfloor management.

For the application of these and also more advanced methods for process optimisation, please refer to the relevant specialist literature, e.g [Dom15]. Especially in the automotive industry, the Toyota production system and further complementary Toyota methods continue to be the benchmark [TUD16]. The corresponding principles and procedures have been adapted by many manufacturers in company-specific systems and been established as a good practice.

6.1 Digitisation Roadmap as Part of Company Planning

Fig. 6.10 Thematic areas of a vision for digitisation [Men16]

The focus of lean management is often in the production area and has resulted in significant improvements there with more efficient processes and procedures. For farther-reaching optimisation, the best practice should be extended to the entire enterprise with the aim of a "Lean Enterprise" [Dom15], [Wie14]. Especially in the indirect areas and in the interfaces between organisations, there is still in-company potential for improvement which can be released using appropriate tools and approaches along entire process and value chains. The initiative and implementation of these overarching projects should be done in agile ways with the focus on quick results. This approach must become part of the corporate culture and is therefore detailed in Chap. 7.

6.1.4 Frame for Digitisation

As explained, all manufacturers have addressed the subject of digitisation at high priority as a field of action and have enclosed the respective objectives in their strategy. Instead of hunting to actionism, driven by the pressure of a hot and forced hype topic, it is necessary to develop a structured approach, which, as part of the company strategy and being integrated into the business model, defines the fields in which digitisation initiatives have to be started with which objective.

In order to help employees and customers understand the objectives, orientation and benefits of digitisation, and to use similar methods and tools even in a large number of projects in big organisations, the definition of a clear, structured vision for digitisation is recommended. The example of a general target image and the digital capabilities required for this are shown in Fig. 6.10. The proposal, based on

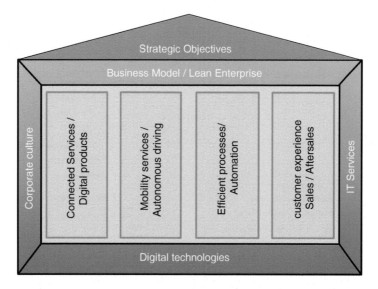

Fig. 6.11 Framework for the digitisation of an automotive company (Source: Author)

the vision of a digital industry evolution, defines concrete topics for individual areas in which the digitisation initiatives should start.

In the area of efficiency, the use of automation tools or a general control of sales channels reduces costs or accelerates processes. For example, the use of monitoring software in the control of supply chains or of communication portals in the handling of procurement processes can significantly reduce the expenditure and thus the process costs.

Also, initiatives are required which support the growth of the company. Through digitisation, new products can be created for instance or existing offers expanded which open up new sales potentials. It is also necessary to identify projects that improve the customer experience in existing structures. For example, the analysis of different data in companies and on the web opens up deeper insights into customer behaviour and their wishes for the creation of tailored offers. In addition to traditional trade structures, new online sales channels can also be established.

Adequate digital skills must be provided or developed to implement these projects. These include technical skills, methods and tools as well as organisational measures and also a structure for innovation management. This is to be designed in such a way that in addition to the established approach for the continuous improvement of processes, methods and technologies, the market is being closely monitored in order to identify disruptive trends as early as possible that could jeopardise the entire business model. The topic of innovation management as a part of the corporate culture is also detailed in Sect. 7.6.

The general goal of a digital vision is now to be transferred to the specific needs of the automotive industry. A proposal by the author is shown in Fig. 6.11.

6.1 Digitisation Roadmap as Part of Company Planning

On the basis of the strategic objectives defined in the company strategy, the overall business model is established with efficient business procedures in the sense of a Lean Enterprise. It encompasses four digitisation fields in which the initiatives are to be implemented. In the automotive industry, these are:

- Connected Services, which also includes new digital products and the resulting sales opportunities
- New business areas with mobility services through a massive impetus from technologies for autonomous driving
- Automation and digitisation of business processes and
- Customer experience in sales and after-sales.

All four digitisation fields use the digital technologies relevant to the automotive industry. The companies must therefore build up competencies, such as 3D Printing, Augmented Reality and Internet of Things, as described in Chap. 4. Furthermore, Big Data, Cloud Computing and Machine Learning IT technologies must be available as a basic service for the digitisation. As a further factor for successful implementations, a corporate culture is to be created which, unaffected by hierarchies, motivates the employees to participate in digital transformation initiatives in the sense of a "start-up mentality" and with thirst for knowledge to open up new domains.

The manufacturers' digitisation efforts can be positioned using the proposed vision frame. This is, for instance, a quotation from the internet presentation of Audi about digitisation [AUDI16].

> Digitisation is changing the everyday life of our customers from the ground up. They expect the linkage of all areas of life – the car becomes part of their interconnected environment. Their requirement is to be "always on" – even and namely in the car.... With digital services for our customers, we are able to tap new business models and sales potentials – in the trade, in the vehicle, and above all with mobility offers beyond the vehicle. The common basis for this is a central platform – a digital ecosystem which is both attractive to customers and partners alike. Artificial intelligence creates faster and most of all self-learning systems. Perspectively, the further development of these systems gives us the opportunity to create outright new customer experiences and processes.... This is why Audi is pursuing its mission to intensify the process of digitisation. We will consistently digitise our processes and create a central platform for integrated, networked premium mobility and digital services.
>
> By 2025, we will achieve:
>
> Integral digitized processes
> Superior user experience
> Substantial contribution to earnings from digital services and mobility offerings
> Scaling myAudi platform.
>
> We want to create a uniform access to the digital Audi world for all our customers and with this platform open up to third-party providers as well as an ecosystem that is equally attractive for customers and partners.

These arguments demonstrate the strategic importance of digitisation for the future of this company. The focal points mentioned, fully fit into the proposed vision frame Fig. 6.11. The link to the overall corporate strategy, the importance of culture as a prerequisite for digitisation and the necessary transformation of IT are not men-

McKinsey	Roland Berger	Capgemini
Be unreasonably aspirational	Think big, then think profit	Understand the threats
Acquire capabilities	Push supply to pull demand	Access your digital maturity
Ring fence and cultivate talent	Build trust in your company	Establish a transformative digital vision lead by senior team
Challenge everything	Interact, integrate and connect with other mobility models	Adopt your business model
Be quick and data driven	Study your customer – then study them more	Strong enterprise level governance
Follow the money	Keep it simple and convenient	Putting organization in motion
Be obsessed with the customer	Build your own ecosystem	Fill skill gaps
	Lobby the authorities right from the start	Quantify and monitor progress
	Think, act and recruit like a start up	
	Harness a jaw-dropping look and feel	

Fig. 6.12 Recommendations of a vision for digitisation (According to [Ola16], [Fre14], [Wes12])

tioned in the Audi presentation, however these are in fact touched upon in the context of Internet representation.

The objectives that Audi intends to achieve by 2025 are, in the author's opinion, of a rather generic nature and difficult to measure. Quantitative targets would be desirable here, for example, which revenue and earnings contribution should be achieved from which business segment. Elsewhere, Audi communicates more aggressive targets [Sch16]. By 2020, half of the annual turnover is to be generated with information technology, software and services based on these related to car driving. The detailing into individual fields is also not done though in this source.

In light of the vehemence with which digitisation will change the industry, it is crucial that the established manufacturers proceed creatively and aggressively in defining a vision for digitisation. This is the only way to counter the newcomers to the industry who are starting from the outset on a higher level of digitisation and thus efficiency based on innovative business models.

The vision for the digitised automotive industry in 2030 as developed in Chap. 5 can serve as benchmark for the initiatives. The forecasts presented therein should be compared against the manufacturers' internal assessment in order to describe on this basis the objectives of digitisation and transformation. Basic recommendations on the approach can also be found in special literature and in different studies. Figure 6.12 exemplarily summarises the advice of three business consultancies on the setting up of digitisation programmes.

Without referring to all the individual points of the list, based on this advice and the author's experiences, the following recommendations for implementation can be summarised:

- Customers to be the focus of all initiatives
- Clear focus on future growth areas and market potentials
- Take up disruptive trends and establish appropriate responses
- All established business models and processes must be scrutinised fundamentally
- Establish an integrated roadmap for digitisation initiatives

- Define measurable goals and key figures for project control and monitor progress
- Implement a comprehensive governance model under senior executive leadership
- Develop the culture of a start-up mentality; create an optimistic atmosphere of departure
- "Think big" in the approach, with simultaneous focus on swift gain
- Speed, speed, speed...

Taking into consideration these recommendations, it is necessary to develop procedural plans for the proposed and selected areas of digitisation which are integrated into an overall plan in order to avail of possible synergies from the projects and to harmonise opposing objectives in the sense of the company strategy.

6.2 Roadmap for Digitisation

In the previous section, Fig. 6.11 showed the business sectors of the automotive industry, on the basis of which digitisation should take place. The following project priorities and procedures can be integrated into a manufacturer-specific roadmap on a time line. For the development of ideas regarding initiatives which should be taken, the relevant technologies for digitisation have to be checked for their possible applications in the enterprise. An overview on these technologies is in Chap. 4. In addition, pilot applications and references in the industry are to be assessed for transferability. As a starting point for this analysis, Chap. 9 describes references within the individual digitisation fields.

In anticipation of the preparation of digitisation initiatives, this is a suggestion for a simple tool to integrally approach the initiatives. It is a matrix which contrasts the selected technologies with the initiatives, Figs. 6.13.

Digital Technologies \ Business Digital Areas	Connected Services Digital Products	Mobility Services Autonomous Driving	More Efficient Processes Automation	Customer Experience Sales / After Sales
	P1		P2	
Cloud	x		x	
Big Data	x			
Mobile	x			
Collaboration			x	
...				
Robotics				
3D Print				
Augmented Reality				
Wearables				

Examples:
→ P1: Service Need on Smartphone
→ P2: Travel Approval within Automated Workflow

Fig. 6.13 The digitisation matrix (Source: Author)

The lines contain relevant digitisation technologies, which are assigned to the individual initiatives in the four business segments. For example, two simple projects are shown: first the display of the service requirement of a vehicle based on diagnostic data in an App, and then the automation of the work process for an in-house travel approval in a workflow solution.

In general, Cloud plays an important role in almost all initiatives and thus is a strategic cross-cutting topic in the company which should be pushed with priority. On the other hand, the subject of 3D Printing is relevant to initiatives in the focus of Industry 4.0 in the transformation of production. 3D Printing also plays an important role in after-sales, in order to produce simple spare parts in future directly in the service organisations on the ground, thus drastically reducing storage for these parts. Since a cross-organisational interest has been identified in this case, it should be considered establishing a joint competence centre for 3D Printing in order to bundle a critical mass and share experiences.

The approach initially serves the integrated view and derivation of initial ideas in order to identify synergy potentials. In the following, the focus areas in the proposed digitisation fields are discussed in more detail.

6.2.1 Roadmap Connected Services and Digital Products

The automobile will evolve from a pure means of transportation and travel into a "driving data centre" which is equipped with considerable computing capacity and networking infrastructure on board. In addition, many vehicles are extensively and constantly linked with different partners. All manufacturers monitor the vehicle condition for instance, and first manufacturers install software updates "over the air". From the driver's smartphone, addresses, music or navigation information are transferred to the infotainment unit, or the next car park provides capacity information into the vehicle. These are typical examples from the Connected Services area. This field of solutions and other new business areas relating to mobility are developing dynamically. Customers expect the vehicles to be equipped with powerful offers that can keep up with the abilities of modern smartphones.

Due to the increasing importance of these business segments, combined with the corresponding growth and earnings potential, all manufacturers as well as many new suppliers are in the process of positioning themselves in this area and developing offers. An intense competition for the customer and his enthusiasm for new mobility models has begun. It is therefore important for the manufacturers to occupy control points in these fields, which enable them to grow successfully in this market. Such an important control point is an integration platform, which acts as a technical hub between the new digital offerings, the vehicles and therefore customers, the manufacturers, and other participants. Figure 6.14 illustrates the approach.

The integration platform is positioned in the centre area of the image, example integration fields are shown in the lower part, and possible solutions and services from the new business fields of the automotive industry are displayed in the upper

6.2 Roadmap for Digitisation

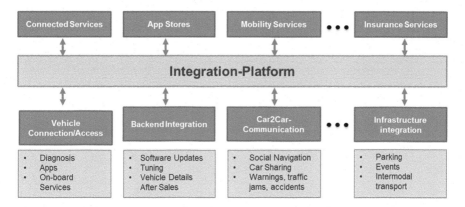

Fig. 6.14 Integration platform as a central solution component for Connected Services (Source: Author)

part. The platform is flexibly composed of IT components in an open, expandable architecture, the details of which are discussed later in this book. It ensures the connection of the vehicles and also the secure dialogue with the IT solutions of the manufacturers, the so-called backend. For example, the car's oil temperature and wear values are detected, evaluated by a diagnostic software, partly in the vehicle or also in the backend, and sent to the driver in the form of recommendations for action.

Apps for navigation or entertainment are loaded into the vehicle, or the operating software receives technical information of the components installed in the vehicle from the manufacturer's after-sales data. Furthermore, the platform can support vehicle-to-vehicle communication so that vehicles not only transmit their current traffic or road condition information to following vehicles yet also obtain additional forecasts or highly precise map information from the backend. In the future, drivers will also communicate with each other in order to flexibly discuss routes through "social navigation" or to arrange car sharing via a mobility service provider, to travel together or to take additional passengers. Information on traffic light control, on a car park occupancy situation, or event information can also be accessed via the platform.

In the implementation of the integration platform, the challenge is to design the architecture in such a way that it is able to follow the dynamic growth of digital applications and users and to enable the various potential uses and accesses of different providers. The platform must be both extensible and flexible in the interfaces, as well as be able to handle large volumes of data at high speed while reliably mastering communication with the vehicles, the backend and other technology partners. The concept of a potential architecture of the integration platform is shown in Fig. 6.15.

In deepening the fundamental considerations about an integration platform, the image shows the technological components of an architectural concept. With the platform, basic functions are implemented as single modules and made available in a Cloud environment. Through messaging and gateways is for instance the vehicle connected via configurable interfaces, which are then integrated with the vehicle IT

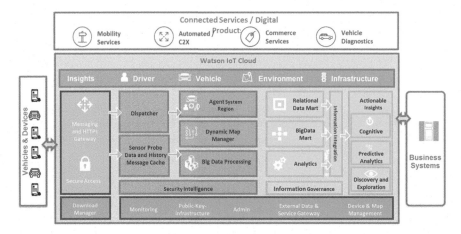

Fig. 6.15 Architecture concept of an integration platform for connected services and digital products (IBM)

(on the left in the picture). Further modules support security services and services for handling and analysing large amounts of data (orange-coloured image area).

A pre-processing of the data through to the rapid implementation of trend analyses and forecasts and subsequent configurable reactions are also based on platform services as well as is the integration with the business systems of the manufacturers and the backend (yellow image area). Further basic services include access management and the linkage of devices and sensors. Application Programming Interfaces (APIs) are provided for the related solutions in the areas of connected services and digital products. In addition, the services of the integration platform can be used, for example, by Apps, mobility services or insurance solutions, which are configured more and more frequently with so-called Microservices, which are small independent IT services [May16]. The topic of mobility services is in more detail subject in Sect. 6.2.2, while Chap. 8 is devoted to more advanced IT aspects such as Microservice architectures and Cloud solutions.

The manufacturers should establish a powerful integration platform as the standard for all vehicles of the company. Via the platform, access to the embedded IT and data of the cars as well as the access to company data is controlled exclusively. As a result, manufacturers obtain control options at least for all solutions that require embedded vehicle or backend data. For these offers it can be decided on how to develop and market connected services or digital products within the company, or with which providers to cooperate. This new business segment in particular is developing dynamically, and Fig. 6.16 shows functional areas for customers in which further Connected Services will be established in future.

The main focus of the offers is in the area of entertainment, driver assistance and safety as well as handling the emergency call obligation. Also, further functional clusters develop in which considerable growth potential is imminent. These are namely the areas of autonomous driving, mobility services and health care offers. The manufacturers will certainly establish themselves with their own solutions in vehicle and mobility-related services, while in the fields of home-based services

6.2 Roadmap for Digitisation

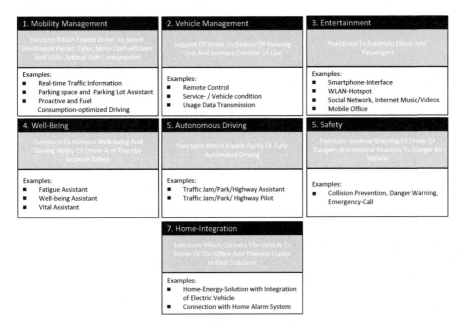

Fig. 6.16 Functional clusters for Connected Services [Bra14]

(home integration), for example safety and heating monitoring and health services e.g. diabetes and fatigue analysis, they are more likely to cooperate with partners or leave the field to independent solution providers. The manufacturers are able to sell the platform usage with services, access options and data via the API layer to interested companies and can thus also participate commercially in this third-party business.

A further requirement for the platform is to design it in such a way that it is open for the integration of established and popular App stores by Apple or Google for instance, so that solutions from this environment can also be downloaded into the vehicle for a complete synchronisation between cars and smartphones.

Mobility services are another important business area that should be built on the platform so that drivers can for example communicate routes and ride sharing offers to mobility providers. Also insurance companies, financial services providers, restaurants, hotels and dealers are users of the platform.

For the successful creation of an open integration platform as the basis of a manufacturer-specific ecosystem, further aspects have to be taken into account: security concept for the access by users from outside the company, API management system for commercial use, and attractive developer environment. These points are briefly discussed below.

Access to a company's IT solutions is controlled by so-called Identity and Access Management (IAM) systems. The system manages the authenticity of the company's employees and controls access to specific applications based on role assignments. These systems are established and the basis of so-called Single Sign-On (SSO) solutions in which with a single password entry applications can be used across the company without having to log on again.

If external users or employees of another company want to access the services of the integration platform via APIs, this access must also be secured, however ideally without re-checking passwords. This avoids effort such as copying of identities or new password prompts. Today's solutions are based on trust relationships between organisations by linking the security solutions of each organisation. The secure exchange of confirmations between the systems that authentications do exist, is then sufficient for access clearing. The identity of the users is in the core system and not duplicated.

This cross-company authentication method is called Federated Single Sign On [May16]. The APIs are addressed at the network level so that for secure communication network or firewall concepts need to be implemented which are secured by the internal company firewalls. For this purpose, so-called API gateways have proved effective. These are upstream computers on which the external API accesses are received. They check the requested API dialogue and direct it through the internal firewalls [May16]. There are proven solutions to solve the security question so that this point should not be an obstacle in implementing an integration platform.

API management systems, which are available on the market in various versions, are used to handle these transactions. It is advisable to integrate a selected standard system into the overall architecture of the integration platform. The system not only handles the dialog between APIs and the applications but also offers functions for version management, system monitoring and load balancing in data traffic, to avoid overloading from old applications with high request volumes, for example. Other important aspects include ensuring consistent detailed documentation for users and developers as well as the provision of analysis and control functions. It also creates a basis for usage quantifications and settlements for commercial models with pricing based on API usage.

The success of the integration platform and the entire business segment will depend to a large extent on whether it is possible to convince as many customers as possible to buy connected services. For broad acceptance, in addition to sensible and powerful Apps, a full integration of the existing smartphone environment and the infotainment unit of the vehicle is a request by customers. At present, the problem is that mobile phone and car electronics are two separate systems which can only be coupled with limited functionality. As a result, even in upmarket vehicles smartphones with a more powerful social navigation solution are used during the trip in tandem with the integrated vehicle navigation system.

The solution is a full replication of the mobile environment, including all Apps, data, images and addresses, complemented by voice and gesture control as well as display on a powerful screen. This integration would lead to high acceptance, so that the first manufacturer offering it could certainly win additional customers for their vehicles. At the same time, this approach facilitates the distribution and handling of payments for the providers of Apps because customary shops could be used.

Also, for external developers the integration means simplification and motivation to develop applications for vehicles. A further prerequisite to attract developers is to disclose the architecture of the integration platform and to provide powerful APIs in an appealing development environment with good documentation, helpful support, and comprehensive assistance and platform service through to access to the vehicle data, if the manufacturer is not keeping this field for themselves. With this "crowd-sourcing" by developers outside the company, it will soon be possible to establish attractive offers on the market and thus further increase the acceptance of the platform. It is also vital to have good earning opportunities for the digital nomads (cf. Sect. 3.6). The basis for this is once again the vehicle stock of the manufacturer with many Connected Services customers.

The most important element for the realisation of Connected Service and digital products is, as was described in detail, the implementation of a company-wide integration platform. Connected services and other digital products from the growing ecosystem of mobility can be installed on this. With the platform, the manufacturers have an important control instrument through the APIs to play an active part in shaping the market for these services and not be overtaken by new vendors. Therefore, important questions and decisions in terms of a roadmap for the implementation of Connected Service and digital products are summarised as follows:

- Decision on the strategic business areas to be addressed as grounded in the company strategy:
  - Which Connected Services should be offered – in-house or with partner solutions?
  - Which further digital products are in focus?
  - Sale of data, commission fees for services – which?
- Development of a business model for the commercial use of these business segments:
  - How to distribute Connected Services?
  - In-house store solution or partnership-based use of established stores?
  - Platform- or API-licensing, or free use?
  - Price models or free use for differentiation and customer retention?
  - Price model of further digital products?
- Development of a solution architecture based on a central and company-wide integration platform for Connected Services and digital products:
  - Base technology to be used... Openstack/Opensource
  - Functionality for the complete integration of smartphones and vehicles
  - Self-programming based on microservices or use of standard components
  - Cloud strategy... Cloud Foundry
- Realisation integration platform
  - Implementation strategy... in-house... partnerships acquisition?
  - Operating strategy
- Implementation Connected Services
  - Agile procedure... Prioritisation according to customer requirements

Servicetype	Weighting Factor	BMW	Daimler	VW	Ford	GM	Geely	PSA	Nissan	Toyota	Honda	Renault	FiatChrysler
Car Rental	1	✓	✓	✓	✓	✓	✓	✓	✓	✓	✓	✓	
Flexible Car Rental	3			✓									
Bikesharing	1		✓			✓			✓				
Carsharing B2B	3	✓	✓	✓				✓	✓				
Carsharing Free-Float	5	✓	✓		✓			✓				✓	✓
Carsharing Micro	3	✓		✓	✓	✓	✓						
Carsharing Peer-to-Peer	5	✓			✓	✓		✓					
Carsharing Station-based	5		✓	✓	✓	✓	✓	✓	✓	✓	✓	✓	✓
Carsharing Valet-Service	5	✓		✓	✓								
Chauffeur Service	3	✓	✓	✓									
E-Chargingstations	1	✓		✓		✓				✓	✓	✓	
E-Chargingstations Navigation	1	✓	✓	✓	✓	✓		✓	✓			✓	
Logistics Services	3			✓	✓		✓						
Multimodal-App	5	✓	✓		✓					✓			
Parking Garage-Finder Realtime	3	✓	✓	✓	✓	✓		✓					
Parkinglot Payment System	1	✓	✓	✓	✓		✓						
Parkinglot-Finder Realtime	1				✓								
Parkinglot-Sharing	1	✓	✓										
Private Drivers	1												
Privat Taxi	5		✓			✓							
Ridesharing	3	✓		✓	✓								
Taxi-Portal	5		✓										

Fig. 6.17 Overview on offered digital services from established manufacturers (Extract from [Bra16])

- Establishing partnerships
  - Technology and development
  - Store or sales channel for connected services and digital products?
  - Third-party providers in the ecosystem, for instance car park operators, cities, toll service providers, insurance companies, intermodal transport partners, network operators,...

Examples for the development of an open ecosystem based on an integration platform with interesting services are solutions from FORD and the PSA group [PSA16], [Nor16]. Both companies are implementing an open platform for their digital services. Through partnerships with technology providers for Cloud and development tools, speed and sustainability of the implementation is achieved. The structure of the architecture and APIs has been published, and the first developer communities and social media communication forums are established to exchange news and experiences. As a result, the motivation of independent developers also improves speed, innovation and range of services. The expansion of the platform towards the role as mobility service provider is pursued at Ford and PSA as well as with other manufacturers as a declared goal. A deepening of these practical examples is given in Chap. 9.

Further to the Connected Services for Driver Assistance, Vehicle Monitoring and Entertainment, the manufacturers are also involved in additional offerings for digital services. An overview is shown in Fig. 6.17.

The range of offerings is wide, from car-sharing and bikesharing options through to the search of electrical charging stations, car parks and chauffeur services. These services are mostly available in the upmarket vehicle segment. Leading by the range of offerings are the German manufacturers BMW, Daimler and Volkswagen, followed by Ford and GM. Interestingly, Tesla Motors, often classified as an innovation leader, currently only services offers in the area of charging stations – probably just the lull before the storm, until this manufacturer as announced will get involved in autonomous driving and presumably also in mobility services? The local Chinese suppliers continue to be absent. But there, too, intensive work is being done on solutions related to the subject of electro mobility. It should also be noted that all of the services offered closely relate to the subjects of vehicle and mobility.

In order to expand the business and its acceptance, manufacturers must be able to use the added value of the vehicle-related data in innovative services before third-party smartphone-based solutions emerge. Typical manufacturer-specific services include, for example, diagnosis and service monitoring, information on driving behaviour and tracking of a driving-specific ecological footprint with carbon dioxide emissions, while third parties offer interaction with an urban infrastructure and car parks, or cooperation with Amazon to handle purchases. The manufacturers could again participate in such a solution by contributing the movement data of the vehicle in return for payment in order to enable location-specific offers.

6.2.2 Roadmap Mobility Services and Autonomous Driving

As shown in Fig. 6.17, some manufacturers offer mobility services, often in cooperation with partners. The customer can choose between different car-sharing models. The services often operate with fixed designated stations, but there are also offers with flexible take-over points for the vehicle. There are also peer-to-peer options, i.e. services offered by vehicle owners, and rideshare offers. As shown in Fig. 6.7, the Daimler and BMW offerings with a corresponding fleet size get noticed on the market, while the mobility services of the other manufacturers are hardly going beyond pilot implementations.

Public opinion and the mobility services business are dominated by a few brand-independent start-ups such as Uber, Lyft, Zipcar and BlaBlaCar as well as Didi Chxing in China. The concepts of these companies differ only in details such as registration procedures or payment processing. They have in common that the services are very easy to use by customers. To start, an App from the chosen vendor is acquired in a single download from one of the usual AppStores and installed with a click on the smartphone. The planned trip is then booked using this App. There are different options available, combined with high transparency about costs and services, for example when and which vehicle is available for a journey. Payment is also made online, whereby the price is based on the service level and the general conditions. Joint trips with other customers are cheaper than exclusive transports and rides at peak times. Also, during rain, the mobility becomes more expensive than with little traffic or good weather. The simplicity of use with easy service handling and attractive prices explain the success of the start-ups. The established manufacturers are hard pressed to compete with these "companies born on the web", which put the customer at the center of their offerings.

As already mentioned in Sect. 5.3.2, the acceptance of mobility services instead of vehicle ownership and thus the market growth of this segment will continue to

Mobility Services	Business Characteristic	Trip Type	Purchasing Criteria
'Just Mobility'	Volume Business	Short distances, Local City Trips	Price
'Seamless Mobility'	Integrator	Long Haul, Intermodal	Simplicity, Security
'Branded Mobility'	Exclusivity	Short / Medium Haul	Image
'Company Mobility'	Reliability	Service trips (medium distance)	Flexibility, Price

Fig. 6.18 Types of offers on B2C mobility services (Source: Author)

grow strongly in the future. In summary, the main reasons are the growing urbanisation with congested infrastructure, modified value systems of the digital natives as customers, and more and more convenient sharing offerings. An additional growth stimulus is generated by autonomous vehicles, which can be used as so-called robotaxis even more flexibly and cost-effectively for mobility services.

A study predicted in 2016 that a share of 30% of all journeys will be carried out with autonomous driving taxis in 2030, and only 45% with private vehicles; currently it is still at 70% [Ber16]. Thus, strong market growth for the mobility service is imminent, which will directly affect the sale of vehicles, since the utilisation levels in car sharing are significantly higher than in private transport. Compensation for losses of revenue due to reduced vehicle sales is the growth in the mobility market. These opportunities are also detected by many competitors, such as the brand-independent service providers already positioned, and, in addition to Google and Apple, other new suppliers in China and from other industries.

In this attractive market, the manufacturers must therefore develop a promising strategy in order to secure business shares. Figure 6.18 shows a simplified comparison of possible mobility offers from manufacturers for customers (B2C, business to consumer) in passenger transport.

Mobility services from the "Just Mobility" segment will occupy the largest share of the business, in which customers are only interested in taking a certain route by car as flexibly as possible. The vehicle brand is of secondary importance in this context, the primary buying criterion is the price. Vehicles of all volume manufacturers are used in this segment, and it can be operated well by Robotaxis. The "Seamless Mobility" service category appeals to customers who wish to travel longer distances intermodally as well. In this case, the change in vehicle type, for example, from the car to a ferry and from there to a bicycle, is entirely organised by the mobility service provider in the background and offered to the customer for selection per "one click shopping". The Branded Mobility is of interest to manufacturers in the luxury and exclusive sports car segment, for who it is important that the brand also encompasses mobility services. This upper segment may also offer golf events or restaurant visits. Here, the mobility becomes a brand-specific experience, and interested customers are thus introduced to the brand. Another special segment in the area of service is "Company Mobility", which includes company mobility services instead of individual company cars.

6.2 Roadmap for Digitisation

Fig. 6.19 Overview on Moovel offers for mobility [Mat16]

Taking into account the target markets and customer segments, manufacturers must decide on their orientation here as well and define an organisational form and a business model. From the viewpoint of the author, the mobility services should be implemented with a new, independent organisation in order to at least mitigate the conflict that the success of the service is at the expense of vehicle sales. In this way, Daimler has organised the subject of mobility in its separate company subsidiary Moovel. Here, as shown in Fig. 6.19, various service offers and partnerships are bundled together.

In addition to the carsharing service under the Car2go brand, bus and bicycle services as well as partnerships with public transport companies and taxi and limousine services are on offer. With these diverse options and partnerships, Daimler is well positioned to succeed in the mobility market with flexible offers through Apps. Complementarily, the "Mercedes me" portal provide connected services, financing services and various sales functions consolidated on the Internet.

For the development and implementation of mobility services, mobility platforms are used. As shown in Fig. 6.20, they are located above the integration platform presented in Fig. 6.14 and use the basic services available there, for example for vehicle connection and backend integration.

Similar to the integration platform, the mobility platform bundles all the functionalities required to process the services. Additional services by third parties, such as intermodal services, alternative means of transport, maps and also weather data, are also integrated via APIs and interface adapters. Furthermore, restaurant bookings or purchases during the journey can be transacted and immediately paid for via

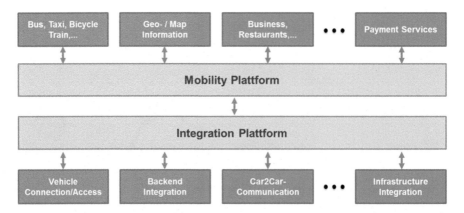

Fig. 6.20 Positioning of a mobility platform based on the integration platform (Source: Author)

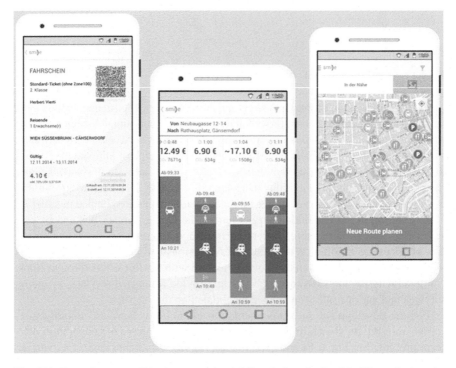

Fig. 6.21 Example screens of the App to use the mobility platform Smile of the Wiener Stadtwerke [Kot15]

connected payment solutions. The services are retrieved using a smartphone App from the mobility platform. An example is shown in Fig. 6.21.

The display examples of the user interface of the mobility platform "Smile" suggest the ease of operation and the functionality. Via the platform, the user arranges intermodal journeys with several means of transport, thereby learning about the prices, the duration and also the CO_2 balance. Required tickets are available online, and the route is visualised in the city map.

The technical details of a mobility platform architecture are not deepened at this point. Similar to the integration platform, open standards, high modularity, extensibility and adaptability as well as API and microservice design are important implementation criteria in terms of profitability and sustainability. For details on architecture and technological components, please refer to a research project of a consortium on open mobility platforms [Bro16].

The Smile platform has been in operation since 2016 in Vienna, and initial experiences with respect to acceptance and the reduction of private transport in favour of bicycle and public transport are positive. Similarly, other cities have also launched projects on mobility platforms, including Singapore, London, Copenhagen and Helsinki [ECO16b]. Next to brand-independent service providers, especially the cities are the drivers of integrated mobility services in order to prevent traffic collapse despite further urbanisation, but also to achieve urgently needed improvements in CO_2 emissions and fine dust pollution.

As a supplier of open mobility platforms, the manufacturers are, with regard to vehicle movement data or mobility offerings, slipping into a passive role. In order to be accepted in the "commodity" area of mobility, perhaps even to achieve the role of an active creator with integrated offers, partnerships or acquisitions are possible options, as many corresponding activities on the market demonstrate. In the areas of special mobility platforms, such as in the upper segment of "Branded Mobility" or "Company Mobility", market niches can still be found and business opportunities are available. Particularly with respect to company cars, mobility services as an alternative to individual vehicles use can open up an interesting opportunity for companies to reduce costs, but at least to do something to support the "green image". The base of the idea is illustrated in Fig. 6.22.

The picture shows the areas to be included into a mobility platform in order to integrate corporate mobility in terms of a corporate car sharing based on fleet vehicles, private cars and public transport services as well as taxi services. The approach is the same as with the public platforms. However, if a company prefers certain manufacturers in its fleet, the company or a consortium of manufacturers could establish the platform to secure vehicle sales, albeit with a reduced volume. Therefore manufacturers should consider positioning themselves in this field.

Another business opportunity in the area of mobility services is the superior optimised control of the vehicles which are accessible to a platform – similar to air traffic control. A "digital mobility shadow" can serve as the basis for such guidance in analogy to the production control. For this purpose, the positions and, if possible, the planned destinations of all vehicles appear in highly precise maps and from there are integrated into a virtual model of the overall traffic situation of a geographical area. Taking into account the destinations and incoming queries from customers, forecasts for traffic development can be derived from this, and the traffic flow can be optimised by routing measures to avoid traffic congestion. With autonomously driving vehicles, the avoidance of congestions can also reduce the risk of

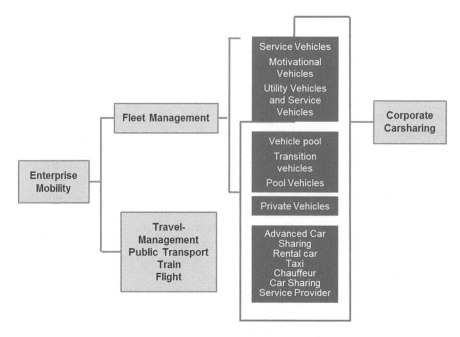

Fig. 6.22 "Company Mobility" as an alternative to company cars [Ren16]

collisions and increase the overall utilisation of the vehicles. Through such superior control, considerable advantages both for individual drivers and also for the overall traffic situation can be achieved. Such a future model could be developed by a manufacturer with a high market share because they have access to all manufacturer-specific vehicle data, and this way a model with the necessary statistical validity is available.

In summary, the importance of mobility services will increase significantly and see additional growth through autonomous vehicles, used as robotaxis. The mobility services are available via platforms, which enable customers to conveniently book intermodal mobility services with the aid of a smartphone App. The market is then essentially dominated by brand-independent companies. Other drivers of mobility platforms are cities and public transport companies.

In order to successfully position themselves in this extended mobility ecosystem and possibly to make up ground that was already lost, strategic partnerships and perhaps acquisitions are important for manufacturers. Further opportunities are in the offering of manufacturer-specific special mobility services, for example in form of a "Company Mobility" or "Branded Mobility". The proposed future project of a superior mobility control also offers an opportunity for market positioning to volume manufacturers. It opens up considerable advantages by avoiding congestion, increasing capacity utilisation and reducing the number of collisions by autonomous vehicles.

6.2 Roadmap for Digitisation

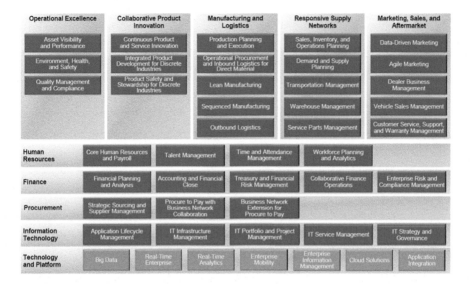

Fig. 6.23 SAP Value Map for the automotive industry [SAP14]

6.2.3 Roadmap Processes and Automation

Digitisation approaches for business processes up to complete automation have the goal to increase the efficiency of same. This is not only about individual business areas, yet about all operating divisions of the company. Along with improvements in the process procedures, further optimisations can be achieved by using new IT technologies such as Big Data, Analytics and especially Cognitive Computing and Machine Learning. To set up an appropriate digitisation program, it is necessary to identify and prioritise in an initial phase those areas that have high need for improvement and thus hold big potential for increasing efficiency.

For this purpose, process-oriented procedures have proved effective which are available from comprehensive studies and the specialist literature and are documented there with reference models for business processes in the automotive industry, e.g [Wed15]. The models are usually described on the basis of standardised languages, and documentation is carried out using special software. The models also provide performance benchmarks for the individual processes, e.g [APQC14]. Furthermore, there are reference models of software producers which also allow a general structuring of processes and data based on the software solutions. Due to the SAP solutions used in many companies, the model of this software producer, the so-called Value Map, is known in the automotive industry. Figure 6.23 shows the upper level.

The Value Map shows the business competences of a company and assigns the essential business process areas thereto. For example, the competence area of a human resources department covers the main business areas of pay, talent development, time recording and employee planning. Other fields of competence include,

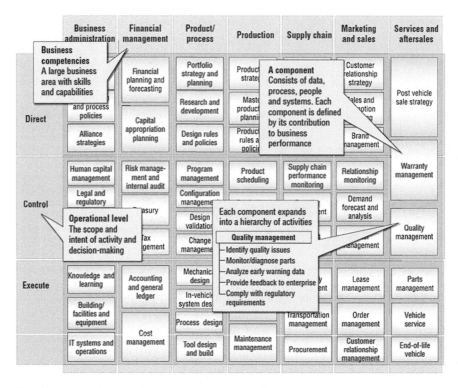

Fig. 6.24 IBM's Component Business Model (CBM) [Kop10]

for example, production and logistics, marketing, sales and after-sales as well as purchasing, each with the associated business areas. With the help of the Value Map a rough corporate structuring and the analysis of the process areas is conducted.

With a similar approach, business competences with associated components can be captured with IBM's Business Component Model (CBM), shown in Fig. 6.24.

The columns show the business competences with the associated components. Components are independent process areas here with their employees, IT solutions and costs. The components are divided into three areas; in the upper third, the direct strategic and planning components (direct), in the middle third the controlling process areas (control) and in the lower third the operational sequences (execute). The CBM-based approach also allows a company and process structuring as the basis for weak point analyses, with the aim of identifying efficiency-enhancing digitisation initiatives and setting them up for implementing a roadmap. For this purpose, the procedure shown in Fig. 6.25 has proved successful.

The picture is a simplified depiction of the steps and supporting methods of this working phase. In the first step, the business processes of the business area to be viewed are recorded and evaluated. Extensive relevant documentation is often available in companies.

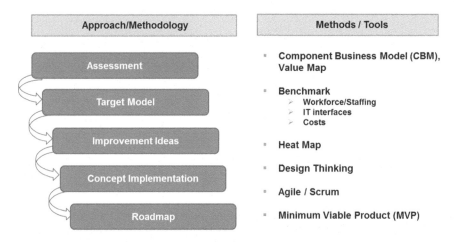

Fig. 6.25 Development of a roadmap for digitisation projects (Source: Author)

For the first rough assessment, it is recommended to summarise the processes of a business area, either in the Value Blocks according to SAP or the CBM method. For the identified process areas, the necessary resources for execution are recorded in respect of personnel, IT systems, and especially interfaces and system breaks are paid attention to. A benchmark with similar business areas assesses between company brands, or even better with comparison values available on the market, the degree of maturity of the respective area and the potential for improvement. Areas with relatively high personnel requirements, with many different IT systems or ruptures between systems and high costs must be addressed with high priority. These are marked in red colour in the CBM representation, the components with medium potential are marked yellow, and the process areas already leading in these evaluation criteria are in green. This creates a so-called Heatmap, which shows at a glance for the company or business areas the fields to be addressed.

The Heatmap then again is the basis for so-called design thinking workshops, in which mixed teams meet with representatives from the fields to be addressed, from adjacent areas and from the IT, in order to develop ideas for improvement measures and concepts in a pragmatic and comprehensive approach. Figure 6.26 shows a tried and tested approach to the development of a roadmap for the implementation of digitisation projects.

For the areas that have a high potential for improvement according to the Heatmap, cross-functional teams develop solutions in workshops and document these in user stories. One business case each evaluates its economic benefits. As a result, projects can be implemented in the sense of "low-hanging fruits" within 8 weeks' time for example, and their costs can be absorbed shortly thereafter through realised savings.

For the selected projects, a first solution is developed within a short time using modern IT tools. The goal is to present a prototype or a so-called Minimum Viable Product (MVP) of the future IT solution, for example as an App on a smartphone, in

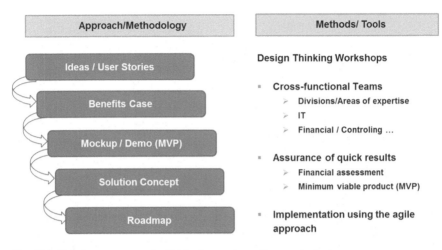

Fig. 6.26 Development steps of a digitisation roadmap (Source: Author)

order to assess at an early stage whether the needs of the users and the project objectives were met appropriately. On the basis of these MVPs, an implementation concept is agreed upon in further agile workshops, and this is then implemented in so-called Scrum courses. Scrum describes an agile development method, which is explained in detail in Sect. 7.2.2. The agreed milestones appear in the roadmap of the overall program.

This contemporary approach to innovative projects in a digitally oriented corporate culture was just briefly outlined here and is deepened in Chap. 7. In the following, process areas are presented, in which digitisation measures often lead to considerable improvements in efficiency and simultaneous quality improvement. The focus is, in addition to administration with finance, human resources and purchasing, also on manufacturing with emphasis on Industry 4.0.

Finance, Personnel, Purchasing

In the management of traditional manufacturers, business processes are often still executed with much manual effort and paperwork. The standard software solutions used, often solutions from SAP, Oracle or Peoplesoft, are accessible on workstation computers installed in fixed office workplaces. Mobile working on tablets or even smartphones is complementary at best.

In larger companies with several brands, in many cases nominally the same software systems are used, however frequently they have been significantly modified and adapted to specific brands or countries. This leads, for example, to the fact that different account system plans are in use, personnel systems are run in different career structures, and purchasing systems use different supplier master data. What's more, special process areas besides the standard software solutions often also use self-developed legacy systems that are connected to the standard packages via complex interfaces. In these heterogeneous environments, overarching analyses in a particular business area are often only possible with considerable manual effort.

This situation leads as an emergency solution to the "Excel madness", i.e. to the manual transfer of data into evaluating spreadsheets, which then can often only be handled by specialists. In such an environment developed over time, targeted digitisation initiatives should be implemented. Areas of action, such as system breaks, comprehensive manual intervention or the lack of integration between business areas, should be identified with the help of the described method (see Fig. 6.25) and documented in a Heatmap.

In order to improve, first the "low hanging fruits" should be picked. Experiences from different projects show that in each area there are some of these quick and easy-to-achieve improvements. Some typical examples from the practice of the author are given here in anonymised form.

- Recording the temperature and humidity of a warehouse with preparation in daily reports; the sensitive hygroscopic material which had been the reason for establishing this process, has not been stored there for a long time though.
  - Immediate removal of the entire process
- Implementation of an SAP-based purchasing solution; the legacy system is still in parallel use even after 2 years.
  - Shutdown within 2 months
- Multiple implementation of functionally identical process sections in different applications – for example receiving incoming goods in logistics, CKD and spare parts.
  - Consolidation – at least for this sub-area as a shared service
- Supplier master data in several systems; manual analyses of business transactions.
  - Analysis using an open, contemporary analysis tool
- In over 30% of the ordering processes on the basis of a catalogue system, manual interventions are necessary, contrary to the original target; several catalogue systems are used in the group.
  - Implementation of the established catalogue products; deviations only with a reasoned approval by the purchasing manager, and take over of the addtional internal costs.

Such examples can in the same or a similar manner certainly be found in any company. The problems are often not addressed as the employees are overburdened by the day-to-day business and become "routine-blinded", and because the improvements can only be implemented across departments or only with IT integration. Furthermore, there is often no incentive to tackle these projects, neither through bonus payments nor under career aspects. In order to improve this situation, a

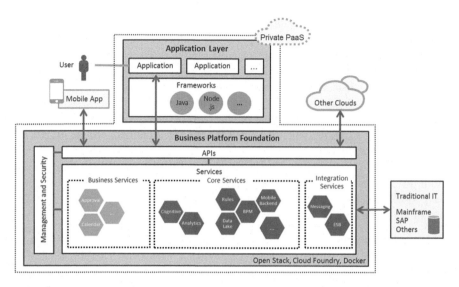

Fig. 6.27 Concept of a business platform for the integration of legacy systems and mobile applications (Image of IBM)

culture must be created that motivates every employee to point out and solve such topics. See Chap. 7 for more information on this.

Furthermore, despite all security concerns, it must be made easier to bring modern software tools, which are often known from the use on a private computer or smartphone, into the companies. An example are chat tools such as WhatsApp or file sharing systems like Dropbox. These solutions are proven to improve collaboration and reduce the communication effort. Although comparable solutions are available to companies in the market, the practical and accepted solutions are not yet used by the manufacturers or in the cooperation between manufacturers and partners. The fast and pragmatic use of contemporary IT tools is certainly another "low hanging fruit" which in addition to improvements in the process sequences also has a motivating effect on the employees.

Implementing the measures in a first wave already leads to improvements and savings. In order to achieve further and sustained progress, integrated holistic solutions are particularly useful for the administrative areas. To this end, a platform approach was proposed in Chap. 5 with the Vision 2030 (cf. Sect. 5.3.10; Fig. 5.21). The basic approach is an in-house business platform to which existing software solutions can be connected and on which new functionalities are built in the form of Apps. This brings together the "old world" of proven IT solutions and the "new world" in the form of Apps which are running on any mobile device. Figure 6.27 shows a refinement of the rough concept proposed in Chap. 5.

The Business Foundation platform provides services to be used jointly, such as Master Data Management (MDM) and DataLake, Analytics and Cognitive Computing. In addition, the integration services secure the integration of the established IT solutions, and business services offer recurring business services, for example, for Blockchain or application releases. Access to the services via flexible mobile devices or web portals is through standardised APIs. The entire solution is based on Cloud structures. The Apps and additional functionalities are configured using Microservices within "Platform as a Service Concepts" (PaaS). The entire concept is based on open architectures and standards and envisages the use of open source software.

So much on the basic concept of an in-house business platform; for further technological IT details, such as operating in hybrid Cloud environments or the use of microservice-based architectures, please see Chap. 8. Following below here is the discussion as to which business areas are eligible for such an approach, what benefits this approach provides, and how it can be set up.

In principle, this concept is useful wherever recurring, standardised procedures prevail which use different data sources and IT solutions. Therefore purchasing, finance and personnel are the first areas to be particularly suited to this approach. For example, in purchasing, the benefit is to access data and applications using simple mobile applications, in order to analyse across the company the purchasing volumes of suppliers or to bundle procurement processes. To this end, purchasers use a new "spending App" with simple dialogue, search and analysis functions. This App accesses the brands' connected purchasing systems through the APIs. The data is automatically consolidated and standardised in the MDM or Data Lake functions. This overarching analysis and bundling allows negotiating of purchase prices without extensive manual evaluations. An additional advantage is the company-wide use of the Apps, so that multiple developments and local special solutions become obsolete. The fast introduction into modern IT solutions and the decoupled renovation of the legacy system landscape are further positive aspects of the approach.

The challenges in the design and implementation of the platform concept are not so much on the technical level. For this, proven concepts, experiences and also powerful IT tools are available. The major challenge rather is the necessary cultural change as the platform should be established and used throughout the company to exploit the potential. If, for example, a common financial platform is the goal, all existing financial applications or adaptions of a joint programme provider in the brands or also in the different markets of a manufacturer have to be linked step by step to the platform. In order to be able to utilise new functionalities in the form of Apps throughout the company, work flows need to be standardised. The acceptability of this centralisation and harmonisation requires a massive change in the traditional self-understanding which is mostly characterised by local solutions with many individual adjustments and "process loops".

In order to successfully implement platform concepts and to achieve the necessary change, it is recommended to start implementing a company-wide Shared Service Centre. For this purpose, the workflows of a business segment are brought together in the current form with the existing IT systems in order to provide the services for the entire company from the centre. After a stabilisation phase, the IT adjustments and connections are implemented step by step in the background, followed by the installation of additional functions. Finally, a mobile application is available to customers to work with the centre. The processes are executed step by step with increasing degree of automation on the basis of cognitive solutions. Thus, through the intermediary of the shared service centre, the goal is achieved to automate process execution in the administrative areas.

An alternative to this sequential approach is the conversion of the applications towards a target system already during the transition phase. This path certainly requires increased change management activities so that employees will support this twofold change in a motivated manner.

Product Development

In product development, IT solutions for product data management, design, calculation, simulation and virtual testing have been used for a long time and are daily practice. However, this area also has considerable potential for increasing efficiency through digitisation projects. Typical subjects are:

- Use of Analytics, for example to increase parts reuse through guided searches and patent researches
- Knowledge management, e-learning
- Assistance of manual work on the basis of cognitive solutions, for example for the identification of potential for improvement in the design through comparative analysis and comparison with "Lessons Learned Databases" and the use of optional materials
- Use of Augmented Reality instead of prototypes for test drives and feasibility tests
- Image search for damage analysis in the error-correction process
- Complete error-correction process; integration of the information flow from the field through development into production up to the supply of parts in after-sales
- Use of collaboration tools across the manufacturer and with external partners
- Crowdsourcing by absorbing customer suggestions through open feedback platforms for conversion into product ideas
- Social media analysis and company-wide analysis of customer-relevant data for the early detection of weak points and customer requirements.

All of the listed approaches have the potential to improve the processes in development through the provision of additional information, the IT-based coupling of adjacent business areas, and targeted support to the engineers. Figure 6.28 summarises with which digitisation technologies which objectives in the area of development are of high potential for improvement.

It shows that the use of Big Data and Analytics tools, in addition to the "self-thinking" cognitive solutions, supports the improvement of performance figures of the development. Also the opening of an internet-based feedback platform for customers, in order to capture ideas and suggestions for the product directly, presents opportunities. The extensive use of collaboration tools to simplify and accelerate

Goals	Technology/Method						
● high Potential	Big Data/ Analytics	Cognitive Solutions	Knowledge-management	Collaboration Tools	Crowd-sourcing	Augmented Reality	Social Media Networking
Increase throughput speed	●	●	●	●			
Decrease effort	●	●	●	●			
Increase quality	●	●	●				
Strengthen Force of Innovation						●	●
Increase customer-orientation	●					●	●
Increase transparency	●			●			
Reduce amount of parts	●	●		●			
Speed up error correction	●	●					

Fig. 6.28 Improving development with digitisation technologies (Source: Author)

collaboration between employees, reduce throughput times and reduce efforts, leads to further improvements.

In addition to measures to improve the process and workflow, from the author's point of view, the development area of the manufacturers should also be responsible for implementing two basic structural adaptations in the vehicle concept, the necessity of which is briefly repeated here. On the one hand this is around embedded IT, and on the other hand around Connected Services.

The current situation of the embedded IT is characterised by long-established heterogeneous structures and a decentralised architecture with a multitude of electronic controllers of different network topologies. The system is prone to error, difficult to control, burdensome to secure, expandable only to a limited extent, and expensive as well. This means that there is a need for action, especially in view of the fact that the new business areas of connected services, autonomous driving, mobility services and digital products all require a powerful data exchange around vehicle data.

In addition to the further standardisation of interfaces and IT technologies, the fundamental transformation of the architecture to a central approach promises significant improvements and the safeguarding of future viability; see Sect. 5.4.5 and [All16]. The central approach takes account of the trend that vehicles are increasingly IT-driven. In the long run, the central processor is becoming the dominant vehicle component around which the "remaining mechanics with wheels" is grouped. This conversion of the architecture is complex and associated with considerable challenges, but from the author's point of view, it is in fact essential. It offers the opportunity, especially in the development of the new electric-drive vehicles, instead of a long-term evolutionary approach, to rely from the outset on the described central architecture.

With the new approach, a second urgent issue is to be addressed with regard to the upcoming e-vehicles. Currently, two "digital worlds" are developing for drivers.

On the one hand it is the Connected Services offered via the infotainment unit, and on the other hand it is the personal smartphone environment. Both areas penetrate with solutions into each other's respective traditional areas.

In addition to the well-known assistance systems, purchasing and booking options for instance will also be offered in the vehicles in future. Via mobile phones, already a constant companion in all private and business matters, powerful navigation solutions are available, and in the future service and diagnosis Apps can be expected.

In order to avoid duplicate work and simplify the use, these two worlds must be brought together conveniently for the customer. Synchronisation solutions are available for this purpose, which, however, only work under functional restrictions. In the author's view, customers expect a seamless integration though (cf. Sect 5.4.2). When the user gets into a vehicle, the familiar smartphone environment should be fully available on a comfortable display in the infotainment unit. The usage should be voice- and gesture-controlled, whilst driving however only with restricted functions. Vehicle-related Apps can also be retrofitted through established App Stores (cf. Sect. 6.2.1). At most, the vehicle-related assistance systems remain embedded in the vehicle.

With this approach, two further points are solved in a pragmatical fashion. The new Apps offer the much sought-after personalisation form the outset, which in the previous on-board approach needs to be established by special software solutions in the first place. Furthermore, the App approach allows short-cycle updates via the smartphone, decoupled from vehicle-related mostly longer innovation cycles. The proposed full integration will certainly lead to an increase in the acceptance of Connected Services and to a greater degree of customer satisfaction.

Manufacturing and Industrie 4.0

In addition to management and development, manufacturing represents a further process area in which digitisation can lead to substantial improvements. The production of the German manufacturers is in this regard guided by the activities, results and recommendations of the platform Industrie 4.0, an initiative under the patronage of the Federal Ministry of Economics and Energy (BMWi). In many committees and working groups, it is continuously pursuing the overall goal of systematically combining digitisation and manufacturing, thus creating the basis for successful international competition for markets and technology leadership [BMWI16]. Concrete recommendations for action and networks between interest groups emerge. Similar initiatives are also being pursued by other countries such as the USA, Japan and China, where producers are similarly supported (cf. Sect. 4.3). The German initiative developed a series of scenarios to specifically describe the future production.

6.2 Roadmap for Digitisation

Order-driven Production
Company-wide production concepts based on automated production market place enable efficient production of customer-specific requests

Adaptable Factory
Production competencies and capacities are being adapted and optimised in a fully automated manner and with the customer request in mind - all within strongly **modularised production facilities**

Self-organized, adaptive logistic
Increase of flexibility and decrease of reaction time of the industrial systems along the whole industrial supply chain based on **comprehensive and automated logistics solutions**

Value Based Services
Virtual platforms consolidate machine and production data and create the basis for need-based maintenance and individual service offers

Transparency and adaptivity of distributed products
Communication-enabled products can be tracked through the entire life-cylce even after leaving the factory and can be updated and adjusted according to the conditions of use of these products

User Support within the Production
Digital Assistence Systems support workers within the production and change content as well as organisation of the work to be performed.

Smart product development for the smart production
Production and product data are being used during the **development of customer-specific solutions** in a holistic and goal-oriented manner

Fig. 6.29 Production scenarios of tomorrow according to platform Industrie 4.0 [And16]

Figure 6.29 shows these production-relevant future scenarios in generic form on which the initiative Industrie 4.0 is based for further work. All addressed areas are also interesting for automotive manufacturers and to be transferred into specific initiatives.

As a starting point, after a process assessment, the production-related Heatmap has to be developed in the proposed approach according to Sect. 6.2.3. In the author's experience, potential initiatives often cover the following project topics:

- Use of 3D printing directly in the line instead of supplying discrete components; e.g. vehicle interior
- robotics solutions for final assembly and body-in-white works to relieve employees
- Superordinate capacity planning of the plants; detail planning of the lines based on this
- Demand-oriented parts storage
- Monitoring the supply and quality situation in the supply chain
- Preventive maintenance of production equipment to avoid downtime
- Early detection of quality problems
- Real-time monitoring of the line supply and initiation of preventive measures to avoid supply shortfalls

Topic	Application Area/ Example
Context-sensitive Assistance System	Employee guidance via Smartwatches e.g. alterting in regard to next step in workflow
Robotic Systems	Relief of employees during heavy-duty work, light-weight robots work hand-in-hand with employee
Simulation und Plant Digitisation	Use of 3D Scanners and high-resolution cameras during measurement of shell construction
Planning and Steering Systems	Capture of quality of parts during production resp. within the flow of goods
Smart Logistics	Capture of status of suppliers in real-time, hedging of supply of parts
Advanced Analytics	Surveillance of screwer device; preventive maintenance to avoid downtime

Fig. 6.30 Digitisation projects in BMW production [SchR15]

- e-learning solutions for the introduction of new employees
- Detailed shift planning using different data sources
- Assistance systems in the control station for the initiation of corrective actions in case of malfunctions
- Predetermination of first-pass situations, i.e. the identification of which vehicles are going through the production process without any problems, or in which sections these are expected, and avoiding failures through preventive measures
- Continuous quality monitoring and early repairs or pull out of vehicles to avoid consequential costs in case of continuation
- Paperless manufacturing; digitised accompanying article card
- Commissioning support; guided parts removal and assembly.

Further typical project examples in the context of the digitisation of the production are also shown in a report by BMW, summarised in Fig. 6.30.

One focus here is on supporting solutions such as employee alerting via Smartwatch in case of deviations in the operations sequence or their physically release by lightweight robots in heavy or tiring works. Analytics technologies are also used to monitor the quality of components as early as possible and to control the delivery situation in the supply chain. In order to avoid downtime, preventive maintenance work is initiated on the basis of trend analyses and forecasts, and the torque monitoring of screwdrivers is also carried out in real time. The examples shown fit into the application scenarios of the platform Industry 4.0 and detail these. Integrational in this regard is the user support through Value Based Services and innovative logistics solutions.

The individual measures pursued in public projects are to be further developed into platforms in terms of an integrative approach. For example, the Industry 4.0 application scenarios envisage a production marketplace for the coupling of customer-specific production structures, a platform for machine and production data as well as a logistics hub. In order to support the intended horizontal and vertical integration of the business processes and for the establishment of the planned data platform, a complete integration of the IT systems at shop floor level is needed. The problem is that the plant IT is usually very heterogeneous and contains a large number of network types, protocols, sensors, adapter types and transmitters. In order to ensure the integration of these different technologies, an approach is the

6.2 Roadmap for Digitisation

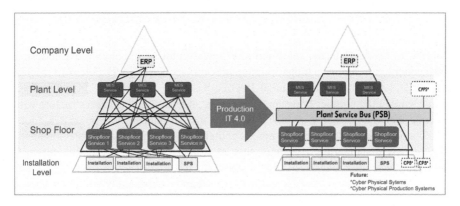

Fig. 6.31 Plant Service Bus for connecting the plant IT in Industry 4.0 Solutions (Volkswagen)

implementation of so-called Shop Floor Integration Layers resp. Plant Service Bus [Die16]. Figure 6.31 shows a proposal by Volkswagen.

On the left, today's heterogeneous world with different IT technologies is pictured, often directly connected to a specific requirement, for example to record the torque of a screwdriver or to provide a robot control with new programs. In the existing structure it is only possible with high effort and special solutions, for example, to register the operating signals of all assembly machines and inserting devices in the production area and to evaluate process trends. The implementation of an overarching monitoring system will be complex as well. Furthermore, any replacement of plant components entails cumbersome adaptation to the IT solution.

The implementation of a Shop Floor Integration Layer addresses this problem and offers, in addition to IT core functionalities, fast adaptation possibilities in order to link different plant ITs. Also, the layer provides various application interfaces (APIs) that can be used in innovative IT solutions for digitisation projects. Such an integration layer decouples the rigid link between technology and applications. Replacement, extension and adaptation of all involved components become easily possible this way. The vertical and horizontal integration required for Industry 4.0 is also feasible [Die16]. For example, order data between processing machines can accompany the operation progress, while the horizontal communication from the development area to the machine ensures a rapid error-cancelling process.

On the Shop Floor Integration Layer, in analogy to the architecture for the mobility platform (cf. Fig. 6.20), an application platform can be set up with Microservices for the Internet of Things (IoT), oriented to production-relevant solutions, for example, for preventive maintenance or for near-real-time quality monitoring. Hence it is important, in addition to the initial projects of digitisation with the aim of rapid use, to observe the strategic orientation and sustainability as well as the conception and implementation of a Shop Floor Integration Layer and an IoT platform. This architecture will in the long term also be the basis for the so-called Digital Shadow (see Sect. 5.4.9). In this virtual image of production, which is supplied with real-time data, future projections and the simulation of action alternatives can be carried out in order to test possible increases in efficiency before the operational deployment.

6.2.4 Roadmap Customer Experience, Sales and After-Sales

Similar to production, the sales and after-sales sector of the automotive industry also undergo an extensive transformation process (see Sect. 5.4.7). Sales of vehicles via dealers will shrink in favour of Internet-based sales platforms. It is therefore expected that the number of dealers will shrink, at least their business structure will change. For the emerging so-called multichannel structures, new processes, tools and cooperation models are to be agreed with the dealers. In addition, sales channels for the new manufacturer offerings, such as connected services, mobility services and vehicle-related data, need to be created. In the area of spare parts, new sales channels, such as shop systems and platforms, can be expected in addition to dealers. Marketing will also change profoundly through the use of digital technologies.

Digitisation initiatives in the main processes of sales, according to the author's experience, include the following topics:

- Sales
    - Intuitive vehicle configurators using customer information from the history and findings from social media contributions; use of the configuration in further customer interactions
    - Augmented Reality for the presentation of the desired vehicle in the virtual space or on high-resolution screens at the dealer or in the private area; virtual test drives in freely selectable environments
    - Online vehicle sales via sales platforms; development of multichannel concepts
    - Cross-brand, continuous lead management from initial contact to vehicle purchase
    - Social CRM (Customer Relationship Management); support of all sales processes taking into account different data from the company and public sources
    - Digitised salesman workstations; assistance systems with needs-based support functions
    - Support in the definition of digital products, the trading of data, the business brokerage in trade via connected services and the use of APIs from manufacturer platforms
    - Definition of sales structures for mobility services and digital products as well as the establishment of suitable online sales platforms
    - Establishment of sales partnerships for the shared use of established marketplaces

- Marketing
    - Customer-specific direct marketing
    - Marketing for new digital products; strategy in line with vehicle business
    - Cross-brand comprehensive customer views, including social media analysis and derivation of a complete view of customers (360° view)
    - Loyalty analysis; early detection of migration trends

- Integration lead management, marketing, sales initiatives
- Transaction-specific customer satisfaction analyses in service
- Tracking of marketing and sales activities with customer-specific supplementary offers for the further development of interest until conclusion of sale
- Offers from partner companies such as retailers, petrol stations and also event providers on the infotainment unit, taking into account customer-specific localisation
- Cross-media and -channel brand management

- Services
  - Continuous vehicle diagnosis; proactive customer-specific service offers
  - Intelligent assistance systems for service reception and maintenance work based on cognitive technologies
  - Mobile devices as an interface for applications in the App format for digitised damage assessment and for customer-specific financing options
  - Software updates "over the air"
  - Harmonisation of warranty management; feed back of trends into the error-correction process
  - Vehicle file online; integrated service and component history
  - Intelligent solutions for workshop capacity planning and disposition of spare parts

- Spare parts
  - 3D printing of spare parts in the local markets
  - Part-specific needs-oriented storage
  - Online parts trading
  - Blockchain methods for part tracking and source protection in the service as copy guard
  - Direct supply of parts from the manufacturer into the service
  - Proactive demand planning based on Big Data and Analytics technologies, using information from sales forecasts, social media, weather data and historical retrievals

- Management and control
  - Rolling sales planning using 360° customer views
  - Continuous usage-oriented post calculation of sales and marketing costs
  - Online reporting with drilldown functions
  - Benchmarking of similar areas between manufacturer brands
  - Efficiency increase in process execution through automation.

This exemplary overview with a multitude of different project themes shows that in sales a fundamental transformation takes place in close connection with the digitisation. Before a roadmap is defined according to the prioritisation of the initiatives, decisions about the orientation and the resolution of target conflicts must be made in the existing sales structure. Here, for example, the following situations must be considered, each with a possible solution approach from the author's point of view:

- The success of mobility services is at the expense of vehicle sales
  - Shift mobility services to independent company organisation; distribution channels and implementation of the mobility concepts with involvement of the dealers
- Direct online distribution of vehicles and parts reduces the dealer business
  - Define a tuned channel concept around central distribution centres for the vehicles ("hubs"); integrate dealers in after-sales services and the distribution of new products
- The majority of importers and dealers do not belong to the manufacturers yet are rather independent companies. The dealers "own" the customer information. The IT infrastructure in the dealer's possession is often outdated and inadequately integrated with the manufacturer. Due to low margins, there are little investment opportunities for innovation and digitisation.
  - Providing open Cloud-based trading platforms for dealers; sharing of customer data between dealers and manufacturers for compensation.
- Distribution channel for mobility services, connected services and digital products
  - Focused sales in the mobility organisation; use of partnerships and established platforms, for example, in the distribution of connected services via App Stores.

Building upon the fundamental decisions on these questions, the digitisation projects must be prioritised on a roadmap. Independent of this, the efficiency of the sales processes is to be continually improved by the use of modern IT solutions in accordance with the approach described in Sect. 6.2.3.

Figure 6.32 shows digitisation initiatives of the automotive industry on a time horizon [Wei16]. The time axis reflects the stability and availability of the required technologies, as well as the readiness for implementation, while the vertical axis indicates the complexity of the solution. Multiple distribution channels are seen here as subject of high priority, followed by other sales channels such as the transformation in after-sales and in the service business, as well as the virtual dealers. In the medium term, the potential of 3D printing will also be qualified as high, besides other initiatives from adjacent areas such as autonomous driving, preventive maintenance and location-related services. The smart factory on the basis of cyberphysical systems is assigned a high complexity here, and it is seen as ready for practice only in the medium term. User-related insurance solutions will also be offered with the help of vehicle data integration.

The digitisation initiatives affect all three levels of the established sales structure, both at the manufacturer, as well as at the importers and the dealers. There are different estimations for the project benefits. An example is shown in Fig. 6.33.

Different effects of digitisation initiatives are shown as examples which, through the established three-level sales structure, affect everybody from the OEM over the

6.2 Roadmap for Digitisation

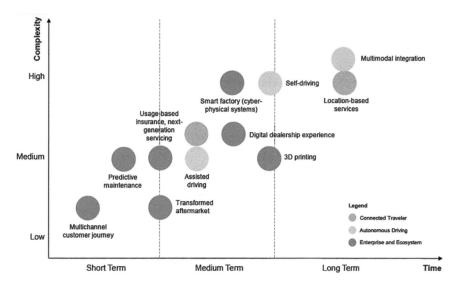

Fig. 6.32 Time horizons of digitisation initiatives [Wei16]

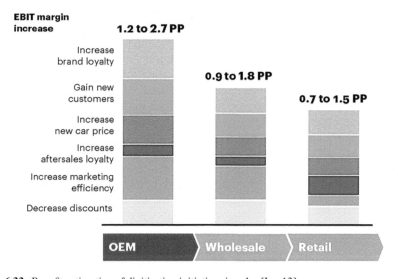

Fig. 6.33 Benefit estimation of digitisation initiatives in sales [Lan13]

importer through to the retail dealer and result in sales and earnings improvement there (values in percentage points). Through more efficient sales and marketing campaigns, more customer interest, so-called leads, is turned into purchases, and loyalty towards the manufacturer is increased by proactive customer support. Customised marketing is used more purposefully and shows more motivating buying effects at reduced effort, so that only minor discounts are required. Overall, sales increases of about 9.5% on the average are expected at each sales stage, while at the same time improving the earnings situation; at the manufacturer it is up to 2.7% and in the sales stages on average about 1.3%. These effects result from the efficiency increase in the existing processes and structures in the vehicle business. New services, structural adjustments and new products are not included in the estimates shown.

6.3 Overview Roadmap and KPIs

As an essential part of a general approach to the development of a digitisation roadmap for a manufacturer-wide digitisation program, a framework concept with four thematic fields and appropriate recommendations for action was proposed in Fig. 6.11. An approach was also developed for the cross-sectional topic of "Increasing the efficiency of the process" through the use of digitisation solutions. This approach was deepened for the areas of administration, development and production with a focus on Industry 4.0. The summarising overview from this is depicted in Fig. 6.34.

The columns show the four digitisation fields and the three most important initiatives for implementing a digitisation strategy. These have been derived from the detailed explanations in this chapter and do therefore not need to be deepened. Rather, the focus is on pointing out the projects and their objectives at a glance. A similar picture should be developed by every manufacturer as the guiding principle of the company-specific digitisation strategy and used as an introduction to communication about the direction. It is important to define concrete measurands for the project objectives and progress on the respective initiatives. The objectives are manufacturer-specific, aligned with the respective initiative, and derived from studies as needed. Information on key figures and methods is for instance in [LeH16], [Col14]. Examples of such figures in digitisation are:

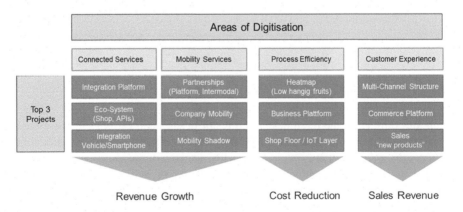

Fig. 6.34 Top digitisation initiatives (Source: Author)

- Strategic orientation
  - Targeted sales, earnings contribution and market share from established vehicle business, mobility services, connected services and digital products with specification at least per the main markets
  - Sales per sales channel – traditionally, but also on the basis of new platforms in the multichannel structure
- Digitisation roadmap
  - Number of employees active to develop new products
  - Investment in digitisation projects
  - Number of strategic partnerships with focus on digitisation
  - Number of Apps developed outside the company (crowdsourcing)
  - Acceptance (active users, followers) of the manufacturer in social media
- Digitisation projects
  - Number of employees and Scrum teams
  - Number of use cases and story points
  - Duration MVP (Minimum Viable Product)
  - APIs, Microservices

The setting of target variables and their communication is part of a culture that promotes and enforces the implementation of the digitisation strategies or the digitisation roadmap in an agile, open approach. The subject of culture as a key success factor in the implementation is discussed next in Chap. 7.

References

[All16] Allmann, C., Broy, M., Conrad, M, et al.: Embedded Systeme in der Automobilindustrie (Embedded systems in the automotive industry), Roadmap 2015–2030, Gesellschaft für Informatik, SafeTRANS, VDA; 2015. http://www.safetrans-de.org/documents/Automotive_Roadmap_ES.pdf. Drawn: 04.11.2016

[And16] Anderl, R., Bauer, K., Diegner, B., et al.: Plattform Industrie 4.0: Aspekte der Forschungsroadmap in den Anwendungsszenarien (Platform industry 4.0: Aspects of the research roadmap in the application scenarios), Bundesministerium für Wirtschaft und Energie (BMWi) (Editor), Result Paper, Berlin April 2016. https://www.plattform-i40.de/I40/Redaktion/DE/Downloads/Publikation/anwendungsszenarien-auf-forschungsroadmap.pdf?__blob=publicationFile&v=14. Drawn: 04.11.2016

[APQC14] APQC: APQC Process Classification Framework Automotive, Standard Business Model, 2014. https://www.apqc.org/knowledge-base/documents/apqc-process- classification-framework-pcf-automotive-members-bpmn-version-6. Drawn: 28.10.2016

[AUDI16] AUDI: Audi company strategy. http://www.audi.com/corporate/de/unternehmen/unternehmensstrategie.html#fullwidthparah. Drawn: 11.10.2016

[Ber16] Bernhart, W., Winterhoff, M., Hasenberg, J., et al.: A CEO's agenda for the ecosystem, Roland Berger GmbH, 2016. https://www.rolandberger.com/publications/publication_pdf/roland_berger_tab_automotive _intransition_20160404.pdf. Drawn: 23.10.2016

[BMWI16] BMWi: Die Digitalisierung der Industrie – Plattform, Industrie 4.0; Fortschrittsbericht (BMWi: The digitisation of the industry platform, industry 4.0; Progress report), editor Bundesministerium für Wirtschaft und Energie (BMWi), Berlin April 2016. https://www.bmwi.de/BMWi/Redaktion/PDF/Publikationen/digitalisierung-derindustrie,property=pdf,bereich=bmwi2012,sprache=de,rwb=true.pdf. Drawn: 04.11.2016

[Bra14] Bratzel, S., Kuhnert, F., Viereckl, R., et al.: Connected car study 2014 strategy &, PwC, CAM, short version 2014. https://www.pwc.de/de/automobilindustrie/assets/automobil-branche-das-vernetzte-fahrzeug-ist-das-grosse-thema-der-zukunft.pdf. Drawn: 03.10.2016

[Bra16] Bratzel, S.: Connected car innovation study 2016, Center of Automotive Management CAM. http://cci.car-it.com/download/CCI_Studie_2016_Web.pdf. Drawn: 06.10.2016

[Bro16] Broy, M., Busch, F., Kemper, A., et al.: Digital mobility platforms and ecosystems, state of the art report, July 2016, Project Consortium Technical University of Munich, Living Lab Connected Mobility. https://mediatum.ub.tum.de/doc/1324021/1324021.pdf. Drawn: 26.10.2016

[Col14] Colas, M., Buvat, J., KVJ, S., et al.: Measure for measure: The difficult art of quantitative return on digital investments. Capgemini Consulting, 2014. https://www.capgemini-consulting.com/resource-file-access/re source / pdf / measuring-digital-investments_0.pdf. Drawn: 08.11.2016

[Chr16] Christensen, C., Dillon, K., Hall, T., et al.: Competing Against Luck: The Story of Innovation and Customer Choice., HarperBusiness, 2016

[Dom15] Dombrowski, U., Mielke, T. (eds.): Ganzheitliche Produktionssysteme – Aktueller Stand und zukünftige Entwicklung (Holistic Production Systems – Current Status and Future Development). Springer-Verlag, Berlin/Heidelberg (2015)

[Die16] Dietel, M., Franken, R.: IBM-Perspektiven auf und Erfahrungen mitIndustrie 4.0 (IBM-Perspectives on and experience with industry 4.0), VDMA Seminar, 15.11.2016. https://zentrum-digitalisierung.bayern/wp-content/uploads/IBM_I40_VDMA_15.11.2106-ext.pdf. Drawn: 06.03.2017

[ECO16a] Economist: Uber – From zero to seventy (billion), The Economist, 03.09.2016. http://www.economist.com/news/briefing/21706249-accelerated-life- and-times-worlds-most-valuable-startup-zero-seventy. Drawn: 06.10.2016

[ECO16b] Economist: It starts with a single app, Economist, 01.10.2016. http://www.economist.com/news/international/21707952-combining-old-and-new-ways-getting-around-will-transform-transportand-cities-too-it. Drawn: 26.10.2016

[Eis17] Eisert, R.: Zetsche bestätigt Milliarden-Umsatzziel von Car2Go (Zetsche confirms billions sales target of Car2Go), Wirtschaftswoche, 21.07.2016. http://www.wiwo.de/unternehmen/auto/daimler-zetsche-bestaetigt-milliarden-umsatzziel-von-car2go/13908452.html. Drawn: 06.10.2016

[Fre14] Freese, C., Schönenberg, T.: Shared Mobility – How new businesses are rewriting the rules of the private transportation game, Roland Berger Study 2014. https://www.rolandberger.com/publications/publication_pdf/roland_berger_tab_shared_mobility_1.pdf. Drawn: 13.10.2016

[Kop10] Koppinger, P., Ban, L., Stanley, B.: Component Business Modeling, IBM Institute for Business Value, 2010. http://www-05.ibm.com/services/bcs/at/industrial/download_ind/a_ge510-3633-00f.pdf. Drawn: 26.10.2016

[Kos16] Koster, A.: Das vernetzte Auto im Zentrum der digitalen Disruption (The networked car at the center of digital disruption), Lecture Automotive Electronics Congress, 14./15.6.2016 Ludwigsburg

[Kot15] Kotrba, D.: Eine App, um alle Verkehrsmittel zu benutzen (One app to use all means of transport), futurezone, 27.07.2015. https://futurezone.at/digital-life/eine-app-um-alle-verkehrsmittel-zu-benutzen/138.423.413.Drawn: 26.10.2016

[Lan13] Landgraf, A., Stolle, W., Wünsch, A., et al.: The new digital hook in automotive, White Paper A.T. Kearney, 2013. https://www.atkearney.de/documents/856314/3032354/BIP+The+

New+Digital+Hook+in+Automotive.pdf/457380e6-78c8-4f09-8612-4a5e33788fb7. Drawn: 08.11.2016

[LeH16] LeHong, H.: Digital business KPIs: Defining and measuring success, Gartner Research Report, March 3, 2016. Drawn 08.11.2016

[Men16] Mennesson, T., Knoess, C., Herbolzheimer, C., et al.: Traditionelle Un-ternehmen in der digitalen Welt – Nachzügler haben das Nachsehen (Traditional companies in the digital world – laggards lose out), study Oliver Wyman, 2016. http://www.oliverwyman.de/content/dam/oliver-yman/europe/germany/de/insights/publications/2016/apr/2016_Oliver_Wyman_Traditionelle_Unternehmen_web.pdf. Drawn: 11.10.2016

[Mor16] Mortsiefer, H.: Strategie 2015: Volkswagen baut radikal um (Strategy 2015: Volkswagen is radically transforming), Der Tagesspiegel, 16.06.2016. http://www.tagesspiegel.de/wirtschaft/strategie-2015-volkswagen-baut-radikal-um/13745924.html. Drawn: 06.10.2016

[Mat16] Matthes, F.: Städtische Mobilität in der digitalisierten Welt. (Urban mobility in the digital world.) Karlsruher Entwicklertag, 15.06.2016. http://fg-arc.gi.de/fileadmin/Architekturen_2016/Matthes-TUMLLCM_Hildesheim.pdf. Drawn: 26.10.2016

[May16] Mayer, M., Mertens, M., Resch, O., et al.: From SOA2WOA, Guidelines of Bitkom eV, 2016-10-21. https://www.bitkom.org/Publikationen/2016/Leitfaden/From-SOA2WOA/160128-FromSOA2WOA-Leitfaden.pdf. Drawn: 21.10.2016

[Nor16] Norton, S.: How Ford is building the connected car, Wall Street Journal 21.02. 2016. http://www.wsj.com/articles/how-ford-is-building-the-connected-car-1456110337. Drawn: 23.10.2016

[Ola16] Olanrewaju, T., Smaje, K., Willmott, P.: The seven traits of effective digital enterprises, McKinsey & Company article, May 2014. http://www.mckinsey.com/business-functions/ganization/our-insights/the-seven-traits-of-effective-digital-enterprises. Drawn: 13.10.2016

[PSA16] PSA: Connected Car; Multi-device connectivity: Develop the technological ecosystem for the car often the future. https://www.groupe-psa.com/en/newsroom/tagged/connected-car. Drawn: 23.10.2016

[Ren16] Renner, T., von Tippelskirch, M. (eds.): Shared E-Fleet – Fahrzeugflotten wirtschaftlich betreiben und gemeinsam nutzen (Shared E-Fleet – Economical operation of vehicle fleets and shared use), Shared E-Fleet consortium, research projects BMWi; Final Report 2016, Fraunhofer-Institut für Arbeitswirtschaft und Organisation IAO. http://shared-e-fleet.de/index.php/de/downloads. Drawn: 26.10.2016

[Röm16] Römer, M., Gaenzle, S., Weiss, C.: How automakers can survive the self-driving era, ATKearney Study 2016. https://www.atkearney.com/documents/10192/8591837/How+Automakers+Can+Survive+the+Self-Driving+Era+%282%29.pdf/1674f48b-9da0-45e8-a970-0dfbd744cc2f. Drawn: 03.10.2016

[SAP14] SAP: SAP roadmap for automotive, 2014. http://www.uniorg.de/images/downloads/leistungen/sap_loesungen/aio/uniorg_sap_aio_automobilzulieferer_en.pdf. Drawn: 26.10.2016

[SchR15] Schmöl, R.: 6 Beispiele für Digitalisierung in der BMW-Produktion (6 examples of digitisation in BMW production); 12/08/2015, CIO of IDG; IDB Business Media GmbH. http://www.cio.de/a/6-beispiele-fuer-digitalisierung-in-der-bmw-produktion,2881919. Drawn: 04.11.2016

[Sch16] Schneider, M.: Audi: Digitalgeschäfte sollen 2020 die Hälfte des Umsatzes ausmachen (Audi: Digital business to account for half of sales in 2020), BILANZ, 04.02.2016. http://www.presseportal.de/pm/114920/3243234. Drawn: 11.10.2016

[Thi15] Thiele, J., Schmidt-Jochmann, C.: Geschäftsmodell der KFZ- Versicherung im Umbruch (Business model of vehicle insurance in upheaval), study Roland Berger GmbH, 2015. https://www.rolandberger.com/publications/publication_pdf/roland_ber ger_kfz_versicherungen_im_umbruch_20151014.pdf. Drawn: 03.10.2016

[TUD16] Technische Universität Darmstadt: 25 years Lean Management, study of Staufen AG and the Institute OTW of the Technical University Darmstadt, 2016. http://www.staufen.ag/fileadmin/hq/survey/STAUFEN.study-25-Jahre-lean-management-2016.pdf. Drawn: 06.10.2016

[Vie16] Viereckl, R., Koster, A., Hirsh, E., et al.: Connected car report 2016 Study Strategy &, PwC. http://www.strategyand.pwc.com/media/file/Connected-car-report-2016.pdf. Drawn: 03.10.2016

[VDI13] VDI: VDI Guideline 2870, Sheet 2 - Ganzheitliche Produktionssysteme Methodenkatalog (Integrated production systems catalogue of methods), Verein Deutsche Ingenieure, March 2013

[Wed15] Wedeniwski, S.: Mobilitätsrevolution in der Automobilindustrie Letzte Ausfahrt digital! (Mobility Revolution in the Automotive Industry Last Exit digital! Springer Berlin/Heidelberg 2015

[Wei16] Weinelt, B. (ed.): World Economic Forum Davos, White paper: Digital transformation automotive industry 2016. https://www.accenture.com/t20160505T044104__w__/us-en/_acnmedia/PDF-16/Accenture-wef-Dti-Automotive-2016.pdf. Drawn: 08.11.2016

[Wes12] Westerman, G., Tannou, M., Bonnet, D.: The Digital Advantage: how digital leaders outperform their peers in every industry, study Capgemini Consulting and MIT Sloan, Management, 2012. https://www.capgemini.com/resource-file-access/resource/pdf/The_Digital_Advantage__How_Digital_Leaders _Outperform_their_Peers_in_Every_Industry.pdf. Drawn: 13.10.2016

[Wie14] Wiendahl, H.-P.: Betriebsorganisation für Ingenieure (Poduction Management for Engineers), 8th edn. Carl Hanser Verlag, Munich (2014)

[Win15] Winterhoff, M., Kahner, C., Ulrich, C., et al.: Zukunft der Mobilität 2020, Die Automobilindustrie im Umbruch (Future of mobility 2020, The automotive industry in upheaval), Study Arthur D Little, 2015.http://www.adlittle.de/uploads/tx_extthoughtleadership/ADL_Zukunft_der_Mobilitaet_2020_Langfassung.pdf. Drawn: 07.08.2016

Chapter 7
Corporate Culture and Organisation

Electrically driven autonomous vehicles are pushing the trend towards mobility services that are available conveniently as required through Apps via smartphones. In the future, a "robotaxi" will be ready for the desired departure date and will drive its customers to the destination and then, while in the background the payment of the services is being processed online, is already on the way to the next customer. The vehicles are linked via Connected Services and guided efficiently through the traffic by a superior traffic control system within the "mobile shadow", evading traffic jams. With these comfortable services, vehicle ownership is at least in the cities of minor importance. Only in the "emerging markets" and also for a few vehicles in the "enthusiast segments", such as sports cars and luxury models, the traditional vehicle business is continuing. This vision of the future automotive industry was explained in Chap. 5 in detail and will be discussed in Chap. 10 in the long-term vision.

Mobility service providers operate different business models. Vehicles for own fleets of providers are to be bought, while in "shared concepts" private or company vehicles are also used. Overall, the demand for vehicles is decreasing as a result of the higher utilisation levels in the service business. In order to be able to participate competently in this changed market environment with new offers to compensate for the drop in revenue from the sale of vehicles, it is imperative for the automotive industry to reinvent itself.

The proven strategy of establishing innovation through powerful engines, elegant bodies, new materials and efficient production technology, will no longer suffice. Manufacturers must establish new strategies and new business models with new offers. The options range from being a "mobility provider with attached vehicle production" up to "Foxconn for mobility services providers or high-performance manufacturers", but also as hybrid forms, depending on the target market. After defining the strategy, manufacturers will have to use comprehensive transformation and digitisation initiatives in order to align their company with the new goals. This was the subject of Chap. 6.

With the impending profound changes, the challenges are not so much the control and availability of the necessary digitisation technologies. It will be much more difficult to motivate all employees to actively participate in transformation and change, and not to stay sceptical and stuck in old behaviour patterns and procedures. Today's corporate culture is often still dominated by hierarchical structures and traditional value systems. A new "digital culture" is required to break up the associated encrustations and create a positive spirit of departure as the basis for the change. This is characterised by curiosity, readiness to change, and flat hierarchies. Speed and agility prevail over formal, slow process flows.

The topic of the adaptation of culture as well as the creation of a start-up mentality as an essential prerequisite for successful transformation driven by digitisation is the subject of this chapter. To this end, new methods for innovation and project management are presented, and organisational proposals are discussed from the viewpoint of digitisation. These are the basis for the discussion as to which organisational adjustments are necessary for the digitisation.

7.1 Communication and Leadership

A corporate culture is reflected in the natural behaviour of its employees among themselves and in the face of customers including the practised values, the climate and the morale. Culture always is also strongly influenced by the company's history and roots, the reputation of its products in the markets, the feedback from customers and the economic environment. It forms the basis for the success of the company, to which the employees contribute purposefully and with motivation to transpose the business strategy [Zel15].

Building on the well over 100 years of history of the automobile and the continuing economic industry success, the culture based on the relevant fundamental values of the company is firmly entrenched in the behaviour of staff. Job outlines and career models have been the same for generations and have often been passed on within families. This established culture must now be sustainably modified in order to successfully implement the digital transformation necessary for the manufacturers to survive.

According to a recent study, company culture is significantly influenced by the following factors [Eil16]:

- Communication
- Leadership
- Flexibility/readiness for change
- Diversity
- Transparency
- Participation

The study evaluated where the surveyed companies see themselves in these areas and their preparedness for transformation projects. The findings are that the companies have on average so far at best only achieved a satisfactory status. As an example, Fig. 7.1 shows the results for the area of communication, which is the most

7.1 Communication and Leadership

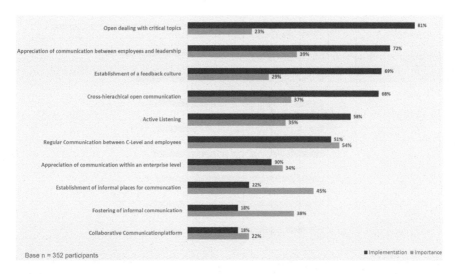

Fig. 7.1 Corporate culture; communication topics and their implementation [Eil16]

important component of cultural influence, closely followed by leadership and willingness to change.

Within the field of communication, the largest gap is in the open handling of critical topics, followed by the feedback culture, the appreciation in the communication of managers with their employees and the communication across hierarchical levels. Unlike that, some of the topics which are easy to realise are well-advanced, albeit of minor importance. Examples of this are the communication of the management with employees, the establishment of places for informal exchange and also the promotion of the formal information exchange. Thus, it is clear in which areas to begin with improvements to the communication. Critical topics must be addressed proactively, even across hierarchical levels, in an appreciative dialogue. These enhancements are clearly a task of the leadership, which cognisantly addresses the relevant implementation gaps and demonstrates exemplary conduct.

Further tasks and challenges for executives, especially in terms of digital transformation, were also assessed in the study. An overview on the results is shown in Fig. 7.2.

Change Management, the handling of growing complexity, the conscious perception of the role model and the creation of transparency in processes are the biggest challenges and show a considerable backlog in the implementation. Other management tasks, such as the creation of a climate to safeguard a work-life bal-

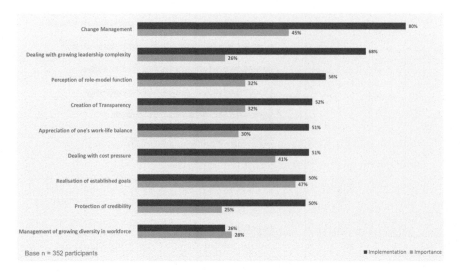

Fig. 7.2 Challenges for executives and their fulfilment [Eil16]

ance, the handling of cost pressures and to maintain credibility, also still show significant deficits in implementation.

In order to increase the ability to transform, executives need to strengthen and retain traditional leadership elements but must at the same time learn to identify new business models and manage the necessary digitisation initiatives. In order to meet these challenges, specialist studies suggest that successful executives act in an "ambidextrous" manner [ORe13]. In the context of innovation and transformation, this means the ability to manage traditional improvement measures in established business areas and alongside these to also develop and drive new disruptive models.

In the past, the manufacturers were almost exclusively focused on the continuous improvement of the existing structures in order, for example, to avoid wasteful production processes by Lean methods or to secure quality through Six Sigma approaches.

However, agility, a willingness to experiment and some degree of risk are typical conditions for the development of ground-breaking ideas. These new skills are to be encouraged, as they determine the cultural change required for digitisation. Previous hierarchical structures and value systems are more and more fading into the background, and project-related forms of organisation, thinking "outside the box", flexibility, willingness to learn, responsibility and an atmosphere of departure are needed. Executives must be able to introduce their employees to this new world, to create degrees of freedom as well as to motivate teams in new organisational forms, to structure complex subjects and to work on the solutions without fixed targets in new approaches. In order to increase understanding and communication across hier-

7.1 Communication and Leadership

archy levels, "reverse mentoring" should be used as an innovative method. In this process, "seasoned executives" regularly meet with a digital native of their company as a personal mentor for coaching on current topics in the field of digitisation. In this way, blockades are removed and an open communication is lived.

As already stated, the basis for this new culture is the vision of the company and, on this basis, the business strategy, which must be communicated internally and externally with clear and measurable goals. Every employee, partner and customer should understand what the company's goals are and how these are to be achieved. To this end, it is to be shown which basic attitude the company expects as a cultural component of its employees. The communicated orientation is then to be deepened in concrete terms through credible example by the executives.

These references to the change in corporate culture are considered to be sufficient for the purpose of this book. For further detail, if needed, please refer to the specialist literature which deals scientifically and extensively with the subject of cultural change [Zel15].

With its "Leadership 2020" program, the German company Daimler is a good example of this clear communication and of a new corporate culture as an authentic role model for digitisation. To illustrate this, here is a quotation from the "Innovation" section of the company's website [Daimler]:

> Digitisation will continue to gain momentum. All the major trends in the industry are already being driven by digitisation nowadays or are driving it forward themselves. We will gradually establish a new innovation culture at Daimler. This is the only way to link the strengths of a global corporation even more closely with the strengths of a Start-up. ... We will continue to be ahead in the digital age as well and create the conditions for an agile and interconnected organisation. In this way, our employees can make full use of their creativity and realise their ideas. ... Our culture is characterised by transparency, trust and flexibility. We place our trust in the abilities of our employees and thus give them the opportunity to actively contribute and shape the "tool" and know-how for the digital future of the Daimler AG. Employee events such as the DigitalLife Days, Open Spaces & Hackathons, the international DigitalLife Roadshows and our internal social network Daimler CONNECT are also available.

The company's focus on digitisation and the necessary changes in corporate culture become apparent here. On the way to an innovation culture, the leadership relies on the creativity and knowledge of the employees and supports them with training courses, tools, innovative projects and open communication based on internal social networks. The reorientation is supported by the authentic appearances of the CEO Dieter Zetsche at the opening of motor shows in jeans and sneakers. Even his speech at a Green Party conference (13.11.2016) in which he predicts that the automotive industry only has a future if it develops emission-free vehicles, reaffirms the new corporate culture. The company's well-known investments and early-stage projects in the area of car sharing, intermodal transport, connected services and assistance systems make the new strategic orientation of the company plausible (cf. Fig. 6.19 in Sect. 6.2.2). To change the culture implies the application and exemplification of new procedures, methods, and tools for project handling and cooperation for instance. Some are therefore presented below.

7.2 Agile Project Management Methods

Many manufacturers rely on agile methods and procedures to achieve fast project successes in cross-divisional teams. Originally, these approaches came from the field of software development but are increasingly being applied to the processing of other work as well. This allows new ideas and approaches to be developed pragmatically, quickly and tested for feasibility and success. These procedures offer an ideal framework for the transformation of the company culture through the consistent application and exemplification.

Agile procedures in software development were already published and applied in the 1990s. The dissemination accelerated considerably through the formulation of the so-called Agile Manifesto in 2001, which was published by 17 well-known software developers and meanwhile has several thousand signatories [Bee01]. The Manifesto defines four values and 12 principles of agile development. A value is, for example, that the cooperation with the customer is more important than the contract negotiation, and the responding to changes has priority over the execution of a plan. Principles are, for example, the delivery of working software in regular short time periods, the almost daily collaboration of the experts, the constant attention to technical excellence and good design, the essential simplicity as well as the self-organisation of the teams in planning and implementation [Bee01]. The values and principles radiate customer orientation, flexibility and dynamism and make it clear why agility is being sought by many companies to promote corporate culture in the sense of digitisation.

Building upon the values and principles, a variety of methods have established, supported by different practices and tools. This context is illustrated in Fig. 7.3.

The picture shows the interrelation between agile terms, from the manifesto and methods to practices, techniques, and tools. In the field of methods, Scrum and Kanban are very common, while as techniques within the methods Story Point processes are popular for describing application scenarios and so-called burn-down charts to show the progress of the creation.

Figure 7.4 shows the results of a study in which six Agile Methods and the classical approach in project management were evaluated by users according to seven evaluation criteria, such as the quality of results, the timing and efficiency. The Agile Methods perform better in each sub-valuation, with one exception, than conventional project management. Of the Agile Methods, Scrum is rated the highest by users. Apart from Scrum, Design Thinking is highly rated particularly in the areas of employee motivation, teamwork and customer orientation. The method is therefore used as a general approach also outside of software production, for example, in innovation workshops and in brainstorming. Punctuality is generally worse in all methods, especially in classic project management, and Scrum is rated best. It should be noted, however, that agile methods often work according to the so-called time-boxing principle, in which the scope of work is variable but the end date is maintained.

7.2 Agile Project Management Methods

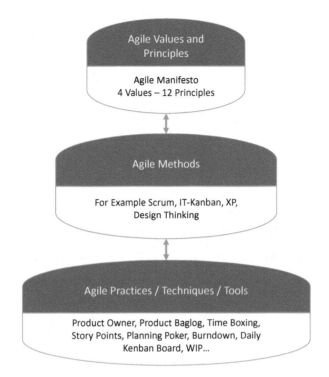

Fig. 7.3 Hierarchy of the concepts of Agile Values, Agile Methods, and Agile Techniques [Kom15]

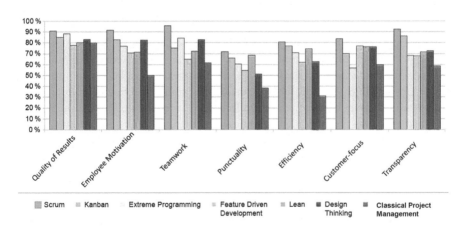

Fig. 7.4 Evaluation of Agile Methods in project use [Kom15]

Design Thinking and Scrum are often used by the manufacturers, at a high percentage in IT projects, but increasingly also in innovation management. Both methods are briefly described below. For a deepening of further agile procedures, relevant studies and technical literature are available, e.g [Rig16], [Now16].

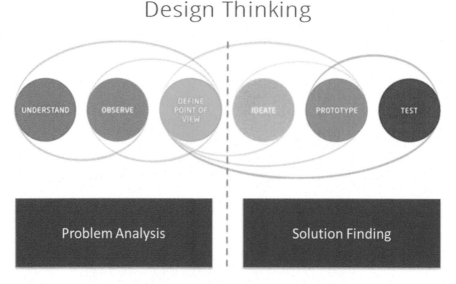

Fig. 7.5 Process steps in the Design Thinking approach [Lob16], [Pla09]

7.2.1 Design Thinking

Design Thinking is an agile approach for solving of problems or developing new ideas. It is important that customer expectations and wishes are the focus of this method, while the technical feasibility and economic aspects only follow in the course of the process. The core of the approach is to develop creative solutions in an iterative approach with interdisciplinary teams in close cooperation with the future customers. Aside from the interdisciplinary groups and the interactive approach, it is crucial for the success that the work is in an appealing atmosphere in rooms which can be designed flexibly, ideally also outside the familiar environment, in order to enable an additional boost to creativity. In order to prepare workshops, the behaviour and processes are recorded at the customer, expectations are determined in dialogue and, situations are documented in typical "a-day-in-the-life" flowcharts for instance [Ger16]. A wide range of other methods and tools are used in the course of the method. There is no standardised approach for the Design Thinking concept, but six process steps have been established, which are shown in Fig. 7.5.

The six working steps are each to be passed iteratively. In repeated runs, new findings or feedback information from the future user can be included. During the process, analytic methods are coupled, for example systematic information evaluations with intuitive methods such as brainstorming and visualisation.

The beginning of a project must be the understanding of the problem situation from the customer's point of view. In doing so, it is important that all team members understand the situation and the task as completely as possible and internalise these

7.2 Agile Project Management Methods

for the next steps. In the next phase of observation, it is important to speak with the customer in depth and to consult with experts on solutions from similar problem areas, for example with assembly workers or logistics companies from the environment in question. Based on this, to conclude the problem analysis, the various information of the team members is consolidated, evaluated together and a position defined in the third working phase. Only then follow three steps to find the solution.

Different creativity techniques are used to develop as many ideas as possible. These are initially just collected and neither evaluated nor commented. Afterwards only, the team evaluates and selects them according to criteria such as technical feasibility, cost-effectiveness and assumed customer acceptance. For the selected favourites, prototypes are created in the fifth step. In software projects, the team is already trying to present the first workflows with a minimal range of functions, for example in the form of Apps, using modern development kits. For representation, Lego or wood models can be used for mechanical problem solutions, and role playing in the team for service processes.

The prototypes are not about the almost perfect implementation of the idea, but rather the illustration of the solution as a basis for the evaluation and development of an improved solution taking into account the requirements. In the final phase, future customers will test the prototype. With the feedback from the tests, the solution is improved in further iterations. In order to confirm the solution and to start the realisation, the technical feasibility has to be confirmed, the economic viability is required, and also all customer requirements must be workable.

The interactive method of the Design Thinking concept is very close to the future customer with a flexible approach and a sound success assessment in an early phase, and it has by now proved itself in many projects [Sch15]. The approach is not only used for the development of new products and business models, but also for the improvement of internal processes and procedures, as well as for the design of more user-friendly software systems.

According to the Hasso Plattner Institute HPI, among the users of Design Thinking are for example companies Airbnb, BMW, DekaBank, DHL, Freeletics, Volkswagen and SAP [Sch15]. Often however, the projects address only individual problems and thus mostly create isolated solutions. The potential of Design Thinking is notably in the area of interdisciplinary work and the creation of creative solutions across departmental boundaries.

In many cases, these possibilities are still to be explored, as a study in this regard shows [Sch15]. Focusing on the customer is at the forefront of the Design-Thinking process and is, besides the analysis and survey of the target group and the personal testing of initial situations, highly important. To try out sequences in role-playing games themselves and to convert these experiences into creative solutions leads to effective team building in the workshops. As a result, this approach, with its open, iterative approach, positively affects corporate culture and communication across departmental borders.

7.2.2 Scrum

Another agile method which is already used by many manufacturers and which offers a comprehensive utilisation potential not only for software development, is Scrum. The term originates from Rugby where it describes an "ordered crowd" to restart the game after minor violation of the rules [Fle14]. This image reflects a dynamic project work with sporadic meetings and an orderly restart. The scrum process also runs iteratively, whereby a project objective is divided into partial steps which the team works on in creation loops, so-called sprints. The sequence of a scrum project, the roles and the essential elements of a Scrum run are illustrated in Fig. 7.6.

The team should comprise between 5 and 10 members and is responsible for the creation of the product in self-organisation without a hierarchy. The method distinguishes three roles. The product owner represents the end customer and his work order and is responsible for the definition and prioritisation of the requirements. He leads the development. The Scrum Master supports the team and is responsible for the organisation and the basic conditions for a smooth work. From the work assignment the team derives the requirements, formulated in the words of users in so-called User Stories, depending on the extent also supported by several Story Points. The User stories form the product backlog which includes a collection of all the

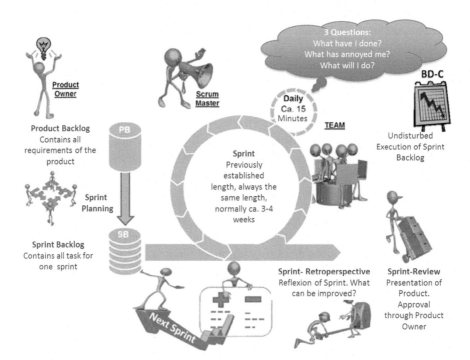

Fig. 7.6 Sequence of the Scrum method [Kom14]

functions and features which the product should have. These descriptions are complemented and refined throughout the project.

From the backlog, the team selects in its sprint planning the tasks to work on in the upcoming work cycle, the sprint. The selected worklist for the team is available in the form of tickets in the sprint backlog. From this stock, the team members take over responsibility for their own tickets to handle. During brief daily stand-up meetings, everyone talks about their work, to include what was perceived as irritating or obstructive, and what to tackle in the next step. The so-called Breakdown Chart (BD-C) documents the status of the actual processing of the work volume compared to the plan.

The duration of the sprints is constant. In the so-called "Time Boxing" procedure, the team adjusts the extent of work, considering priorities. At the end of the sprint, the flows and results are analysed in the retrospective, and necessary improvements are made for the upcoming runs. The work results of the sprints, the partial solution of the product, is given to the end customer, represented by the product owner, for testing. Taking into account the feedback, adjustments of the work results or also additions to requirements are made.

A characteristic feature of the Scrum method is the iterative approach in fixed cycles with self-organising teams. The close integration of the end customer into the assessment of interim results from the sprints as well as the direct implementation of feedback and the flexible adaptation to changed requirements ensure that the final product meets the expectations. The transparency of the worklists, the work statuses, the daily communication and the open teamwork create a motivating work atmosphere and quickly lead to desired work results.

In all, the use of agile methods is highly recommended. However, it should be borne in mind that these do not work in all situations, and there are quite a few areas which are better approached using proven waterfall concepts. Based on the author's experience, Fig. 7.7 simplifies recommendations as for which framework conditions which method is appropriate.

The traditional approach of the waterfall method from outline into details is always preferable when it is necessary to implement established processes with clear requirements using existing tools in tried and tested procedures. Even though

Waterfall method	Agile Methods
➢ **Project focus:** static processes and requirements	➢ **Project Focus:** Innovative solution for requirements are fixed
➢ Established procedure based on existing tools	➢ Cross-divisional objective with high need for co-ordination
➢ Limited availability of know-how and decision-making authority in the team	➢ Know-how and decision-making powers available in the team
➢ Test the project result only as a complete product	➢ Permanent end customer integration possible
	➢ Iterative testing possible

Fig. 7.7 Method selection in project management (Source: Author)

in the implementation phase end users with overlapping process knowledge and decision-making authority are available only to a limited extent, or the project cannot be tested iteratively, the waterfall method makes sense. Whenever innovative solutions with not yet fully defined requirements are to be found and interdisciplinary teams with know-how and decision-making mandates are available, agile methods are the preferred option.

7.3 Entrepreneurship

As a result of the increased use of agile project management methods, the project members are developing new attitudes and habits which are key to the envisaged "digital culture". The basic approaches of these methods are:

- Trust and confidence in the employees
- Independence
- Use of group intelligence
- Learning by experiment
- Acceptance of errors and failure
- Give feedback and reflect
- Result orientation in iterative phases
- Self-organisation

Working together in interdisciplinary teams encourages creative, independent work in the development of new ideas and handling of projects and is then more and more becoming part of the corporate culture. Many of these elements do not conform to the well-established traditional leadership elements of classic Command & Control and Micromanagement, which have promoted the lack of independence and working in fixed hierarchical structures. But it is important to alter just this approach in the future in order to enable innovation, creative thinking, cross-sectoral work and own initiative.

In this transformation, the executives have an important role to play. It is necessary to integrate the corresponding elements into their own style of leadership and to live them. It should be relied on open, team-oriented coaching, and workers need to be encouraged to take risks and responsibilities, but also to look beyond the boundaries of their own organisation. This is the biggest challenge for the organisation, and it is precisely this which, according to the author's experience, is lacking with some established manufacturers. In many cases, the focus is almost exclusively on optimising one's own department.

Ultimately, this thinking must be changed, as well as the breaking of existing structures through more entrepreneurship must become a further part of the enterprise culture. There is no single definition of this term, which has been used for many years and is also the subject of research projects. In the sense of the intended digital culture, the explanations of the online Lexikon der Gründerszene (lexicon of the founders' scene) describe in the distinction between manager and entrepreneur well what is needed in the future [Grü16]:

The present understanding of entrepreneurship was essentially characterised by the work of the economist Joseph Schumpeter, who differentiates between the entrepreneur and the manager. According to Schumpeter, the manager is a company administrator who has little to do with creation of the new. It is the entrepreneur who acts as an innovator and takes up and establishes new ideas. In the context of the innovation process, the entrepreneur first destroys existing structures before creating new and better structures. He is also heavily involved in the development of these.... An entrepreneur does not invent, yet transforms what is presently there through successful restructuring and an analytical understanding of the market into innovations.... Entrepreneurship is characterised by the ability to recognise and take up market opportunities and to realise these in pursuit of profit. This includes the coordinated use of resources as well as the calculated assumption of risks. Entrepreneurship is therefore a three-step process consisting of the identification of market opportunities, the development of business ideas, and their implementation.

This definition vividly describes how a corporate culture should evolve towards more entrepreneurship. It is necessary to establish more start-up mentality coupled with a positive atmosphere of departure and innovation. The employees should be encouraged to engage in the transformation process like an entrepreneur [Fal11]. An example of this is a well-known founder in the USA, who was unable to complete the purchase of a vehicle entirely online. He recognised the opportunity and founded the start-up "Drive Motors", which now offers eCommerce solutions for car dealers, among other things the "Buy Online" option in the dealer's Internet platform. After a short time, already more than 150 dealers use this function, and sales sharply increase [Som16]. This is, on the one hand, exemplified entrepreneurship which the manufacturers urgently need, but on the other hand, it also is a missed opportunity for manufacturers to move forward with their own creative ideas.

7.4 Resourcing for Digitisation

Further to the change in corporate culture and the resulting employee behaviour, another important prerequisite for successful implementation of digitisation projects is the availability of employees with the necessary knowledge and experience. There are different options to secure the appropriate capacity:

- In-company training of permanent staff
- Hiring of digital natives and leaders with relevant knowledge and experience
- Cooperation and partnerships with companies from the technological field involved

These options will be discussed in the following.

7.4.1 E-Learning as Basis for Digital Education

In order to implement digitisation initiatives, the employees must be trained in the new IT technologies and also in agile project management methods in order to actively participate in the project implementation and to be prepared for the new type of work. There are different ways of training, such as frontal teaching, internships, online courses, or self-study. These traditional routes are not detailed within the framework of this book, nor are didactic questions and general adaptations, which are necessary in the context of the digitisation in vocational education and study. Rather, the focus is on how digital solutions for the training can be used up-to-date in order to provide learning contents and to offer them to the learner as required, flexibly and individually within the course of e-learning.

E-learning is understood as all forms of learning that use digital solutions for the presentation of learning material and the dialogues between learners and teachers [Ker12]. Since the 1980s, the so-called Computer Based Training (CBT) has first established, in which multimedial learning contents are often available on CD-ROM or even DVDs, in addition to traditional training formats. With the CBT, learning means self-study with no possibilities of interactive communication with a tutor or other learners.

With the spread of the Internet and Intranets in the companies, so-called web-based learning has established as an advanced development. Thereby, the distribution of the learning contents takes place online via the network. These are available on workstation computers via webbrowser or portal solutions, and more and more also via learning Apps on mobile end devices such as tablets or smartphones. Parallel to the study of the learning contents, the learner can have online dialogues with trainers and also co-learners. As a basis, companies use special IT solutions, often as so-called Learning Content Management Systems (LCMS). Figure 7.8 shows an example.

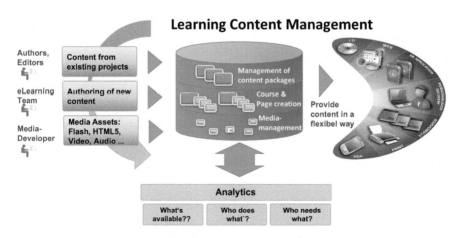

Fig. 7.8 Learning Content Management System (IBM image)

The core of these learning platforms are management systems for the provision of learning contents. They are hold in configurable, small modules for course management and also for the administration of media data. The learning contents are generated with authoring systems, whereby documentation from projects, such as user manuals or image material and information from forums and chats can also be integrated. The learning material is available at the learner's option on different systems. A general use by instructors in classroom teaching with large displays is possible as is a fully individualised e-learning based on a smartphone at any time of the day. In this case, the learning contents to cover certain learning needs for a particular activity, or taking into account the knowledge profile and the previous experience of the learner, are customised accordingly. The individualisation is based on the analysis functions of the learning systems, which evaluate learning histories, personal training data and activity profiles of the learner for instance.

These integrated systems thus enable the flexible provision of learning contents from different authoring systems on different workstations and the integration of different information sources. In this way, they support the organisation of learning events and ensure online communication between learners and trainers. The systems are web-based, so users only need an Internet connection and a web browser on their workstations. For mobile devices, Apps are normally available. Thus, learning is possible at any time and in a flexible manner. Due to the flexibility and the cost advantages, by now many manufacturers utilise e-learning for the further training of their staff.

As a basis, central learning platforms are recommended for all organisational units as a Cloud solution. The provision of learning content should also be carried out throughout the company on the basis of the same processes and tools, in order to ensure the most comprehensive use of the learning contents as well as to keep the content up-to-date with as little effort as possible. Then, for example, the pictures and the video of a learning module for machine operation can be used internationally. For local application, the learning system is easily adapted to the specific branding within the group, and the texts are translated into the user's language, while the structure and basic materials are the same for locations all over the world.

7.4.2 New Ways of Learning

E-learning is often used in combination with traditional learning methods such as external seminars or in-company classroom teaching. This so-called Blended Learning combines the advantages of both methods with didactically synchronised contents [Eic13]. The advantages of e-learning are in the efficiency and the flexible potential uses, while the presence education strengthens the social and team-building elements by the personal contacts, and only in this way the practising of activities is possible. Both learning methods complement each other, namely when accompanying social media such as Chat, Wiki, eBooks, interactive White Boards and file sharing are available. In this combination, Blended Learning has established itself as a standard procedure in companies, but also in schools and universities.

The widespread use of smartphones as a terminal for e-learning is accompanied by a clear trend towards so-called Microlearning. In this, learning contents are subdivided into small modules and retrieved "just-in-time" as needed, for example to provide specific support in the pursuit of a new work situation. This flexible use corresponds to the behaviour patterns of the digital natives with their always-on mentality (cf. Sect. 3.1). Current activities do not have to be interrupted for longer than necessary, and through this flexible "learning by doing" swift learning results are achieved.

Microlearning uses existing learning contents from the LCMS archive. During the preparation of the learning contents, this possibility should be considered when cutting the learning modules into small units, so-called learning nuggets. Current Microlearning solutions offer the user-controlled retrieval of learning contents. Self-learning cognitive solutions are expected in the future. These systems accompany the user in the background in their work, know their training and experience horizons, and offer proactive learning contents even before the need is recognised or retrieved.

Further opportunities for innovative learning include the integration of Virtual Reality technologies into learning solutions. For example, in a virtual environment, service sequences or machine control can be practised. In the final assembly, the operator can see the execution instructions displayed directly at the machine in Augmented Reality glasses. The support could be made available in a chat by an experienced colleague at a similar assembly station in another plant and called up as required. Further trends in the development of learning, which will also be used by companies in the future:

- Massive Open Online Course (MOOC) [Gab16]

- MOOC means free, open online learning offers with very large numbers of participants. Different course forms are available. For instance, universities such as Harvard, MIT or Udacity offer entire such courses. In contrast to this, in the interactive format, the participants develop their own learning material based on specifications. Such procedures could also be used across a manufacturer and be regarded as the signs of a new culture.
- Social Learning
- In a simplified way, this method involves shared learning in social media, for example, in thematic Blogs or Wikis with or without tutorial participation orchestrating the process.
- Cognitive Learning

- In the context of the increasing individualisation of learning opportunities, Big Data methods analyse relevant data of the learner in order to proactively configure learning opportunities and offer these in the preferred way.

This short outlook on further trends in learning highlights the fact that new learning needs, as well as new learning opportunities, are emerging as a result of digitisation. Taking these opportunities and the digitisation roadmap into consideration, manufacturers should develop an education planning system and implement it in learning offerings to reach out to both the digital natives and the older employees. The performance-oriented young employees use smartphones "naturally" and accept a smooth transition between leisure and work. Cognitive Microlearning on demand accommodates them, whereas older employees tend to organised Blended Learning offerings. The inclusion of older employees is important in order to use their experience in the project work and on the other hand in order to counteract the expected shortage of skilled workers through a prolonged working life of this employee group.

7.4.3 Knowledge Management

Especially in the context of growing complexity, quicker processes and the increasing employee fluctuation, the storage and the easy accessibility of knowledge is of great importance. Due to the demographic change and the upcoming retirement of a high proportion of long-term employees from important company functions, knowledge management is a subject of interest. It is important to systematically archive the experiences and long-term know-how, which is often unstructured in various documents such as presentations, emails and drawings, or even just in the minds of the employees, in order to make it available for subsequent use. Furthermore, the knowledge of employees, who are active at different points in a work process, has to be merged along the process chain in order to lay the foundation for overarching digitisation projects. It is also useful to electronically record and store the knowledge which is in different places in the company partly for similar tasks and ideally proactively suggest it as best practices for future work. Knowledge management thus becomes part of continuing training in the work life.

Efficient knowledge management gives companies clear benefits. The importance of knowledge as a differentiating resource is also shown by the recommendations of the ISO 9001:2015 standard [Bre16]. It calls for organisations to acquire and preserve knowledge in order to sustainably improve the quality of products and services. In order to meet this requirement, the necessary knowledge is to be identified, kept up-to-date and further developed in order to become effective in the company. The norm thus suggests the bridge between knowledge and learning. The recommendations of the standard imply also the implementation of powerful application solutions for knowledge management.

Today, a large number of standard software for knowledge management is available. The selection should focus on the user experience for end users in order to achieve a high degree of acceptance and motivation for sharing knowledge and to shape it as an element of corporate culture. The knowledge should be available on mobile devices with short response times. Powerful search algorithms and flexible integration in workflows are also important. Instead of isolated applications, standard packages across different sites and the whole organisation should be used wherever possible. The software should be intuitive and easy to use, without the need for additional learning, and facilitating a dialogue between experts and searchers for feedback and queries.

Conventional systems, unlike today's Web 2.0-oriented solutions, often do not offer these options. Since the new tools are mostly familiar from the private areas of life, employees expect them as solution in the company too. Typical software modules for knowledge management are:

- Knowledge databases
- Document management systems
- Search Engines, textmining
- Workflow solutions
- Collaboration tools
- Sharing systems
- Wikis, Blogs,

A large number of standard solutions are offered in these areas [Sie17]. Successful projects for knowledge management, however, are not primarily a question of software selection, but rather a matter of change management and cultural change in which the dimensions of technology, organisation and human beings are to be brought into line under a joint objective.

7.4.4 Hiring

Digitisation also advances in the personnel sector. Against the background of necessary hiring to expedite and safeguard the digital transformation, the following is a description of the change within the framework of the search for talents and new appointments in times of digitisation in order to give advice as to what should be considered in this area in the future.

The times of paper-bound procedures with job advertisements and applications by post are gone, at least in the areas relevant for digitisation. In addition to the active, often computer-based search for potential candidates on online platforms such as Xing or LinkedIn, job vacancies appear on the online channels of the manufacturer webpage, the industry or even the press. The digital applications are prefiltered using publicly available information on the applicant as per the search criteria. Suitable candidates receive, after having successfully completed an online test in the first selection step, an invitation to a first interview with a recruiter via webcam. If this selection stage is also mastered, an invitation to an assessment centre and deepening discussions is made. In this final selection phase, personality and soft skills are the focus of the decision.

The briefly described hiring process largely runs with digital support and can be, for instance in the active search for candidates and also in the analysis of applications, fully automated. However, recruiters' experience is required in assessing if candidates fit in with the company culture and the team structure. The described

7.4 Resourcing for Digitisation

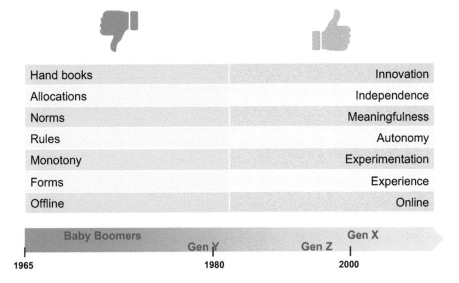

Fig. 7.9 Motivation elements of digital natives (Source: Author)

procedure assumes that the open position and also the searching company are interesting enough to motivate a large number of prospective applicants. However, demographic change and the growing need for digitisation experience will lead to fewer applicants for vacant positions. Therefore, in the future new hires will be based on an active search and approaching potential candidates on the social web.

With the help of intelligent search engines, suitable candidates are to be tracked and based on their development possibly followed in a kind of "external talent pool" in order to encourage them to change job through personal contacting them when the time is right. It must be borne in mind that, particularly in the case of digital natives, the motivation for a job change cannot be influenced solely via the income level. Instead, a challenging task, an interesting innovative work environment and the image of the seeking company, characterised by its products and its culture, play an important role. It should also be considered that the new generation of workers are more willing to change their company.

The parent generation's decades-long loyalty to a company is a matter of the past. In the future every company must keep securing the commitment of its employees. Similar to the above-mentioned factors of career move motivations, development potentialities and also the basic freedom to change the place of work and the organisation are influencing factors as well. It is important that the direct executives continuously take up, address and as far as possible meet the needs of their co-workers. Fig. 7.9 shows the motivational elements of digital natives in summary.

The work generations currently employed in the companies are assigned to their respective birth cohorts on the timeline. The post-war generation of the baby boomers are followed by the generations X, Y and Z, which were socialised more and more frequently with the Internet and smartphone. In the right half of the figure, motivation elements for the younger generations are shown in as implified manner.

Everything which is of a regulating and structured nature is perceived rather as an obstacle, whereas innovation, dynamism, flexibility and meaningfulness create a drive. In keeping with these elements in mind, the work environment has to be designed, as well as development perspectives and project contents be aligned, in order to get good work results with high motivation and to achieve a commitment of the younger generation to the companies.

7.5 Cooperation Forms

Another way to provide the necessary resources for digital transformation and at the same time to positively influence the corporate culture through cooperation with external employees are partnerships and alliances. If the task is simply to expand the capacity and to involve additional employees with project-relevant knowledge and pertinent experience, cooperation with service providers or close cooperation models with sourcing platforms are useful.

The use of "liquid workforce" platforms (cf. Sect. 3.6.2) enables the global search for appropriate resources so that this procurement approach will become established, and manufacturers should consider presenting themselves as preferred partners on such platforms. Furthermore, for more complex digitisation programmes, they should consider engaging in partnership-based framework agreements with a selection of already established sourcing partners. A long-term bond offers economic advantages and can also minimise the manufacturer-specific loss of know-how from employees leaving the company.

In addition to pure sourcing, strategic partnerships are an important means to increase the availability of technology expertise, to secure access to certain technologies, and to influence the orientation of developments in a manufacturer-specific approach [Str16]. This kind of partnership has yet to be practiced by today's manufacturers in the digitisation environment. Oftentimes, IT is still purchased as an "indirect material" by the same procurement department and with the same approach as for oils, catering, fire extinguishers and gardening services. The decision is based solely on the price of the service requested. An analysis and review in terms of Total Cost of Ownership (TCO) is only done rarely.

This approach has led to very heterogeneous IT landscapes with various technologies and a high number of supplier relationships. As IT is set to become the core element of vehicles, rethinking must begin. It is necessary to enter into partnerships under strategic aspects, and not to solely judge them by the purchasing price. On this occasion, the manufacturers should assess the often mentioned risk of a "vendor lock-in", i.e. the dependency on just one partner. In the author's view, this topic is often overstated in an unreflected manner, thus blocking the view on opportunities. Derived from the transformation of the industries and the constantly expanding value chain, strategic partnerships and alliances with focus on digitisation are conceivable in the following domains for instance:

7.5 Cooperation Forms

- IT technology – e.g. Cloud, embedded IT, software
- Development and operation of mobility platforms
- Utilisation of shop solutions
- Cooperation of several manufacturers – e.g. common platform for mobility services or set-up of charging stations
- Research partnerships – e.g. for battery technology
- Development of solutions in Crowdsourcing (cf. Sect. 9.1) – e.g. for Connected Services
- Software development with alliances and contributions in Open Source, Open Stack
- Service chains; after-sales services
- Telecommunications
- Electric providers
- Charging infrastructure
- Car park operators
- Cities, toll operators
- Content provider – e.g. weather information, stock exchange data
- Payment processing
- Insurance, retail trade, hotel chains
- Intermodal transport – route partners

Manufacturers need to decide in which areas they want partnerships in order to create market access, to gain speed, or to share development risks. For reasons of know-how security or in order to differentiate themself with own solutions on the market, a company may alternatively rely on purely company-internal developments.

Strategic partnerships are based on long-term joint objectives. This could be the development and subsequent operation of a mobility platform or the joint further development of payment solutions for connected services. The joint development of a platform for the integration of the manufacturing IT in the plants for Industry 4.0 solutions in order to implement and operate them worldwide could be another field.

In any case, with such partnership agreements, besides technological aspects and questions concerning the intellectual property of the joint developments, commercial aspects are to be treated as well. So-called risk sharing models have proved their value in which both partners jointly bear the economic risk arising from the uncertainty of a market success, and on the other hand share the earnings. This approach emphasises mutual trust in the partnership. As a basic principle, it is advisable to define a specific objective in the agreement and to agree upon a timeframe and an agile approach in order to achieve the necessary flexibility in the implementation.

Due to the complexity and extent of the digital transformation, partnerships are an important success criterion. It is therefore important for the manufacturers to address this subject and to learn of how to work in strategic partnerships. It is certainly possible to use existing experiences from the traditional manufacturer environment, for example, from joint development partnerships for components. These must be complemented though by the specific aspects of digitisation, such as start-up mentality, agile approach, speed and globalisation.

7.6 Open Innovation

Through partnerships, for example with research facilities or leading technology partners, the innovative capacity of companies can also be increased. Traditionally, the focus and the strength of the manufacturers are in innovations in vehicle-related areas such as new materials, production processes and drives. It is now necessary to achieve a similar innovative energy in the field of digitisation. To this end, the concept of so-called "open innovation", i.e. the inclusion of external sources in the established innovation process, should be integrated as part of the corporate culture. This approach differentiates three core processes according to Fig. 7.10.

The outside-in process involves the knowledge of external sources, such as customers, suppliers and research facilities. For this purpose, a targeted scouting is to be organised to identify possible creative ideas, approaches and suitable partners. Specific innovation and scouting units are recommended, but at least cooperation with start-ups in the innovation centers of digitisation such as Silicon Valley, Israel, India or Munich and London. In this way, trends and shifts, especially "disruptive forces", are identifiable at an early stage for the own business model, and they must be taken up in the innovation process.

The inside-out process is about publishing own ideas, testing them in interest groups or cooperations and enriching them. This also involves the participation in open-source developer communities, such as the Open Automotive Alliance, which is driving the use of Android in vehicles, or the Open Stack Community which devel-

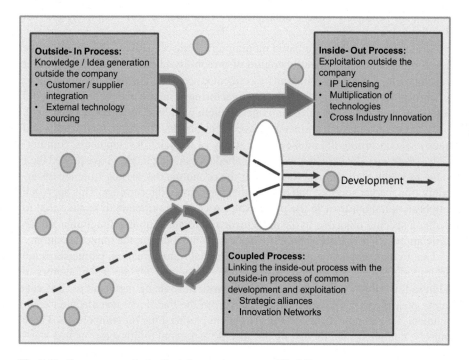

Fig. 7.10 Core processes in the Open Innovation concept [Gas06]

7.6 Open Innovation

ops the open software for Cloud environments [OAA], [Ope16]. Such collaborations provide impulses for own ideas and are also influencing the company culture.

The Coupled Process, with a combination of the two directions of innovation, is aimed at accelerating the implementation of ideas in alliances, joint ventures or partnerships and quickly commercialising them. In this context, the cooperation with so-called incubators is also to be mentioned, which provide start-up assistance namely to young companies through founder know-how and capital. Speed is the key competitive factor in times of digitisation. In new business models, the first will always have an advantage if successful and occupy a market where followers will find it hard to partake.

In order to exploit the potential of the concept, the manufacturers should practise work in Open Innovation initiatives and make it a part of their corporate culture. The widespread "not-invented-here" attitude, which often leads to scepticism through to rejection in the implementation of these concepts, can be counteracted by setting examples. Many references are known in this regard, mostly in pilots or also in defined areas. For example, many manufacturers rely on so-called co-creation labs or on innovation competitions with defined objectives. Audi for instance has successfully revised an infotainment unit in a Virtual Lab as part of two open-ended Crowdsourcing projects [Tho15].

A further example of the consistent and sustainable application of Open Innovation and Crowdsourcing concepts is Local Motors, a so-called Open Source automobile manufacturer. The development of the vehicles is through open communities on the Internet. Local Motors has quasi moved its development department in the sense of Crowdsourcing to an online community. Roughly 1400 designers around the world work together to develop vehicles [Buh16]. The designers, engineers and technicians often have an industry background but are not employed by Local Motors. All the more they are thrilled in the agile projects to develop individual vehicles, which are then manufactured to customer specifications. Many components are created in 3D printing processes and the assembly of the vehicles takes place largely under customer involvement locally at the dealer. The vehicle details are open for inspection by anyone and can therefore be easily used for further vehicle developments. The development times for vehicles are approx. 18 months, and 12 months is the target, while traditional manufacturers currently need up to 5 years. The examples illustrate the potentials of crowdsourcing initiatives.

It seems promising to transfer the approach of Open Innovation to the transformation to digitisation. Here, too, are first examples in which manufacturers and suppliers have carried out so-called Hackathons. Reports on this can be found under this search word on the Internet. One example is the Campus-Hackathon, which defines its goals as follows: "The Campus Hackathon is not a programming competition in the traditional sense, rather a programming festival! It is a collective, social experience and has nothing to do with work". And it continues: "At the Campus Hackathon, you will be able to develop ideas as part of a team and convert them into codes within one weekend" [Eco16].

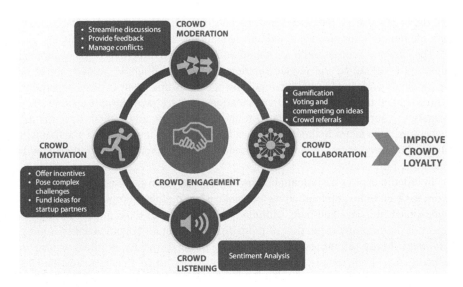

Fig. 7.11 Methodology for successful Crowdsourcing [Cha16]

Such events are organised by companies under one theme, and students and interested Digtial Natives are invited to programme an app within a limited period of maximum 2 days to solve problems in the given topic area. The IT infrastructure and catering are provided by the organiser. Several teams form and compete with their ideas. In the end, there is a cash price and the pursuing of the idea in the organising company with inclusion of the participants [Lec12]. The results of these hackathons are often promising and bring new impulses into the companies. Further options are idea contests and laboratory events from the IT environment.

Overall, from the author's point of view there is considerable potential to gain speed in digital transformation via Crowdsourcing, integrated into a structured innovation process. The basic method is shown in Fig. 7.11.

The projects should start in a limited thematic area under a clear objective. First of all, the participants need to get to understand each other, for example on the basis of a so-called sentiment analysis. The evaluation of social media data relating to the topic field is used for this. Then the motivation of the participants needs to be stimulated by price money or start-up financing. A moderation of the workflow is the prerequisite for channelling ideas in the desired direction and to take up new trends in the dialogue. The coaching also promotes motivation in the work process. Agile project management methods and the use of innovative methods such as gaming or mutual assessment (votings) of the group results provide motivation to continue the collaboration even after the conclusion of, for example, a hackathon or development project. The loyalty of the participants to a project should also be maintained over longer development phases, for example in vehicle projects at Local Motors.

By and large, Open Innovation is a process for increasing the innovative capacity which also offers high potentials especially for digitisation programmes. First pilots

7.7 Organisational Aspects of Digitisation

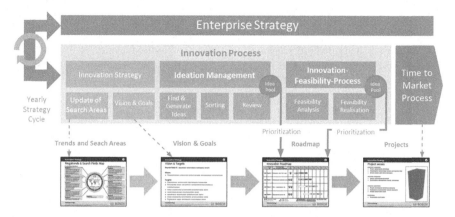

Fig. 7.12 Innovation process in conjunction with the company strategy [Kno11]

support this assessment and surprise with good results. The manufacturers should expand such pilot projects into complete programmes and make greater use of the method as part of the overall innovation process. The results and new trends and findings from the scouting of the outside-in processes should be discussed at least once a year, in light of the dynamics of digitisation preferably every 6 months, in the company management and compared with the digitisation roadmap (cf. Sect. 6.1). This aspect and also a structured innovation process are shown in Fig. 7.12.

The picture puts the elements of the innovation process into context. All elements should be implemented by the manufacturers in a similar way, with key figures in the proposed digitisation frame (cf. Fig. 6.11). Within the framework of the innovation strategy and in line with the vision for digitisation, concerning Connected Services for instance, the search fields are defined for which subsequently ideas are created with the help of Open Innovation procedures. These are filtered and evaluated, prioritised in an ideas pool and held in a roadmap. In the feasibility process, this is followed by their evaluation and the implementation of selected ideas, especially under time-to-market aspects. The illustrated structured process, applied to each digitisation field, ensures that ideas are consistently being developed, tracked and implemented without losing topics or "dynamically" changing previously set priorities.

7.7 Organisational Aspects of Digitisation

For the development of the digitisation strategy, the development of an integral roadmap and its implementation along with the innovation process, the responsibility must be clearly defined and reflected in the company organisation. The organisational structures are in many cases sedate and hierarchically oriented which is why they obstruct matrix-oriented project structures. They often do not correspond to the desired new culture and start-up mentality, which is required in the implementation of digital transformation. Hence the organisation has to be adapted and its hierarchy be made flatter than nowadays to facilitate quick decisions. This creates new career outlines, and career models will change. Some of these organisational aspects are deepened in the following section.

7.7.1 Chief Digital Officer (CDO)

The digital transformation is not about the one-time implementation of new software or purchasing new IT technology, it rather is about the survival of the company through the change of business models and the introduction of completely new products. This requires change in the entire company, organisation and culture. This task is not delegable, yet must be borne by all employees. The overall responsibility is with the CEO and in certain parts with the other members of the Management Board. At the levels below, all managers and all employees are to be involved. In order to tackle this enormous task in a structured manner, it is essential to clearly define and communicate the responsibility for the development of a digitisation roadmap, the cascading of the tasks and projects to be derived from it, and the monitoring of the implementation.

Many companies install a so-called Chief Digital Officer (CDO) or alternatively put the responsibility for digitisation on other executives, such as the Chief Information Officer (CIO), the development or sales executive. From the author's viewpoint, a new cross-sectional organisation under the direction of a CDO, at least until an advanced digital maturity level is achieved, is recommended because of the breadth of the task which became clear in the previous chapters. The CDO should report to the CEO, who is still in charge of the overall responsibility and must over and over again highlight the importance of the digital transformation and also personally represent the cultural change by exemplifying the transformation of traditional behaviour patterns.

The willingness to remodel the company, to change culture and the definite commitment to digitisation are the basis for successful work of the CDO. Multi-brand manufacturers should consider using one specific CDO for each brand which then report to the Group's CDO. The strategic goals, the relevant digitisation framework and the methods and standards are agreed upon at corporate level. The implementation is then carried out in the respective brands.

The task of the CDO is to work together with the CEO to promote the digital transformation. In accordance with the principles described in Chap. 6, Fig. 7.13 shows a proposal for the division of the digitisation fields which need to be addressed.

The columns show the digitisation fields, and the lines the company management areas. The individual areas of the company management are assigned a main responsibility and partial responsibility for one or more digitisation fields. The overall responsibility for the transformation is borne by the CEO in close collaboration

7.7 Organisational Aspects of Digitisation

		Digitalisation Fields						
	● Responsibility ◐ Partial responsibility	Business Model	Culture	Connected Services; Digital Products	Mobility Services; Autonomous Driving	Efficient Processes	Sales / Aftersales	IT Services
Management	CEO/CDO	●	◐	◐	◐	●	◐	◐
	Finance	◐				◐		
	Personnel		●			◐		
	Marketing	◐		●	●	◐	●	
	Development			◐ Lead Autonomous Driving & Integration-Plattform	◐	◐		
	Production					◐ Lead: Industry 4.0		◐
	Procurement	◐				◐		
	IT	◐	◐	◐	◐	◐	◐	●

Fig. 7.13 Assignment of responsibility for the implementation of digitisation fields (Source: Author)

with the CDO reporting to him [Wes14]. He ensures the precise coordination between business sectors when digitisation fields in the implementation concern several business sections, as is the case with Connected Services and digital products, for example. In terms of customer orientation and market knowledge, the head of sales assumes the main responsibility, while the head of development provides a modern embedded IT architecture and a powerful integration platform, and the CIO provides efficient Cloud-based IT platforms.

A similar division of responsibilities can be found in the mobility services, which are experiencing a special boost with autonomous driving. For the technical implementation of this technology as well as the realisation of the integration platform (see Sect. 6.2.1), the lead responsibility should be with the development department. Establishing efficient digitised processes is a responsibility which all company divisions in their organisation assume, for example, the lead for the implementation of Industrie 4.0 is with the head of production. The CEO and the CDO must then again be responsible for the overall process programme.

Crucial for the success is whether a profound digital transformation can be implemented within the existing organisation. In the past, manufacturers have achieved innovations with incremental improvements to vehicles, manufacturing processes and procedures. With the expected disruptive changes in the business model, especially in the areas of Connected Services, Digital Products and Mobility Services, it remains to be seen whether the frequently cited "innovator's dilemma", which means that established companies cannot manage these erratic innovations, can be solved [Chr11].

In order to gain independence and agility, many manufacturers are moving these new business areas into smaller, separate organisations. It remains to be proven that this route will achieve its objectives. According to the author's experience, agility is gained on the one hand, but a possible synergy is weakened on the other, namely if individual organisations within the brands build separate units such as innovation and mobility labs. This model of internal start-ups or spin-offs which are operating near the existing organisation, is called Intrapreneuership. The work of these units is certainly important and pioneering by exploring the possibilities of new technologies and agile work, which also allows for failure and still creates innovative solutions.

However, the backflow of these "lighthouse developments" from the labs into the manufacturer's organisation proves to be problematic, and the impact on the productive day-to-day business is rather modest [Küh00]. Nevertheless, the route via separate units offers great opportunities and should therefore be organised across the group and with joined forces. In order to achieve a clear development- and separation boost, this step should be done in cooperation with a strong partner who is already established in the new business segments and brings his business models, the start-up mentality and the necessary assets into the collaboration.

Regardless of this organisational question, the CDO should have a strong team to support the company divisions in the digitisation initiatives in order to provide kickstart assistance to their colleagues, for example through digital competencies in the relevant digitisation technologies, and the organisation of design thinking workshops in process assessments and for the development of new IT solutions. At this, IT provides the necessary tools and test fields and implements the new solutions in an agile approach.

7.7.2 Adaptation of the IT Organisation

In order to accelerate the digital transformation in the corporate divisions, the internal organisation of the enterprise IT must be made more efficient. In the past, IT was also intensively involved in the automation of business processes by implementing software solutions. The approach was rather dominated by technology. In this sense, IT is still often broken down into three areas, according to the IT life phases. The "Plan" phase implements the projects and organises the process, in the "Build" phase, the projects are implemented in software solutions, and in the final "Run" phase, the solutions are mostly operated in the own computing centres of the manufacturer as safely as possible.

In the Plan and Build phases, the organisation is usually roughly organised according to the process areas of development, sales, customer order fulfilment and administration. In addition, there are other IT teams to support the component plants, while the Run phase is usually grouped into technologies, such as mainframe computers, Linux servers, and networks. This organisation with mature heterogeneous processes and structures often proves too inflexible for the requirements of digitisation [Hel16]. The situation which is typical, also by the author's experience, in the cooperation between IT and specialist areas is illustrated in Fig. 7.14.

The cooperation between the corporate divisions is often characterised by reservations. "The IT" is regarded as too slow and little responsive, too bureaucratic and

7.7 Organisational Aspects of Digitisation

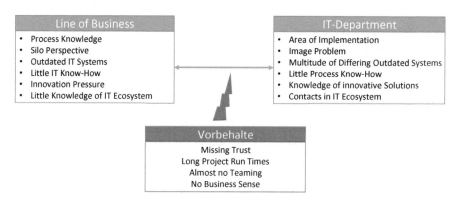

Fig. 7.14 Situation in the cooperation between department and IT (Source: Author)

with little business understanding. The application systems used are considered outdated, hardly expandable and rigid. On the other hand, the IT division sees in the specialist divisions a lack of IT knowledge, no understanding for their concerns and no cross-departmental thinking.

These views are often based on experiences from previously common large-scale implementation projects under IT management, which partly stretched over a number of years. Starting from a so-called blueprint phase which documented the requirements, and the subsequent implementation and test phase, the process was concluded in the rollout phase. This project approach often led to results that surprised the specialist divisions, which in turn resulted in extensive adjustments and reworking. In large manufacturers, several such projects frequently ran in parallel, which led to isolated solutions, today's heterogeneous application landscapes. This history has put a lot of pressure on the relationship between the areas, and against the background of the upcoming digitisation urgance it needs to be put on a new basis.

As a reaction or probably just emergency solution, specialist divisions had begun in parts to implement their own IT solutions in their organisation and thus create a kind of "shadow IT" that runs past all security precautions in public Cloud environments or on "servers under the desk". This definitely is not an advisable route. A further solution is the concept of so-called bimodal IT proposed by Gartner, which means an organisation of two different speeds [Lai16]. The development and operation of the traditional application environments is often referred to as "IT 1.0", and the agile new digitisation projects, frequently developed in so-called innovation labs, are called "IT 2.0". However, this approach is also viewed critically as it does not achieve enough sustainability and breadth in the business divisions, while the "IT 1.0 employees" are little motivated and just uncouple themselves. A "two-class society" emerges, with too few synergies [Vas16]. Hence this route is also rather questionable.

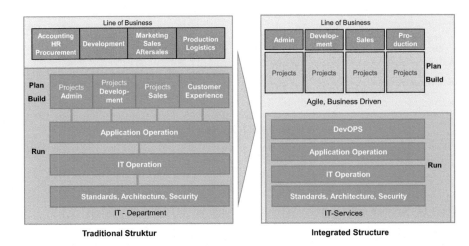

Fig. 7.15 Traditional and integrated application development (Source: Author)

It is indisputable that divisions and IT need to move closer together in order to jointly cope with the challenges of digital transformation. The upcoming projects are of a very different nature than the previous technologically oriented projects. Together, transformation measures must be agreed upon in order to implement the first digitisation solutions in an agile manner close to the business, which gradually grow in short cycles or are stopped again if not useful. For this new approach, however, the business divisions often lack the necessary solution-related IT know-how.

From the author's point of view, a profound organisational change is promising, with a proposal shown in Fig. 7.15.

On the left of the picture the current common organisation of IT in the automotive industry is shown, as was already briefly described at the beginning of the chapter. In the planning and build phases, application development is made according to the structure of the divisions with all of the described acceptability and coordination problems. In order to create more proximity to business, processes and demand, all IT staff, as shown in the picture on the right hand side, should switch from their previous organisation to the divisions. The close integration of process and IT knowledge in the business leads to considerable synergies, for example in order to address the digitisation of process chains in Design Thinking workshops and to test them quickly with a first prototype.

The consolidation of the application IT in the specialist division thus follows the overall trend that the added value of a car will in the future be defined to a large extent by IT. In this respect, it is only logical to sustainably strengthen the IT competence of the divisions by integrating the IT staff from the respective plan and build areas. The IT area can then focus on the provision of IT services. The coaching of the approach and methodological competence must be provided by the CDO organisation, and the IT architectures and technology standards should be supported by

7.7 Organisational Aspects of Digitisation

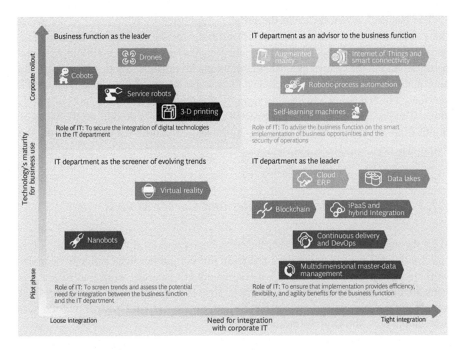

Fig. 7.16 Role of IT in digital transformation [Gum16]

guidelines. This enables all divisions to address the desired digital transformation on the basis of the business processes.

In doing so, the CIO assumes an important role with his organisation. It must ensure that highly efficient and secure IT structures are available for the solutions of the specialist divisions. On the one hand, these are implemented as hybrid cloud architectures and on the other hand include basic technologies such as Middleware, Big Data tools and Microservice environments. Furthermore, the IT has to provide so-called DevOps concepts in order to enable the rapid use of new applications and, based on this, of a "smartphone-oriented" release cycle. DevOps is a word combination of Development and Operations and aims to improve the collaboration between software developers and IT. For details on this and further technological IT tasks, please see Chap. 8. In addition to these tasks, IT is also a consultant to the specialist divisions and a scout for new technologies. This positioning is shown in Fig. 7.16.

The picture illustrates the different roles and responsibilities of IT in the context of digital transformation. In addition to the briefly described responsibility of IT to provide secure service environments and the operation of basic applications such as Enterprise Resource Planning (ERP), IT acts as a consultant and supporter of the specialist divisions. New technological trends such as Nanobots for instance are to be evaluated at an early stage in order to take account of the IT requirements, to consider them in the solutions and, if necessary, to ensure integration and operation.

The business divisions are in the projects supported by IT in order to use solutions such as drones for logistics or 3D printing for the production of parts. The responsible division and IT work closely together in the further rollout of these solutions and in the case of extensions, for example, towards robotics-based process automation and self-learning machines. IT is responsible for efficient and secure IT services, which quickly and flexibly support new requirements of the specialist divisions. The process integration and restructuring of business models on the basis of digitisation solutions are the in the proposed structure responsibility of the divisions.

7.7.3 New Occupational Profiles and Career Models

With the new distribution of tasks, the roles and scopes of duties of both the divisions and the IT are changing. It is crucial in this comprehensive change to keep the workers highly motivated through modern human resources work and to accompany them in the transformation journey. To this end, offers are required which appeal to the entire work force, in fact to both the new Digital Natives and the veteran employees, the Digital Immigrants.

Through the possibilities of digitisation, agile methods, the changing value system and new forms of cooperation, previously proven principles of human resources work become obsolete, and new models are needed. The following trends must be considered at this:

- New occupational profiles, e.g. computer linguists
- Longer working life
- Work in virtual teams and loose networks, e.g. in partnerships, Open Innovation
- Changed value pattern, fading of status symbols, e.g. time vs. money
- New career models in "mosaic processes", e.g. personnel management alternating with specialist career phases and occasional sabbaticals [vRu15]
- Flexible work time models
- Specialist careers of equal status as executive positions
- Life-long learning
- Diversity as a matter of course
- Growing importance of feedback and coaching
- Emphasis on degrees of freedom, autonomy and self-realisation
- Growing project orientation and work in temporary tasks.

These trends require new, much more flexible career and work time models, combined with other compensation and incentive systems, including organisation of time and training opportunities. The new models must also appeal to current employees and thus contribute to the motivation and the change of corporate culture. Offers must be designed in such a way as to convince applicants to join the company. Due to the foreseeable shortage of qualified workers, it is necessary to be different in the "war for talents". A reversal is looming in the way to start a career: Companies are then applying to highly qualified career beginners and experts.

7.7 Organisational Aspects of Digitisation

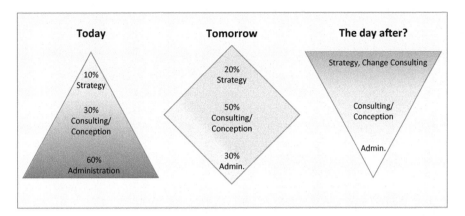

Fig. 7.17 Change of human resources work [Gor13]

The leaders play a decisive role in the new world of work. They are direct reference persons and thus point of contact in order to fill the new models with life and to promote the change of corporate culture. In continuous feedback and coaching conversations talents should be tracked and developed, feedback from the employees got and be returned to the Human Recource department. At the same time, the persons responsible for human resources also have to change their task weighting, as shown in Fig. 7.17.

Coming from today's high administrative proportion, soon consultancy and conceptual work on the development of new structures and models will increase through to strategic consulting and change management. The change reflects the implementation of the new requirements by the digitisation, the demographic change and the changing value patterns of the future employees.

The outlined change in human resources work can be significantly supported by the digitisation and automation of administrative processes in the personnel area. The establishment of a shared service centre and the introduction of a platform solution according to Sect. 6.2.3.1 also help to create open spaces in administrative work. Based on employee-focused human resources work, regular cooperative coaching between direct leader and their employees with open feedback on both sides leads to high employee satisfaction, which in turn is the basis for good work results.

7.7.4 Change Management

Apart from attractive working models and human resources offerings, an innovative work environment is another important element in employee motivation. To the Digital Natives the size of the office or the number of office windows is not their concern. It is much more important to create an appealing, agile work atmosphere by using modern working tools and open communication. This also includes the

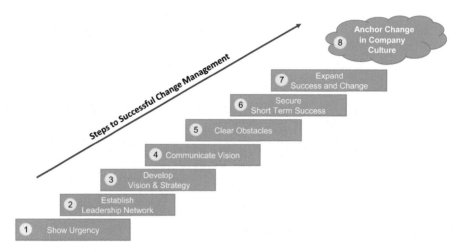

Fig. 7.18 8-step model for change management (According to [Kot11])

free spatial arrangement of the workplaces and the layout of a work environment by a team on a large floor area without walls. Complemented by the elements mentioned in this chapter, a corporate culture develops in this manner which is open to changes and transformations.

These motivation elements form the basis for the successful implementation of digitisation, which must be accompanied by focused change management. It is important to integrate all employees into the change processes at an early stage in order to avoid potential open or hidden resistance. Unlike previous project-related changes, the digital transformation affects almost all areas of the company and will be a long-term process. Consequently, continuous change management has to become an integral part of the company culture. To reiterate, the change must be lived by the board and be credible by authentic behaviour.

The overall responsibility for this comprehensive change process lies with the CEO and as a change leader with the CDO. In turn, it must motivate all executives to become change agents in order to implement the change in the conversations with employees and the daily work. Not the technology, yet to win all of the employees over and get them enthusiastic about the change is the key to success. On the basis of this challenge, the approach model in Fig. 7.18 was developed.

The picture shows an eight-step approach to successful change management. First of all, it is important to clearly show all executives and employees the necessity of a change based on scenarios and to generate motivation for change. Then a network of people is to be set up, who continuously maintain communication during the change process. In the next steps, a vision is to be developed along with a comprehensible strategy on how these goals are to be achieved. In the fifth step, obstacles in the structures or work processes and deficits in the project equipment need to be eliminated.

7.7 Organisational Aspects of Digitisation

Shaping Corporate Change (Doppler / Lauterburg)	Change Management Study (Capgemini Consulting)
Awaken Energy And Building Up Trust	Clearly Formulate Goals
Thinking in Terms of Processes Instead of Structures	Establish Action Plan
Organise Outside -Inwards	Secure Resources
Align the Company to its Environment	Develop Skills
Networking through Communication	Create Incentives
Learn to Ensure	Ensure Good Communication

Fig. 7.19 Core elements on successful Change Management [Dop14], [Boh15]

At the beginning, successes should be achieved as quickly as possible in order to strengthen communication and to make further changes in the next step with new ideas and complementary measures. In the final step, it must be ensured that the changes are permanently embedded in the company and that change has become part of the corporate culture. A relapse into old structures and behavioural patterns must be prevented in any case.

This approach focuses on the communication and the integration of people based on the digitisation vision and strategy. Other methods which have been tried and tested as well have somewhat different focuses. For the sake of completeness, Fig. 7.19 shows the core elements of the procedures according to Doppler/Lauterburg, authors of a standard work on change management, and comparatively the recommendations of a consulting firm, derived from a study.

The key factors for a successful change management, shown on left hand side of the picture and having been tried and tested for many years, should largely be incorporated into the eight-step approach as per Fig. 7.18 in the type of communication, the development of the vision and the strategy as well as the implementation. The authentic orientation of the company towards its environment and to ensure learning, are accompanying measures of the new corporate culture.

The recommendations shown on the right of the picture ensure the availability of action plans, resources and abilities. The considerations on incentive systems are also worthy of note, since they additionally secure the motivation of the workers. It is important that the Change Management, led by the CDO, is set up at an early stage, the employees are respected and continuously involved.

7.8 Case Study: Transformation IBM

It became clear that the change in corporate culture towards more agility, entrepreneurship, start-up behaviour and the willingness to assume overarching responsibility, as well as to actively pursue things and to implement them quickly, are the most important prerequisites for a successful digital transformation from a automotive company into a mobility company. As a reference for a successful transformation associated with a fundamental change in corporate culture, the company IBM can be considered. The change from a hardware manufacturer to a Cloud-based service provider is an analogy to the outlined transformation path of the automotive industry. The transformation of IBM is therefore presented below.

IBM is an international corporation, office-based in the U.S., employs several hundred thousand people, has been for more than 100 years in the market, and has a long-standing position worldwide with very many large and medium-sized companies of all industries as their IT partner. With many inventions and innovations, IBM has time and again substantially shaped the computer industry. One of the most important innovations are the universal computers, the so-called Mainframes, developed in the 1960s. These were unchallenged market leaders in the 1970s and even in the 1980s, and are still today the central computers of many companies and institutions.

This resulted in a considerable economic success, which led to saturation in the organisation at that time, which in turn led to lethargy and let uncontrolled growth of internal bureaucracy happen. The quite admirable and at that time common term of "Big Blue" suggested to the employees an inviolability and resulted in an image that was partly perceived as arrogant. This was certainly justified in a situation unimaginable today. The mainframes were partially "allocated" to the customers; there were long waiting lists for the latest systems, and the customers were falling over themselves to improve their ranking in the list for the supply [Mus10]. This success also led to a kind of tunnel vision. Changes in the market were detected late or just not considered relevant.

In the 1970s the personal computers (PCs) emerged, among others from the companies Apple, Commodore, Tandy and Atari. These were increasingly used by companies, but also privately. After IBM had ignored this trend for a long time, the company launched its first own PC in 1981. It consisted of components that could easily be configured into different systems. Contrary to the previous business strategy, essential components were purchased from other companies. Intel supplied the processors and Microsoft the operating system.

The system was a huge success with high market attention, and within a short time the PCs also spread in private households. Sure, IBM's success at the time can also be summed up in the common term "IBM compatible". However it was at the same time a Pyrrhic victory as well which ultimately weakened the IBM position. Since the modules were freely available for the PCs, many providers entered the market, both large companies and many microcompanies, so-called "no names", which were engaged in fierce price wars in this market. IBM essentially contributed to the success of the PCs through the open component construction, but thereby the company had no means of control in order to adequately participate in the growth. Even worse, the PCs drove the then-common IBM typewriters with word processing solutions out of the market and with growing performance and networking partly became an alternative to the Mainframe.

The cumbersome nature, bureaucratisation and the retention of old structures and product lines resulted in economic problems within a short time and to high losses in the early 1990s. It was considered to divide IBM into independent companies to ensure speed and at least a survival of the healthy parts. In 1993 the net loss was $8 billion. Then Louis V.Gerstner joined the company as CEO, the first time in the history of IBM that a board chairman came from outside the corporate. The choice proved to be a lucky find for IBM. After a short analysis, Gerstner made fundamental decisions and consistently implemented them, so that they still have an effect on IBM today. The main parameters were [Ger03]:

- Maintain IBM and focus on integration instead of fragmentation
- Expansion of the service business
- Network centric solutions (afterwards: "eBusiness")
- Worldwide standardisation of business processes
- Reduction of bureaucracy and streamlining of procedures
- Consolidation and harmonisation of internal IT
- Expansion of the software business also through acquisitions
- Open Innovation
- Transformation of the corporate culture.

These decisions have led to a complete change in IBM from a hardware manufacturer to an integration service provider, who, in addition to services, also offers leading software and hardware technology in its portfolio. The challenging task of this transformation was in Gerstner's perception the change in corporate culture, with respectful inclusion of all staff. The author is still familiar with a few catchphrases of the time, such as "customer first" and "execute". It was precisely this customer orientation, and at the same time the focussed execution of projects instead of renewed discussions, which marked these departure years. The successful adaptation of the employee behaviour fits well into the world of some automotive manufacturers at the beginning of digital transformation.

Gerstner's term ended in 2002 and at that time, the culture of IBM had changed this far that the initial "must-change" had become an attitude of "wish-to-change". This is an important behavioural change in order to tackle transitions even in economically healthy times. The disintegration of the "old mainframe culture" succeeded due to the demonstrated leadership of Gerstner combined with an intensive communication and inclusion of workers, yet certainly also since the impending insolvency had left no alternative to the need for change. This makes it all the more remarkable that the newly established corporate culture induced further changes in economically successful times too. The situation is illustrated in Fig. 7.20 with the distribution of sales of IBM in the main business areas of Hardware, Services, Cloud and the relatively new field of Cognitive Solutions since the 1960s.

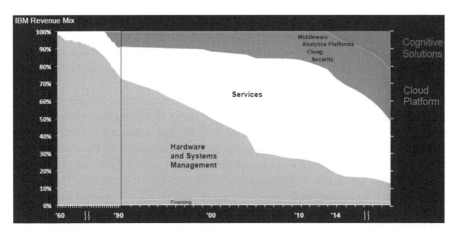

Fig. 7.20 IBM distribution of sales (IBM image)

The dominance of the hardware proportion in the 1960s and 1970s and its subsequent continuous erosion are clearly visible. The decrease is due to the exponential growth in the hardware performance on the one hand, and price decline on the other; later on also through the sale of "commodity" hardware business, such as the PC division to Lenovo in 2004. The growth and the increase in the significance of the services, of the integrated solution business, and the development of the software proportion based on Gerstner's trend-setting decisions are clearly visible.

In 2017 too, IBM is well positioned in the high-performance hardware market, with great importance to many customers and the company, although this business proportion has fallen significantly to below 20%. An increasing part of the computer and storage capacity the companies get, at the expense of their own hardware, in the form of Cloud services, is a newly established strategic growth area of IBM. The new Cognitive Solutions division expects to achieve above-average growth. It is regarded as an important component of the digital transformation of companies in the automation of business processes and the development of smart "thinking" assistance systems (cf. Chaps. 6 and 9). The incessant change in the IT industry and the need for IBM to focus always on new business areas requires even after Gerstner's departure as CEO to continue with change and transformation.

One way of doing this are for example innovation jams. These are temporary online brainstorming events, organised on the intranet, open to all employees. In these sessions selected topics for discussion and participation are on the agenda. The communication process is moderated by senior executives and directed by specific comments and contributions towards the desired directions. As an example, in such a 72-hour event, the company values, the relevance of the company and the future orientation and working areas were debated quite controversially by the employees. The conclusions resulted in a new, clearer definition of the company's objectives and basic values, from which the so-called practices were derived as

7.8 Case Study: Transformation IBM

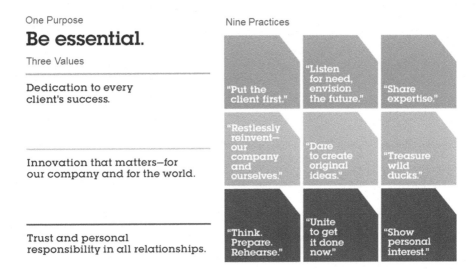

Fig. 7.21 IBM corporate values and leadership principles (Image IBM/)

management guidelines. This structure, which is presented under the term "1–3–9 ibm" also on the Internet, for example in YouTube, is shown in Fig. 7.21.

IBM has set itself the superordinate objective of making an important difference to the world and to the customers, and to be "essential". The three basic values shown on the left specify the goal with the focus on customer success, innovation and personal responsibility in all relationships. These basic values are assigned to nine so-called practices as management principles, for example to share experiences or to accept exceptional opinions and procedures, characterised by the term "treasure wild ducks".

This structure today characterises IBM's corporate culture which, however, continues to change continuously and flexibly, driven, among other things, by ever-changing market requirements. As part of the change, processes were also changed and structural adjustments made. Here are some major measures:

- Reduction of hierarchical levels from nine to a maximum of seven
- Quarterly instead of annual employee-target interviews
- Equalisation of specialist and management career
- Expansion of online learning (at least 40 hours per year)
- Introducing of senior executives from outside
- Mobile workstations, shared desk
- Modernisation of the IT landscape

As a result, IBM has very successfully undergone a comprehensive transformation and will keep changing continuously. The now firmly established readiness to change has become a fundamental part of the corporate culture. A similar motivation for the mandatory change must also be created in all employees of the automotive industry. The transformation of culture, hitherto shaped by a business model that has been successful for more than a hundred years, is the most important prerequisite for a successful digital transformation. In this sense, the development and experience of IBM provide opportunities to learn for this path.

References

[Bee01] Beedle, M., van Bennekum, A, Cockburn, A., et al.: Manifesto for Agile Software Development. (2001). http://agilemanifesto.org/. Drawn: 11.11.2016

[Boh15] Bohn, U., Crummenert, C., Graeber, F.: Superkräfte oder Superteam? (Super powers or Superteam?) Wie Führungskräfte ihre Welt wirklich verändern können. (How Leaders Can Really Change Their World.) Change Management Study 2015; Capgemini Consulting. https://www.de.capgemini-consulting.com/resource-file-access/resource/pdf/change-Management-Study-2015_4.pdf. Drawn. 08.12.2016

[Bre16] Brecht, A., Bornemann, M., Hartmann, G., et al.: Wissensmanagement in der Norm ISO 9001:2015; Praktische Orientierung für Qualitätsmanagementverantwortliche (Knowledge Management in the Standard ISO 9001: 2015; Practical Orientation for Quality Management Responsibles), Deutsche Gesellschaft für Qualität (DGQ), (2016). http://www.gfwm.de/wp-content/uploads/2016/05/Praktische_Orientierung_fuer_Qualitaetsmanagementverantwortliche_GfWM_DGQ. Drawn: 24.11.2016

[Buh16] Buhse, W., Reppesgaard, L., Henkel, S.: Der Case Local Motors: Co-Creation und Collaboration in der Automotive-Industrie; Case Study. https://www.tsystems.de/umn/uti/782372_2/blobBinary/LocalMotors.pdf%3Fts_layoutId%3D760278. Drawn: 30.11.2016

[Cha16] Chatterjee, S., Khandekar, P., Kumar, V.: Reimaging Enterprise Innovation through Crowdsourcing; White Paper Tata Consultancy services. http://www.tcs.com/resources/white_papers/Pages/Reimagining-Enter-prise-Innovation-Crowdsourcing.aspx. Drawn: 02.12.2016

[Chr11] Christensen, C., Matzler, K., von den Eichen, S.: The Innovator's dilemma: Warum etablierte Unternehmen den Wettbewerb um bahnbrechende Innovationen verlieren (Why Established Companies Lose The Competition for Groundbreaking Innovations); (Business Essentials); Vahlen (2011)

[Daimler] Internet presence Daimler AG; div. pages in the section "Innovation". https://www.daimler.com/innovation/digitalisierung/digitallife/mitarbeiter-im-mittelpunkt/. Drawn: 11.11.2016

[Dop14] Doppler, K., Lauterburg, C.: Change Management; Den Unternehmenswandel gestalten (Shape The Company Transformation); Campus Verlag; 13th edn. (2014)

[Eco16] Campus Hackathon; eco – Verband der Internetwirtschaft e.V. (Association of the Internet Industry). https://www.eco.de/2016/veranstaltungen/branchentermine/campus-hackathon.html. Drawn: 03.03.2017

[Eic13] Eichler, S., Katzky, U., Kraemer, W., et al.: Vom E-Learning zu Learning Solutions (From E-Learning to Learning Solutions); Positionspapier AK Learning Solutions; Ed. BITCOM (2013). https://www.bitkom.org/noindex/Publikationen/2016/Sonstiges/E-Learning-Studie/Positionspapier-Learning-Solutions-2013.pdf. Drawn; 24.11.2016

[Eil16] Eilers, S., Moeckel, K., Rump, J., et al.: HR Report 2015/2016; Focus on culture. https://www.hays.de/documents/10192/118775/hays-studie-hr-report-2015-2016.pdf/8cf5aee3-4b99-44b5-b9a9-2ac6460005da. Drawn: 11.11.2016

[Fal11] Faltin, G.: Kopf schlägt Kapital. (Head beats capital.) Die ganz andere Art, ein Unternehmen zu gründen. (The Very Different Way to Start A Business.) Vor der Lust, ein Entrepreneur zu sein (Before the Desire to Be An Entrepreneur); Hanser, (2011)

[Fle14] Fleig, J.: Agiles Projektmanagement – So funktioniert Scrum (Agile Project Management – How Scrum Works); business-wissen.de; b-wise GmbH; 27.06.2014. http://www.business-wissen.de/artikel/agiles-projektmanagement-so-funktioniert-scrum/. Drawn: 16.11.2016

[Gab16] Gabler Wirtschaftslexikon: keyword: MOOC, Springer Gabler Verlag. http://wirtschaftslexikon.gabler.de/Archiv/688938794/mooc-v4.html. Drawn: 24.11.2016

[Gas06] Gassmann, O., Enkel, E.: Open Innovation – Die Öffnung des Innovationsprozesses erhöht das Innovationspotential (Open Innovation – Opening Up The Innovation Process Increases The Potential for Innovation); zfo wissen 3/2006. http://drbader.ch/doc/open%20innovation%20zfo%202006.pdf. Drawn: 30.11.2016

[Ger03] Gerstner, L.: Who Says Elephants Can't Dance? Inside IBM's Historic Turnaround. Harpercollins, New York (2003)

[Ger16] Gerstbach, I.: Design Thinking im Unternehmen: Ein Workbook für die Einführung von Design Thinking (Design Thinking in the Company: A Workbook for the Introduction of Design Thinking); Gabal, (2016)

[Gor13] Gora, W., Jentsch, P., Erben, S.: Innovative Human Resource Management White Paper, 2013; Cisar – consulting and solutions GmbH. http://www.walter-gora.de/media/e34a849ad27e8a45ffff805effffff0.pdf. Drawn: 08.12.2016

[Grü16] Gründerszene: Was ist Entrepreneurship (Founder Scene: What is Entrepreneurship); Online Lexikon, gründerszene.net. http://www.gruenderszene.de/lexikon/begriffe/entrepreneurship. Drawn: 24.11.2016

[Gum16] Gumsheimer, T., Felden, F., Schmid, C.: Recasting IT for the Digital Age; BCG Technology Advantage, April 2016. http://media-publications.bcg.com/BCG-Technology-Advantage-Apr-016.pdf. Drawn: 05.12.2016

[Hel16] Helmke, S., Uebel, M.: Managementorientiertes IT-Controlling und IT-Governance (Management-oriented IT Controlling and IT Governance); Springer Gabler, (2016)

[Ker12] Kerres, M.: Mediendidaktik: Konzeption und Entwicklung mediengestützter Lernangebote (Mediendicactics: Conception and Development of Media-based Learning Offers); De Gruyter Oldenbourg, 3rd edn. (2012)

[Kno11] Knospe, B., Warschat, J., Slama, A., et al..: Innovationsprozesse managen, Fit für Innovation (Managing Innovation Processes, Fit for Innovation); Report of the Working Group 1 in the Collaborative Project "Schnelle Technologieadaption" ("Rapid Technology Adaptation"), 2011 Funding by BMBF and ESF. http://www.fitfuerinnovation.de/wp-content/uploads/2011/07/Fit_Fuer_Innovation_AK1.pdf. Drawn: 02.12.2016

[Kom14] Komus, A., Kamlowski, W.: Gemeinsamkeiten und Unterschiede von Lean Management und agilen Methoden (Similarities and Differences Between Lean Management and Agile Methods); Working Paper BPM-Labors HS Koblenz, (2014). https://www.hskoblenz.de/fileadmin/media/fb_wirtschaftswissenschaften/Forschung_Projekte/Forschungsprojekte/BPMLabor/BPM-Lab-WP-Lean-vs-Agile-v1.0.pdf

[Kom15] Komus, A., Kuberg, M.: Status Quo Agile; Studie zu Verbreitung und Nutzen agiler Methoden (Status Quo Agile; Study on the Spread and Use of Agile Methods9). https://www.gpm-ipma.de/fileadmin/user_upload/Know_How/studien/Studie_Agiles-</g>PM_web. Drawn: 11.11.2016

[Kot11] Kotter, J.: Leading Change: Wie Sie Ihr Unternehmen in acht Schritten erfolgreich verändern (Leading Change: How to Successfully Change Your Business in Eight Steps), Vahlen, 1st edn. (2011)

[Küh00] Kühl, S..: Grenzen der Vermarktlichung – Die Mythen um unternehmerisch handelnde Mitarbeiter (Limitations of Marketization – The Myths of Enterprising Employees); WSI Mitteilungen 12/2000 Hans Böckler Stiftung. http://www.boeckler.de/pdf/wsimit_2000_12_kuehl.pdf. Drawn: 5.12.2016

[Lai16] Laitenberger, O.: Bimodale IT – Fluch oder Segen? (Bimodale IT – Curse or Blessing?) CIO od IDG Media Business Media GmbH, 07.03.2016. http://www.cio.de/a/bimodale-it-fluch-oder-segen, 3253885. Drawn: 05.12.2016

[Lec12] Leckart, S.: The Hackathon is on: Pitching and Programming the next Killer App; Wired Magazine 17.02.2012. https://www.wired.com/2012/02/ff_hackathons/all/1. Drawn: 02.12.2016

[Lob16]   Lobacher, P.: Innovationstreiber Design Thinking (Innovation Driver Design Thinking); Informatik aktuell,12.01.2016. https://www.informatik-aktuell.de/management-und-recht/projektmanagement/innovationstreiber-design-thinking.html. Drawn: 11.11.2016

[Mus10]   Mustermann, M.: Ändere das Spiel. (Change the game.) Die] Transformation der IBM in Deutschland und was wir daraus lernen können (The Transformation of IBM in Germany and What We Can Learn From It); Murmann Verlag GmbH, (2010)

[Now16]   Nowotny, V.: Agile Unternehmen – Fokussiert, schnell, flexibel: Nur was sich bewegt, kann sich verbessern (Agile company – Focused, Fast, Flexible: Only What Moves Can Improve); BusinessVollage, (2016)

[OAA]   Open Automotive Alliance: Introducing the Open Automotive Alliance. http://www.openautoalliance.net/#about. Drawn: 30.11.2016

[Ope16]   OpenStack: Open source software for creating private and public clouds. http://www.openstack.org/. Drawn: 30.11.2016

[ORe13]   O'Reilly, C., Tushman, M.: Organizational Ambidexterity: Past, Present and Future; 11.05.2013 Graduate School of Business Stanford University Havard Business School. http://www.hbs.edu/faculty/Publication%20Files/O'Reilly%20andTushman%20AMP%20Ms%20051413_c66b0c53-5fcd-46d5-aa16-943eab6aa4a1.pdf. Drawn: 11.11.2016

[Pla09]   Plattner, H., Meinel, C., Weinberg, U.: Design-Thinking. Innovation lernen – Ideenwelten öffnen (Learning Innovation – Opening Up Worlds of Ideas); FinanzBuch Verlag, (2009)

[Rig16]   Rigfby, D., Sutherland, J., Takeuchi, H.: Innovation: Embracing Agile; Harward Business Review, May 2016. https://hbr.org/2016/05/embracing-agile. Drawn: 11.11.2016

[Sch15]   Schiedgen, J., Rhinow, H., Köppen, E.: Without a whole – The current State of Design Thinking Practice in Organizations; Study Report, 2015 Hasso-Plattner-Institut Potsdam. https://hpi-academy.de/fileadmin/hpi-academy/buchempfehlungen/Parts_Without_A_Whole_-_Download_Version.pdf. Drawn: 11.11.2016

[Sie17]   Siegl, J.: Wissensmanagement im Vergleich – Die besten Wissens-management Anbieter im Test 2017 (Knowledge Management in Comparison – The Best Knowledge Management Provider in Test 2017); trusted GmbH, Munich 01/2017. https://trusted.de/wissensmanagement. Drawn: 04.03.2017

[Som16]   Summer, C.: How This 30-something Entrepreneur Is Giving The 100 Year-Old Automotive Industry a Tune-up; Forbes online, 28.11.2016. http://www.forbes.com/sites/carisommer/2016/11/28/how-this-30-some thing-entrepreneur-is-giving-the-100-year-old-automotive-industry-a-tune-up /#7bbd3c981ae8. Drawn: 30.11.2016

[Str16]   Strelow, M., Wussmann, M.: Digitalisierung in der Automobilindustrie –Wer gewinnt das Rennen (Digitalisation in the Automotive Industry -Who Wins the Race); Study Iskander Business Partner GmbH (2016). http://i-b-partner.com/wp-content/uploads/2016/08/2016-09-06-Iskander-RZ-Whitepaper-Digitalisierung-in-der-Automobilindustrie-DIGITAL.pdf. Drawn: 24.11.2016

[Tho15]   Thomas, R.; Kass, A, Davarzani, L.: Experimental Design with Audi's Virtual Lab; Accenture Case Study (2015). https://www.accenture.com/t20150825T041248_w_/us-en/_acnmedia/Accenture/Conversion-Assets/DotCom/Documents/Global/PDF/Dualpub_20/Accenture-Impact Of-Tech-Audi.pdf. Drawn: 30.11.2016

[Vas16]   Vaske, H.: Forresters Abgesang auf die bimodale IT (Forrester's Farewell to the Bimodal IT); CP of IDG Media Business, Media GmbH, 29.042016. http://www.channelpartner.de/a/forresters-abgesang-auf-die-bimodale-welt,3246894. Drawn: 05.12.2016

[vRu15]   von Rundstedt, S.: Lebenslauf-Mosaik & neue Karrierereformen: 4 Tipps für Unternehmen (Curriculum Vitae Mosaic & New Career Forms: 4 Tips for Companies); Blog in berufebilder. de, 22.09.2015. http://berufebilder.de/2015/lebenslauf-mosaik-neue-karriere-formen-4-unternehmen/. Drawn: 05.12.2016

[Wes14]   Westerman, G., Bonnet, D., McAffee, A.: Leading Digital – Turning Technology into Business Transformation. Harward Business Review Press, Boston (2014)

References

[Zel15]   Zelesniack, E., Grolman, F.: Unternehmenskultur: Die wichtigsten Modelle zur Analyse und Veränderung der Unternehmenskultur im Überblick (Corporate Culture: the Most Important Models for Analysing and Changing Corporate Culture as an Overview); article initio Organisationsberatung, (2015). https://organisationsberatung.net/unternehmenskultur-kulturwandel-in-unternehmen-organisationen/#_ftn1. Drawn: 11.11.2016

Chapter 8
Information Technology as an Enabler of Digitisation

Information technology as a company function plays a key role in digital transformation. It is not only required as an advisor and supporter of all specialist departments in the implementation, but also has to operate the existing technical IT environment in a cost-effective and secure manner and to develop itself comprehensively at the same time. The established architectures and technologies have constantly to be modernised or replaced by new solutions in order to meet the requirements of the specialist departments flexibly and efficiently. But not only new architectures and infrastructures have to be implemented, yet also the organisation and the internal processes of IT must change, along with the development of the knowledge and the behaviour of the IT staff.

This situation is covered in the following with a focus on the business IT, while the IT installed in the car with control devices for various functions, such as air conditioning or distance control as well as the factory IT with its field bus systems, plant and robot controls, are left out at this point. Firstly, the challenges of IT as a supporter and enabler of digital transformation are explained, followed by the approach to the development of a holistic IT strategy based on the digitisation roadmap and, finally and briefly, a method for assessing the usefulness of IT for the business processes. The detailed elaboration of an IT strategy is not the subject of this book. Extensive approaches to this can be found in the respective specialist literature and standards such as COBIT, ITIL and ISO, e.g. [Joh14], [Cox16], [ITG08]. Instead, the focus is on typical initiatives and thematic areas, which based on the author's experience, are upcoming in the IT organisation of the manufacturers and must be implemented urgently. Two examples finally support the discussion.

8.1 IT Transformation Strategy

The IT organisation is in a balancing act between the traditional application landscapes and computing centres on the one hand, and the new world of digital natives, characterised by Apps and smartphones on the other. For decades, computers in the automotive industry have been used for a wide range of tasks, and specific solutions have been implemented for the specialist departments. For example, the vehicle developers use systems for drafting and parts list management, the commercial sector uses so-called ERP systems, and the production works with control stations for monitoring the production lines. The systems are either functionally specific in-house programming solutions, so-called legacy systems, or standard packages such as CATIA from Dassault Systemes or SAP. They are partly more than 30 years old and continue to support a large part of the business processes with high reliability.

The usually highly customised and sophisticated applications are widely regarded as business-critical and require high quality of service. Due to the programme scope and the complex, often poorly documented process/technology integration, an adaptation of the processes and the IT systems is very complex. Also the complete replacement of these solutions by Apps is not foreseeable. At the same time however, the specialist departments demand modern Apps which run on mobile devices. As used to in the private life, these applications should be available quickly and flexibly. Thus, there will be a long-term coexistence of both approaches.

The current situation in the companies and the required orientation of IT is summarised in Fig. 8.1. The business processes are to be reviewed and made lean, and the IT systems and the IT infrastructure have to be transformed and consolidated hand-in-hand. The established systems are often referred to as monolithic, because the applications are created as a closed block in a technology, aligned to an application area and adapted to new business processes only with great effort. The solutions run on dedicated infrastructures, whereby the technology is defined by the applications.

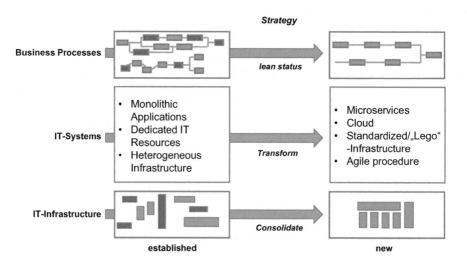

Fig. 8.1 Integrated IT strategy (Source: author)

As a result, a tremendous infrastructure diversity, a "technology circus", has emerged in the computing centres, which can only be operated safely with great effort. According to new studies, the operation and the application support of these areas often require more than 70% of the available IT budget, while for innovations only 14% are available [Kur16]. Standardisations and consolidations help to improve this cost situation, but there is still a significant budget ratio of the legacy systems.

The development of the "new IT world" is characterised by Microservices as solution elements for applications as well as by standardised hardware components whose capacity is easily expandable by connecting "Lego blocks" which are connected in Cloud architectures. Projects are characterised by agile procedures. The necessary investments must at least be partially financed through savings in the established sphere in order to be able to provide the required modern applications in support of the transformed business processes.

8.2 Building Blocks of an IT Strategy

In order to tackle the new challenges in a purposeful way, an IT strategy is needed that combines old and new technology areas and is sustainably geared towards achieving the business objectives. Transformation initiatives and optimisation projects build on this, which lead to programme planning. It shows the individual projects on a time axis with implementation priorities. The strategy describes, for example, the sourcing strategy, the target architectures, the technological IT standards and the programme to implement the required transformation.

The classic approach to strategy development is based on corporate objectives and the derived business processes, and subsequently defines the goals and basic architectures of the IT. This approach is proven in relatively stable processes and established IT environments with an existing process and application landscape. At the same time it is important though to take up the selected initiatives of the digitisation roadmap in close consultation with the specialist departements and to include them in the IT strategy. The particular challenge is that disruptive new business models and structures may become necessary in the short term. IT needs to be prepared for that with flexible structures that can quickly meet the new requirements. For example, the platform concepts proposed in Chaps. 5 and 6 for mobility services, administrative functions and sales management as well as for new digital products should be covered by the IT strategy in order to provide the necessary services quickly and efficiently.

When formulating the strategy, it must also be taken into account that the role of IT as a powerful enabler of digital transformation must change dramatically in the future. In the past, primarily technical expertise was required in order to develop applications, build the necessary infrastructure and operate it securely. The IT experts often were like within "a world of their own", partially isolated also by the use of technical terms, which created reserve in the users. Business processes were described in the definition phase of implementation projects according to the technological requirements. A continuous exchange with the specialist departements rarely took place.

However, this situation is changing fundamentally now. Through the use of modern IT technologies also in the private sector and the rejuvenation of the work force through the entry of the digital natives, the knowledge about the fields of application and use of IT is growing in the specialist department. There are hardly any more fears of contact, and the expectation towards the internal IT services is increasing to provide modern solutions quickly and flexibly. In return, IT is called upon to build more process knowledge and business orientation in order to expedite the ideas and projects of digital transformation in future together with the specialist departments, for example, in design thinking workshops. IT must also build this knowledge in order to be an accepted advisor and driver in the digitisation. Should this fail, there is a risk that the specialist departments will decouple and independently create own Cloud solutions and App developments. This "shadow IT" is to be avoided in any case since such isolated solutions certainly generate additional costs and often also constitute a security risk.

An IT strategy must therefore illustrate how the jointly developed objectives are to be achieved, taking into account the initial situation. The path should be clearly described, responsibilities should be assigned, and clear measurement points and key figures must be defined for progress monitoring in the work packages. The following topics are to be dealt with in an IT strategy:

- Architectures/standards
    - Architectures derived from corporate architecture
    - Standard applications/development planning
    - Concepts for handling of data and acquisition of information
    - Technology standards

- Applications/Microservices
    - Strategic services: Big Data, Analytics, Cognitive Computing
    - Application strategy
    - Software development … Tools … Methods … Open Source Software
    - Concept: Microservices/PaaS … API management
    - DevOps

- Infrastructure
    - Platform strategy, operating systems, integration technology
    - Cloud strategy
    - Communication technology, network concepts
    - Operating strategy … SLAs

- Sourcing
    - Core business vs. Commodity
    - In-house vs. Outtasking/Outsourcing

- Nearshore/offshore
- In-house factory concepts: software development, testing
- Partner concept
- Supplier strategy/-consolidation

- Mobile devices ... "Bring your own device"-concept
- Security concept
- Innovation management
- Computing centre concept ... site consolidation
- Training planning
- Organisation, internal processes, governance
- Investment planning, personnel planning, controlling.

For a detailed elaboration of an IT strategy, as already mentioned, please refer to the relevant specialist literature. It is important that the strategy integrates both IT worlds holistically. It should be noted that the development of an IT strategy is a dynamic process, especially with regard to rapidly evolving technologies and the changes in the requirements which accompany the transformation. The strategy and the implementation programme derived from it should therefore, together with the specialist departments, be validated and revised about biannually in parallel with the review of the digitisation strategy. A regular external benchmark is also useful to identify where new measures may be needed.

8.3 Cost and Benefit Transparency

An IT strategy will also have to be measured against the envisaged and achieved efficiency by clearly defined goals. In the past, the automotive industry evaluated its IT organisation at quasi natural service quality almost exclusively based on costs, while the benefit of IT support was not an issue in the execution of business processes. The goal was usually to achieve the lowest possible value in pure cost benchmarks. For this purpose, the ratio of the summarized IT costs to the company turnover as a percentage value has established itself as a standard parameter. As a general rule in the industry, volume manufacturers in Europe should achieve values below 2%, and in higher-quality vehicle segments, up to 4% are considered acceptable due to the smaller vehicle volumes. To be able to use this characteristic value for meaningful comparisons, it is necessary to define exactly which cost types of IT are involved. For example, there is a need to define the extent to which the IT costs are included in the production lines, the extent to which the costs of the in-car IT are taken into account, and which communication costs to include.

The main disadvantage of the exclusive cost focus is that the benefits generated by the IT are not assessed, and also the development of the digital maturity of the companies and thus the development of the competitiveness is not measured. It is therefore recommended to define together with the specialist departements parameters which, in addition to the costs, also describe the intended business benefit and

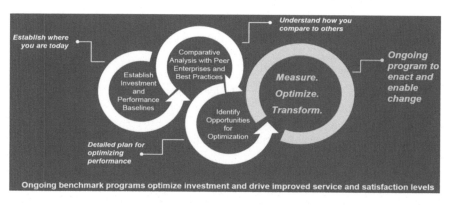

Fig. 8.2 Initial situation and direction of an IT strategy [Gue16]

the support of digital transformation, without neglecting measures to increase efficiency. A procedure for setting such characteristic values is shown in Fig. 8.2.

Firstly, the current IT costs must be recorded with an adequate level of detail. This includes the costs for personnel, hardware, software, delivered services, communication and also buildings. This rough view is then to be refined functionally, and as a basis for the progress control as well as external benchmarks performance, key figures are to be added, such as the costs per terabyte memory by service class, Linux costs per instance or also MIPS costs (million instructions per second) in the mainframe computer area.

On the basis of this classification, it is possible to identify IT areas in which improvements are possible and optimisation measures are useful in a cost comparison between the brand organisations of the manufacturers or also by comparison with benchmark indices of a comparable industrial sector. This procedure has been established, described many times and is therefore not to be deepened here [Tie11], [GadA16].

Furthermore, it is recommended to allocate IT costs as much as possible according to their causes to the main business processes of the specialist departments. A suitable basis for this is, for example, the segmentation described in Sect. 6.2.3 as per the SAP Value Map or the Component Business Model. Together, process areas with potential for improvement can be identified, also considering IT costs, and transformation initiatives be established.

The allocation of costs in line with their origin is relatively simple for specific application solutions, such as CAD program licences, whereas distribution keys are to be specified for cross-company IT services such as firewall solutions or server operation. This should be done in coordination with the specialist departements in order to create a common basis also for future cost tracking. The specialist functions should then assign benefits to these IT costs, for example the IT costs for the execution of a sales transaction, or the share of the IT costs in the construction of a vehicle component. Although the method proposed here is initially rather effortful, it is recommended in order to bridge the gap between IT and special departments, since this approach is particularly useful in digital transformation [Fre16].

8.4 Transformation Projects

On the basis of the corporate objectives, the corporate organisation is defined with the necessary business processes, and the objectives and the strategic requirements for IT are derived from it, which in turn becomes the basis for the IT overall architecture to be defined as the framework for action. It encompasses the thematic areas listed in Sect. 8.2, such as applications, data, security and technology. In addition to the requirements and needs, namely technological trends and innovations as well as resulting utilisation potentials must be included in order to ensure the sustainability of the IT solutions. In the following, essential aspects of the IT transformation are deepened.

8.4.1 Development in the Status Quo

Modern IT applications should be cost-effective, secure, and scalable, even with strong demand fluctuations, yet also easily adaptable, expandable when business needs change, and also function with adjacent applications. Furthermore, they should provide App-oriented appealing user interfaces that allow intuitive operation, and control should be via smartphones, voice, or gestures.

When migrating the existing applications to the target scenario, three options are available, summarised in Fig. 8.3.

The upper part of the picture shows the requirements for modern IT applications which have already been covered in the introductory explanations. In the implemen-

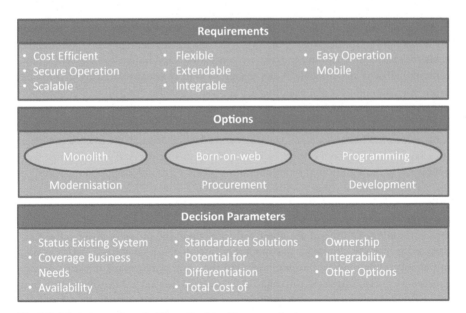

Fig. 8.3 Migration options for IT applications (Source: author)

tation of the development planning, the options listed in the centre of the figure range from the modernisation of existing applications, and the procurement of software solutions based on the latest technologies, effectively "born on the web", through to new development using innovative technologies and methods. In the lower part of the image, the key decision parameters are listed to choose between the options. The most important influencing factors are the status of the legacy system with regard to security and operability, the coverage of the business needs by the chosen solution and the total cost of ownership.

With stable business needs and safe and economically viable applications, modernisation is a good option. If an old system is to be replaced for lack of functionality or permanent instability, it is recommended to get a system according to modern standards. If this is not available in the required functionality on the basis of new technologies and also operable in Cloud environments, a new development is the remaining option, using innovative development tools and implementation methods.

The existing and traditionally developed applications include all the necessary programme modules, libraries and interfaces required for smooth operation. Because of this architecture, even small adjustments are already very complex, since comprehensive tests, compilations and deployment of the overall application are necessary. This explains why modernisation, in order to reconfigure these applications for example for operation in Cloud environments or to create new user interfaces, is very complex. The extension of functions and the transition to access via mobile devices require extensive works.

In order to improve this situation and, in particular, the agility of the application in the long term, a replacement of the application or a fundamental renovation is recommended with the modularisation and conversion to a modern, open architecture [Old15]. In order to support this work, methods and tools are available and make the process economical if migration factories in offshore centres are used. With this option, further to the pure project costs, the implementation risks and the costs for testing and rollout of productive operation need to be assessed. The transformation risk is manageable if the legacy system is encapsulated by a software layer, for example, for the handling of accesses and the control of the sequence integration, an architecture based on the concept of so-called strangler patterns. These allow a mixed use and the stepwise transition to the new system [New15]. A further option is to connect the legacy system to be preserved to the integration layer of a business platform in order to enable through this an App based, up-to-date operation, the integration of new functions in form of mobile function modules, or the use of innovative solutions for data analysis (cf. Fig. 6.27 in Sect. 6.2.3.1).

In addition to the individually programmed legacy systems, many manufacturers are using standard applications. SAP solutions have been very common for years. A major advantage of this off-the-shelf software is that the provider ensures the continuous further development and modernisation of the application. For example, in-memory technologies (programmes and data are in the main memory during execution) provide high-performance data analysis, the operability in Cloud, and latest SAP releases offer high modularisation, a new user interface and mobile access options. In this regard, a challenge for automotive manufacturers is that the

8.4 Transformation Projects

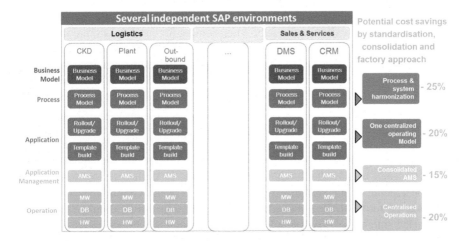

Fig. 8.4 Example of a developed SAP environment (Image IBM, author's experience)

standard software in many cases has been changed by parametrisation, customising or complex additional programming. Thus being far away from the standard, the release updates often cause considerable efforts and expenses.

Furthermore, many manufacturers are using not only one single SAP system, but rather special systems in different organisational units, which have grown as isolated solutions. These changes are often maintained as sector-specific standards, so-called templates, in order to achieve a reuse during the rollout into comparable organisational units at home and abroad. To do this, the required local adaptations in the templates must however be maintained up-to-date as a standard, also a complex task. In addition, SAP systems consist of three independent individual systems each for development, test and production, and operation respectively. Figure 8.4 illustrates this situation.

The picture shows the established SAP landscape of a manufacturer. The columns each represent an organisational area, for example, in logistics the sections CKD (exports in the Completely Knocked Down concept), the internal plant logistics, and the outbound logistics for the supply of parts. Also, in Sales & Service there are the units DMS (Dealer Management Systems) and CRM (Customer Relationship Management). The areas are shown here in part. There are further fields, for example, in the financial environment or procurement.

The lines are for the different project and operating phases and include the three-level system structure (development, test, production) of an "SAP island", e.g. the plant logistics. To start an implementation project, the business units define the system requirements for IT support in the form of a business and process model. The customising of the system is performed in the development system and then checked in the test system by the specialist departements. In rollouts, local extensions of the template are preferably added as standard to the "template build", often even by special development and test systems.

The systems are further developed and maintained in the application management area where so-called Application Management Services (AMS) are provided. In the IT operation, alongside the SAP applications also the middleware (MW), databases (DB) and the hardware (HW) are running, in addition to other technologies, for network integration and security for instance. The image illustrates the complexity of a developed SAP environment which in large companies may well cover several hundred different SAP systems.

Such IT landscapes offer significant savings potential if the systems are first returned as close as possible to the software standard. This must be done in close cooperation with the specialist departments since an absolute prerequisite is the adaptation and cross-organisational standardisation of business processes, which must be adhered to in a disciplined way on the basis of strict requirements management also in future changes and extensions. In addition, efforts should be made to reduce the number of SAP templates and to consolidate these across all areas. Depending on the company size, an ambitious consolidation goal is to be defined in order to significantly reduce the number of systems without too much restricting the agility of rollouts and extensions due to overly large application monoliths.

As a result of the reduction and standardisation of the applications, the number of hardware systems and through the unification of the application also maintenance and operating costs decrease. The standardisation also reduces the effort required for release changes, so that innovations are made available to the specialist departments substantially faster. Also, development and operational tasks can be carried out with considerable synergy gains by teams across the organisation, set up in several regions in 7 × 24 h modus operandi "follow the sun". These teams should be uniformly organised globally and collaboratively according to "factory concepts", using modern tools for problem analysis, knowledge management and automation in order to further reduce costs while at the same time improving service. Against this backdrop, the savings potentials shown on the right in Fig. 8.4 between 15% and 20% annually are a rather conservative estimate and often much higher, depending on the initial situation.

In many cases, the application inventory of legacy and SAP systems thus has significant potential for improving service, increasing the ability to innovate, and especially achieving savings. The latter should be used to further standardise applications and develop them towards the target architecture in order to finance innovation projects.

8.4.2 Microservice-Based Application Development

With increasing digitisation, the need for software continues to rise. IT becomes the core element of business processes and products. As a result, the importance of software development in companies is growing and becoming an integral part of the core business. The expectations of the specialist departments towards the IT to provide fast and flexibly scalable applications which can be further developed in

Monolith Architecture vs. Microservice Architecture

	Monolith Architecture	Microservice Architecture
Architecture	A monolith architecture represents a single logical program unit. All functions, libraries and dependencies are located within one "application block".	A microservice architecture consists of a series of small services that work independently and communicate with each other. Each service can be used in more than one application.
Scalability	The entire application scales horizontal behind a load balancer.	Each service scales independently and on demand.
Agility	Changes to the system lead to the compilation, testing and deployment of the entire application.	Each service can be changed independently.
Development	The development based on a single programming language.	Each service can be developed in a different programming language. The integration happens via a defined API.
Maintenance	Very long and confusing source code.	Many pieces of source code that are easier to administrate.

Source: Crisp Research AG, 2015

crisp

Fig. 8.5 Comparison of monolithic and microservice-based application architectures [Büs15]

short cycles, can hardly be fulfilled with traditional software development methods, technologies and the architectures for monolithic applications.

As an alternative, so-called Microservices are becoming increasingly important [Ste15]. As an evolutionary development of object-oriented programming or SOA (service-oriented architectures), microservices are small, standalone functional modules that can be created in different technologies. More extensive application programmes are created by coupling many microservices. The individual objects are executable independent of each other, tested as individual module and scalable, so they can easily be adapted to changing requirements. This concept is called a Microservice Architecture and offers many advantages over monolithic applications [Fow15], Fig. 8.5.

The first line of the listing compares the architectural approaches. Closed programmes stand as monoliths in contrast to a connection of independent individual modules which are also usable in other applications. The scalability of the microservice-based applications is very high, since heavily loaded modules can be given additional computer, memory or data transfer capacities. The availability of the entire system is also high as the failure of a single microservice does not necessarily cause the failure of the entire application. Agility and maintainability of the microservices are very high as well since only the modules concerned are to be tested for the deployment of adaptations and extensions, rather than the entire programme.

The development of the modules can be carried out by different developer teams, and interconnection of the modules is via predefined APIs (Application Programme Interfaces). Due to the parallelisation, the concept is well suited to work in Scrum teams. The low test effort and the flexible possibilities for the implementation facilitate rapid response to the demands of the specialist departements. These advantages of the microservice architectures can be enhanced by the use of further technologies and concepts such as Container and DevOps.

In logistics today, standardised containers are a common means of transport for all kinds of goods. The standardisation of the containers gives considerable advantages in the transport process since cranes and vehicles worldwide are adapted to the handling in ports, railway stations or transfer points. In line with this concept, the IT container technology encapsulates Microservices together with the required operating system services and runtime services by a surrounding software, thus quasi packing them into a container that can be run on any kind of infrastructure and operating system environment [Pre15]. The implementation of this idea takes place for instance via the Open Source project Docker and has enjoyed increasing popularity since then [Kri16]. The "packaged microservices" which can run on any server enable a very convenient infrastructure utilisation, and the transfer of the application, for example between development teams and computing centres, becomes very flexible. For traditional applications, extensive effort was needed to ensure that the infrastructure, including the system software, was set up completely identical between the systems.

Multiple Docker containers can run independently on the servers, without the need for further installations or virtualisation measures. The necessary resources are assigned to each container individually. The growing complexity from increasing container numbers can be mastered in this way even in running IT operation. Figure 8.6 shows an overview on the use of microservices with additional services.

In the centre of the figure are, placed on the IT infrastructure, containers C_1 to C_n, each of which encloses Microservices (MS_1 to MS_n) with the individually required libraries (Lib) and operating system components (OS). The orchestration of the microservices to applications is done via a separate MS management system [MSV16]. Resource allocation is done through container registration, and this level also manages releases and areas of use of the microservices. Monitoring and logging functions which log internal processes for example, support the operating teams. This means that comprehensive tools are available throughout the entire lifecycle of microservices-based applications. These are offered by different firms, so that the use of the architecture across the corporate is secured.

Apart from the considerable benefits, a number of challenges must be considered when deciding on this architecture. In addition to technology knowledge, develop-

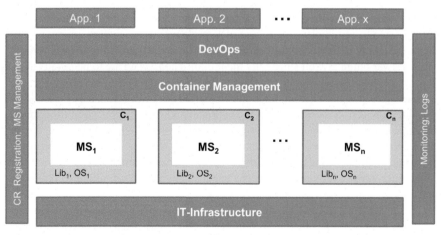

App. Application, C Container, CT Container-Registration, Lib Library, Logs Log-Functions
MS Micro-service, OS operating system component

Fig. 8.6 Orchestration and operation of microservices in containers (Source: Author)

ers also need Cloud know-how in order to avail of the advantages of hybrid hardware architectures, which are discussed in Sect. 8.4.4. It is in this context that latencies, i.e. delays in data communication, have to be taken into account in the operation of the distributed services, which must be absorbed by powerful communication concepts. Furthermore the developers should have knowledge in the topic fields of DevOps and APIs since both are vital for the full use of microservice-based architectures.

DevOps models are methods and tools which bring the areas of development (Dev_elopment) and operation (Op_eration) closer together in order to enable fast release cycles and short implementation times [Bos15].

What is important here is that a culture of close, open cooperation also establishes itself through the sharing across locations and organisations. It is important to create in an agile project approach as efficient processes as possible together, which are also increasingly automated in certain phases, such as testing and production roll-out. The use of containers, in particular, reduces the effort required since no application-specific infrastructure environment is needed any longer.

Another advantage of microservice-based architectures is the ability to easily also integrate external microservices. Especially in the Open Source environment, there are a variety of offerings, in many cases free of charge on sharing platforms provided by developer communities. Services from other companies or external suppliers can also be used. It is also possible of course to provide one's own microservices, programmes or even data via APIs for use by interested parties outside the company. For example, vehicle-related data could be sold to insurance companies or marketing organisations, or providers of services can be given access to the vehicle for displaying offers on the infotainment unit (cf. Sects. 5.3.2 and 6.2.1). To pursue this business strategy, an API management system is to be established not only in the microservices environment yet also for interaction with other application technologies.

8.4.3 Data Lakes

"Explosions" of data volumes in the enterprises, social networks and the Internet of Things open up opportunities for greater transparency, new insights and additional business. For this purpose, future-oriented data architectures and the targeted utilisation of the information from these data through the use of high-performance technologies are essential. Previous evaluation and reporting applications worked with a defined temporal grid via hard-coded interfaces with access to data in also defined structures. Reading the data from different applications and preparing the reports takes a lot of time. Ad hoc queries were not possible, and deviations from the report structure or the integration of further data sources were complex.

To avoid these disadvantages, so-called Data Warehouses (DWH) have been developed. A DWH imports data from different source systems, transfers them into a target data structure and stores them in the DWH. Reports and evaluations are fed from the target data of the DWH, while the output data in the source systems are overwritten [Dit16]. In situations with precisely defined analyses and stability of the data sources to be integrated, this concept is still reliable and widely used.

Aside from the effort for processing and storage, the main drawbacks of data warehouse architectures are the limited flexibility in reacting to spontaneous queries, the fixed data structure and the loss of the raw data after storage in the target system; also, evaluation directions must be specified in advance. It is not possible to gain new insights through modified queries or a new combination of raw data. However, modern solutions for the processing of information must offer flexible evaluation possibilities of most diverse data formats and efficiently work "near real time".

In response to these demands, the concept of the so-called Data Lakes has proven itself [San15]. The approach is to store all types of raw data without further processing in a flexible system at low cost. This may be, for example, structured data from established application systems, data from vehicles, machine-generated information, social media data or also audio and video files. The summary in Fig. 8.7 contrasts DWH and Data Lake solutions according to various criteria.

First, data storage is compared in both concepts. While in Data Lake the raw data remain unchanged, including the link to the source data, and are acquired in almost "near time", various technologies are available for data storage, whereas in DWH the data are stored in a target structure without the raw data being kept. Relational databases are used, and the data is transferred in a fixed rhythm, often at the end of a working day (EOD, end of day).

		Data lake	"traditional" DWH
Data storage		• Technical 1:1 data storage to original system • Technical mapping can be reconstructed in original system	• Technically harmonized and technically standardized data storage • No availability of original data in persisted target data model
		• Long-term persisted source data • Logical integration of source data possible	• Temporarily persisted input data ("staging area")
		• diverse technologies for user-oriented provision (e.g. Hadoop, NoSQL-DB)	• usually relational mapping
		• Near-time data transfer possible	• Data transfer usually EOD
Query		• Individual access rights (e.g. HiveQL, JAVA, PHP, SQL); partly complex	• Simply structured query option via SQL
		• Preparation of data (e.g. harmonization of field versions) "on the fly"	• Preparation / standardization / harmonization already completed in target data model
Performance / scaling		• Available technologies developed for very large data amounts ("big data") • Linear scaling based on "standard" hardware	• Relational DBMS usually designed for large data volumes • Scaling partly only possible based on special hardware
Development		• agile / iterative procedure • Environment under development (div. technologies, models under setup)	• target image-oriented procedure • established environment (tools, models, etc.)

Fig. 8.7 Comparison of data warehouse and data lake [San15]

Retrievals in Data Lake are more complex, since the data processing only takes place in the course of the query. Date Lake solutions scale their need for large amounts of data by using flexible technologies and based on standard hardware better than DWH solutions, which often require special hardware. For standardised queries with predefined evaluation direction and regular consistent reports, DWH's continue to have their strengths. Data Lake concepts, on the other hand, offer flexible analysis options and fit well into the agile, digital world because of the possible gain of new knowledge. Big Data technologies from different providers are available for the technological implementation of the Data Lakes as well as for the definition and processing of the queries [Gad16]. Through the use of flexible tools, specialist departements can independently carry out all kinds of evaluations without the involvement of IT and therefore welcome the new flexibility and independence.

Both concepts can also be combined, for example, if a Data Lake is connected to a DWH as a front-end system [Mar15]. Therefore, both DWH and Data Lake should be integrated as a core component in data architectures.

The following trends and technologies in the field of Big Data are to be considered in the architecture definition:

- Data Stream Management Systems (DSMS)

- DSMS systems process data streams that occur continuously and at short intervals [Ara04]. Search algorithms are permanently extracting desired results from the data stream and make them available for processing. Examples are vehicle motion data or data from camera systems during autonomous driving.
- In-Memory Data Management
- With in-memory data management, data is stored in the main memory of servers rather than on separate storage media and thus is available to processing in a high-performance manner. To efficiently use the available memory bandwidth, the data are read sequentially in the flow. This technology is often used when analyses of current data are required. Examples of use are complex reports, evaluation of sensor data and also real-time evaluation of social media data [Pla16].
- Appliances

- Appliances are integrated turn-key systems, optimised for a specific application. In a casing there are servers, memory, system software including visualisation, and partially software for data management as well. Examples of use are high-performance Big Data analyses via a distributed infrastructure and applications.

In order to deepen the technologies, which are only briefly mentioned here, and which are gaining in importance in the automotive industry in times of growing data volumes, for example from vehicle motion sensors and the Internet of Things of Industry 4.0, reference is made to further literature [GSM16], [Mar15]. Overall, a growing number of products are available for information processing.

8.4.4 Mobile Strategy

As was explained in detail, mobile devices and Apps are gaining ever more importance. In continuance of the private habits, employees, business partners and also customers expect new applications such as from the area of Big Data and Analytics to be available as Apps on mobile devices such as smartphones and tablets. This modern operation should also be available for established in-house applications. It places the use of workstation PCs in the background and will namely be for "power users" or programmers and testers. The use of mobile applications should be done through appealing graphical surfaces that are intuitive and easy to use without training.

Furthermore, users expect at any time and from anywhere access to the company applications used by them and the relevant data. In order to be able to address the rapidly growing needs in a structured manner, clear specifications for mobile terminals and mobile applications must be defined. This mobile strategy is particularly important because it defines the interface to digitisation and the move towards innovative usage models for all employees and customers. Also the agile handling of requirements from the field and the fast provision of solutions must be possible by powerful technologies, defined in the strategy.

Firstly, the device standard for company-internal equipment must be defined and also decided whether to offer the employees a BYOD option (bring your own device) to allow them the use of their private devices in the company as well. Next to commercial considerations on the basis of a comprehensive TCO (total cost of ownership), the safety aspects play an important role in this (see Sect. 8.4.7). Due to the worldwide customer acceptance and the resulting market shares, with Apple and Android devices two essential technology directions are dominant. For both platforms, architectures and solutions for secure integration into the enterprise IT are available so that many companies support both platforms.

Based on the device standards and the safety architecture, the application strategy has to be defined. Particular challenges arise from the fact that mobile devices are subject to very short innovation cycles, and the applications must therefore be able to run on several device and system software generations. The integration of mobile applications into the existing enterprise applications, the so-called back-end integration, also requires a viable architecture and specifications in order to generate synergies in the development and to facilitate operation and backup.

Also, criteria are to be defined in order to decide in a comprehensible manner on the type of implementation of the applications. There are three options. Native Apps work directly on the device operating system and use device functions such as camera, sensors and communication interfaces with high efficiency and safety. Programming, maintenance and operation in this case are more complex compared to the second option, the web applications. At this, the application development is independent of the device, and their use is done via a web browser. This means that special device options are not used, so that the user-friendliness often is diminished. The development and operating costs are lower though. A further option are so-called hybrid applications, which are a mixed form. All three routes are proven and have their benefits of use.

Namely the field of App development is particularly well suited to bring together specialist departments and IT in digitisation initiatives as a team and to make joint decisions on design and implementation. The App is wrongly much-underestimated and partly dismissed as just "a few colourful pages for mobile phones". These applications are however just the right tool to drive the transformation of business processes and to make them easily accessible and tangible to the users. They are, in a sense, the user interface of the digitisation. In this way, IT can position itself as an innovative, agile organisation, which also understands the business processes and realises fast implementations and adaptations. IT should avail of this opportunity and not only handle the technology and security aspects of Apps. A potential sequence of this cooperation is shown in Fig. 8.8.

In Design Thinking workshops, the team develops initial ideas by clarifying the business situation of the area, analysing disruptive trends and elaborating potentials for process improvements through Apps. The prioritised ideas are collected and, in the following concept phase, the processes (user story) have to be mapped and interfaces (user interface) are discussed, in order to then agree upon the scope of a first prototype (MVP, minimum viable product) (see also Sect. 6.2.3 and 7.2.1). This also includes an accompanying requirements management.

Development, testing and deployment should be based on developmental environments, the so-called mobile development platforms MDP. These support different device types and implementation paths also with the integration of microservices. The finished Apps are made available on company-owned App Stores for easy download.

The devices are managed using an MDM (Mobile Device Management) solution, with reliable architectures being available for safe integration regarding connectivity (network capability) and security. These technology kits support agile procedures especially in the test phase, so that feedback can always be given to the

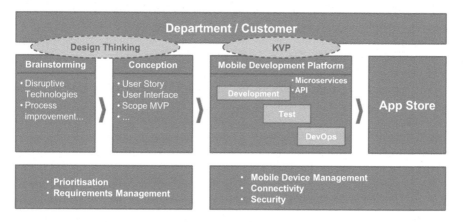

Fig. 8.8 Agile App development (Source: author)

specialist departments in order to continually improve the product and further releases. DevOps methods ensure fast release changes. The procedures and requirements for the user interface design as well as colour concepts to ensure a certain branding or "look and feel" should be described in addition to the mobile strategy in an online manual. Examples can be found in the relevant specialist literature and also on the Internet [DHS16].

A structured description contradicts at first glance the agile approach required in the field of mobile applications. Nevertheless, some rules are needed to avoid a sometimes occurring uncontrolled development in this rapidly growing area of solutions, and in order to achieve high quality and operational efficiency. On the basis of a well thought-out mobile strategy, the IT should therefore create the technical and organisational preconditions to surprise the specialist departments by implementation speed, flexibility and quality and thus be accepted as a partner in the development of mobile solutions. Implementations outside the IT which unfortunately happen frequently in this area, can thus be avoided.

8.4.5 Infrastructure Flexibilisation Through Software Defined Environment

The preconditions for high reactivity and efficiency are created in the computing centres. Challenges such as in the area of applications the integration of legacy standard applications and container-based microservices in a single target architecture, are similar in the IT infrastructure. In this context, infrastructure means a combination of all the technical facilities required to run applications, store data and transfer them for utilisation.

In the past, for each major application special hardware packages were installed, often with specific system software. In this way, complex heterogeneous infrastructure environments have developed over the years. System operation was in technology clusters such as Linux servers, storage systems and networks. These often ran without any reference to the applications, let alone any customer reference. The provision of system environments for new applications or projects lasted months. Deployment of solutions or the provision of minor software updates required long-term planning and special maintenance time windows in the business process.

This structure and serviceability were sufficient in times of stable business processes, stationary workplace systems, without Internet and at "manageable" data volumes. In this day and age, isolated server, storage and network structures have served their time. In times of digitisation with demanding requirements in the areas of mobility services, autonomous driving, Big Data, IoT, blockchain and social media, new solutions are imperative.

As a vision and a sustainable response to the requirements of digital transformation, the concept of the Software Defined Environment (SDE) has established itself with intermediate steps through consolidation, virtualisation and partial automation

[Qui15], [Bec16]. In this approach, the entire technical infrastructure is controlled by a special software without human intervention or changes to the hardware. There is a control layer above the computer nodes (i.e. computer units without further subsystems such as I/O units and power supply) and the storage and network units. Their software recognises the system requirements of the applications and automatically implements them through adaptations in the connected infrastructure, avoiding the previous complex manual work. The hardware technology fades into the background. This concept is further characterised by the following features:

- Automatic real-time adaptation of the technical infrastructure to the needs of the applications
- Automatic deployment initiated by software request
- Continuous dynamic optimisation of the configurations to achieve the definable target service levels and resource utilisation
- Highly scalable and "breathable" at load fluctuations and modification of operating parameters
- Fail-safe and self-healing if needed
- Hardware-independent
- Modular, open concept without "vendor lock-in".

The challenge on IT is to define a target architecture that ensures full coverage of the business requirements, takes into account the existing infrastructure, and allows a coexistence of the established and new world. For example, the following will have to coexist for several years:

- Monolithic applications/Microservices
- Virtualisation/Container
- Commercial software/Open Source
- Hard drives, tape drives/flash memory
- Mainframe technology/standard server ("lego bricks")
- Dedicated own hardware/multi-Cloud solutions

For this purpose, a transition road map towards the Software Defined Environment is to be specified and implemented. Figure 8.9 gives an overview on a target scenario with the technological components and their coupling to the applications.

A precondition for implementing the SDE concept is a complete virtualisation of the technological infrastructure consisting of computers, storage and network [Men16]. This refers to the complete logical representation of the hardware in software-based logical units. Virtual servers, storage systems and network components, so-called images, are managed in a way completely decoupled from the details of the hardware versions. A further element of SDE concepts is the integration of Cloud services in both manufacturer-specific (private) and publicly distributed (public) solutions; cf. Sect. 4.1.1).

To implement the so-called hybrid Cloud architectures, many companies rely on OpenStack technology. This is a comprehensive software portfolio for building open Cloud solutions, developed by the OpenStack Foundation and made available as an Open Source solution. The open community includes more than 600 compa-

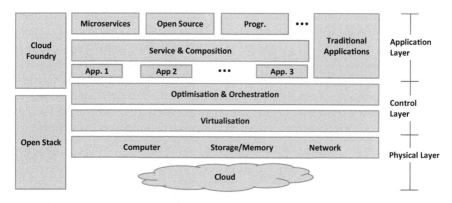

Fig. 8.9 Overview on the software defined environment (Source: author)

nies and its over 50,000 members are from more than 180 countries [Buc16]. A fast-growing number of large companies such as Walmart, Disney, Volkswagen and SAP rely on this technology to build and use Cloud environments. The manufacturer-independence, the openness of the architecture and the rapid further development of the technology with the opportunity to contribute their own input and to influence trends, all are motivators. Due to the active designing in the open environment, it is also possible – by contrast with closed, manufacturer-specific solutions – to ensure the control over own data.

Building on full virtualisation, based on the detected infrastructure requirements of the applications, the control layers automatically adopt the orchestration of the resource allocation and its optimisation. For this purpose, cognitive solutions are integrated. The avoidance of failures is based on preventive maintenance technologies, which in turn require a continuous analysis of the operating parameters and the operating logs (log-on and log-off).

The software layer in the SDE concept also adopts the automatic deployment of both traditional standard packages and new applications which are traditionally programmed, based on Open Source software or configured as microservices. As a development environment in this setting, CloudFoundry has established itself with considerable growth. This is a Cloud-based software environment for developers, a so-called PaaS (Platform as a Service) solution, for example, with services for development, production, and test, provided by the CloudFoundry Foundation as an Open Source. The foundation includes many major industry players, such as SAP, IBM and Cisco, which help to establish the solution as a standard [Sch16]. The full compatibility with Open Stack and thus the possibility to integrate into Software Defined Environments support a decision to also include this technology in a mission.

In SDE concepts, IT technology loses much of its importance. While previously specific performance and configuration parameters of servers or storage systems were important in order to run certain applications as efficiently and securely as possible, the SDE software now recognises the requirements and assigns the required number of virtual resources. The applications then run distributed on any server and are optionally saved on storage systems to enable the desired service level. The physical IT layer consists of standardised units, which are sufficiently available in a manner comparable to a Lego concept. In the event of a failure, the respective units are automatically discarded. In order to replace or extend the capacity, the installation of such "Lego bricks" is implemented, which are adapted by the virtualisation layer.

Apart from these hardware-independent applications in Software Defined Environments, there still remain a few islands that require special technology.

These include, as briefly described in Sect. 8.4.3, so-called appliances, which efficiently process very large amounts of data through high parallelisation of computer units. Furthermore, neuromorphic chips can be foreseen, which map neuronal structures directly in silicon circuits (cf. Sect. 2.6.4). This technology is particularly strong in the recognition of patterns and images. Another emerging trend is quantum computers. After stabilising the above-mentioned technologies and increase in the business requirements, the control software of the SDE architecture is expanded so that these systems can also be integrated in the future.

8.4.6 Computing Centre Consolidation

With transformation projects in the area of application and infrastructure, it is in both cases necessary to establish a target scenario as part of the IT strategy and to transfer the current inventory step by step. By consolidating and optimising the growing heterogeneous environments, costs are saved which are available for innovations. Similarly, the consolidation of the existing computing centre structure provides significant savings potentials. Large manufacturers often have well over a hundred computing centres, ranging from smaller computer rooms at importers and dealers, special IT rooms in the plants, to computing centres in the brand organisations. As recently as in the 1970s, local requirements necessitated special server and operating software installed there because the networks at that time did not have the bandwidth and security to hold the required infrastructure in a central computing centre. Also, hosting and Cloud solutions were not available. As a consequence, at many manufacturers a heterogeneous computing centre structure has developed over many years, Fig. 8.10.

The left side of the figure shows the typical computing centre structure of a manufacturer, as was already described briefly. In addition to the large separate computing centres each per brand or financial sector, there are often smaller computing centres (CC) in the plants and also at the importers, at least shielded server rooms, and even the dealers have their own IT areas. In many cases, local operations teams run the respective infrastructure with local applications. The utilisation factor of the infrastructure was estimated by the author to be less than 50%. The high energy consumption due to this low utilisation, but also due to the partially aged computing centres, certainly holds improvement potential.

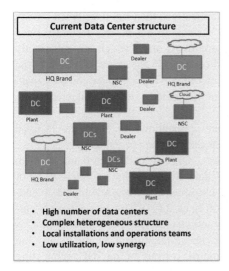 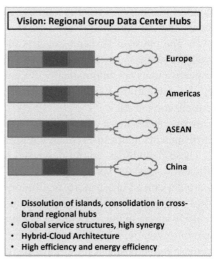

Fig. 8.10 Consolidation of a computing centre structure (Source: author)

In addition to this current situation, a possible target scenario is shown in the image on the right. In order to achieve significant savings and reduce the complexity of the structure, the primary goal should be to minimise the number of computing centres and maximise the use of the Cloud instead. Today, the technological developments in the network sector and the infrastructure allow very ambitious consolidation targets [KriS16].

For this reason, the author suggests that manufacturers completely refrain from local computing centres and rather build regional CC hubs, for example in Europe, for the Americas and for Asia. Due to the specific security issues, a separate hub in China makes sense, as well as in Africa if there is a market presence. The hubs then serve brand organisations, plants and also sales organisations of the respective region on the basis of hybrid Cloud architectures. These are implemented in the hubs as private Cloud environments, which in turn are linked to public Clouds in compliance with security concepts. Clear goals should be implemented to massively expand Cloud components outside the manufacturer environment in order to keep the previously used productive CC area as constant as possible despite the significant increases in demand due to the foreseeable digitisation and the growing data and storage requirements. The operation of the infrastructure is organised globally and is based on standardised processes. The hubs are interlinked and safeguarded by backup and emergency solutions so that in the event of a disaster situation, a region can jump in for another region and continue the operation of this failed region.

The implementation of such scenarios is technologically manageable, and the advantages in terms of costs, complexity reduction, securing options and flexibilisation are considerable. The road to implementation is quite demanding, and many manufacturers are moving in significantly smaller steps, with no overriding big goal in mind. However, this should be defined as part of the IT strategy and approached according to the concept "think big, start small, move fast" [Low16]. Prerequisites for implementation are, for instance:

- Powerful networks with high service levels
- Regional computing centres for the hubs with sufficient space availability and high energy efficiency
- Strong global governance and enforcement opportunities for the dissolution of the local and organisational "kingdoms"
- Global integrated service structures
- Global IT strategy with future-oriented goals and transformation targets for applications (especially: standardisation, "Cloudification", microservices) and infrastructure (especially: software defined environment, virtualisation, standardisation)
- Partnerships for Cloud services and project implementation
- Leadership and entrepreneurship

The above emphasises that it is not a lack of technological possibilities which bars the way to ambitious consolidation, rather the challenges are in the breakup of customary procedures and the establishment of global, overarching organisational structures.

8.4.7 Business-Oriented Security Strategy

Another very important topic of the holistic IT strategy is security. Almost daily there are reports on hacker attacks, the theft of company data and the intrusion into company software by viruses, partly dormant in Trojans as a latent risk. The integration of nearly all business processes with application solutions which is further increasing in the digitisation, the penetration of the cars with IT, the direct coupling of dealer applications with the manufacturer's backend, or the implementation of Industry 4.0 initiatives along with the "Internet of Things" at the plants, give more and more opportunities for attacks. Also the open agile project processing methods with temporary integration of experts, partly involving badly secured Internet communication, increase the risks.

This is why many companies have established CISOs (Chief Information Security Officer) who are in charge of security in the company. However, in many cases their area of responsibility does not include the plant and vehicle IT, which is carried out independently by the specialist functions. The security of the dealer and service network remains a topic of minor importance even in companies which belong to the manufacturer, and it is addressed by independent dealers even without manufacturer integration. Security measures are primarily regarded as technology projects there, at best loosely linked to business objectives and specific security requirements, such as protection from intruders or restarting a plant.

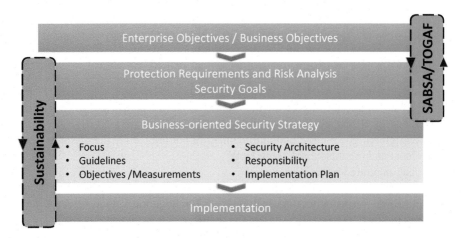

SABSA: Sherwood Applied Business Security Architecture
TOGAF: The Open Group Architecture Framework

Fig. 8.11 Development of a business-oriented security strategy (Source: author)

These exemplary problems illustrate that in many cases there is a need for action to approach the issue of security with the appropriate focus, sustainability and business orientation. The recommended major development steps of a business-oriented security strategy for traditional IT are shown in Fig. 8.11.

It is recommended that IT and specialist departements jointly define and implement the security strategy in order to achieve acceptance of this topic, which is often seen as a nuisance. The starting point is the company strategy and the business targets derived from it. On the basis of protection requirements and risk analyses, the team identifies the potential risks and defines the specific security objectives. This should be done in a pragmatic way so that IT security is not perceived as a hindrance to business and the acceptance of the subsequent measures is already developing at this stage.

To ensure the completeness and for the structuring of the cooperation, it is recommended to use the SABSA and TOGAF methods proven in the industry in the areas of security and corporate architecture [Kni14]. SABSA (Sherwood Applied Business Security Architecture) is a framework for the structured recording of security requirements as well as for the development and implementation of security architectures. The focus here is on the security aspect, but there is no link to the business processes. This gap is bridged through the use of TOGAF (The Open Group Architecture Framework). The combination of the two approaches provides a suitable approach to record safety requirements from a risk and business perspective in a structured manner and to transform them into a target frame for a business-related security strategy.

In the security strategy, the essence of security is described for instance, and the general guidelines for achieving security are defined with objectives, standards and

Infrastructure	IT Operation
Vulnerability Management	Information Security Process
Patch Management	Access and Authorization
System Hardening	Asset Management
Remote Access	Employee Security
Software Development	OPS-Demands: Malware Protection, Logs, Back-Up, Network
Cryptographic Solutions	Change Process
Documentation	Security Incident Management
Reporting Security Breaches	Physical Security, Access Control
Non-technical Security	Security Outsourcing Process

Fig. 8.12 Work areas information security (Source: author, following [KRIT16])

responsibilities. Concrete targets and key figures are the basis for the implementation plan. In line with sustainability, security planning should be continuously monitored, and a change in the business objectives be reflected in the adaptation of the protection requirements. Within the scope of the implementation, there are a number of topics. Some typical work areas, derived from "best practices" experience, are shown in Fig. 8.12.

The areas are grouped according to infrastructure and IT operations. Vulnerability management, for example, is the continuous surveying of all components of the infrastructure for vulnerabilities as well as the identification and removal of detected problems. Firewall gaps may occur or a lack of protection at the server BIOS level for instance. These are eliminated within the framework of patch management, which covers all software components of the infrastructure. Further areas of work are the use of encryption techniques and namely also remote access for technicians and service staff from outside the company. In IT operations, many fields concern the secure handling of processes, such as change management and incident management, which includes problem management within the service management, but also the regulation of access rights and personnel security.

A further detailing of the implementation of the strategy for information security of the business IT is left out here, as well as an explanation and deepening of the large number of relevant standards, norms and guidelines of the German Federal Office for Information Security. Rather, reference is made hereby to extensive literature and sources as well as to explanations on new requirements derived from Cloud services and open architectures, e.g. [BSI16], [BSI09], [NIS17].

It is important to understand that the issue of IT security cannot be managed and implemented by a CISO or the security department's staff alone. The topic is rather to treat it top-down by all business areas with appropriate care. In the company's management meetings the subject has to be on the agenda time and again, the more since various laws provide for the personal liability of executives in case of failures or negligence.

8.4.8 Security of the Factory IT and Embedded IT

The procedures and security measures described so far relate to securing of the business IT. In addition though, all manufacturers are required to also secure the IT directly at the production lines in the plants as well as the embedded IT in the vehicles. Both fields require special measures, which are briefly discussed below.

During the implementation of Industry 4.0 projects, a massive penetration of relevant business processes with IT is taking place. The production and assembly lines are equipped with additional sensors and the line sections are linked via IT solutions based on fieldbus systems in order, for example, to assign individual measured values from the work operations to orders during the manufacturing process, or to flexibly control subsequent stations. The production planning system forwards order data to the lines via the Shop Floor system, so that pick-and-place robots automatically change the correct gripping device using their programmable logic controller (PLC). Data from the line monitoring is compared with the information from lines of other plants in order to derive proposals for preventive maintenance measures. The examples illustrate that traditional IT and plant IT are more and more converging and thus the security risk is increasing. This also explains why the plant IT is to be integrated into a holistic security strategy and should therefore be included in the responsibility of the CISO.

The IT security procedures presented above do in many cases however not meet the requirements of the plant IT. For example, the use of anti-virus software in control computers is problematic since the scan process consumes computing power and thus results in performance losses in the near-real-time sequence control. Also, common measures of the anti-virus software when tracing a virus, such as setting up in quarantine and shutting down the computer, are contraproductive to the desired uninterrupted production operation, and so are regular updates of the signature database with corresponding runtime as well. Therefore, if the use of anti-virus software is not possible, other security measures must be taken, such as swapping out the unsecured components into a special network segment protected by an additional firewall. Detailed advice on securing information in the Industry 4.0 environment can be found in the relevant specialist literature [Bac16].

Just as with the plant IT, special procedures are applied to secure information of the embedded vehicle IT, different from the usual measures of information security. Although, the protection of the vehicle IT also has other aspects with respect to the security objectives. At this, the focus is not "only" on corporate objectives, but on the interests of many persons and parties concerned. The security risk of the vehicle IT is increasing with the rapid expansion of the IT in the cars, indicated by the rise of computer units, the so-called embedded control units (ECUs) (cf. Sects. 5.3.3 and 5.3.5). In the premium vehicle segment, more than one hundred ECUs are often used to control procedures and driver assistance systems. Details on this were also given in Sects. 5.4.5 and 6.2.1. To implement these functions and also as part of the future mobility ecosystem, the vehicle has to become part of a comprehensive communication, as shown in Fig. 8.13.

8.4 Transformation Projects

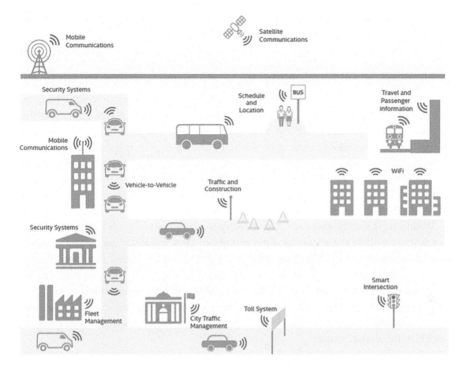

Fig. 8.13 Communication in the ecosystem of mobility [Bro16]

In intermodal transport, cars are used in order to reconcile the passenger changeover points with other mobility services such as regular buses and other means of public transport. "Vehicle-to-Vehicle" communication, also including sensory data from the infrastructure, is used, for example, to warn each other of dangerous situations in the course of the road. Mobility providers record vehicle information to increase the utilisation of their fleets or to dynamically adjust routes as needed. Traffic control in cities and the online handling of toll payments also requires a vehicle connection. Beyond the examples shown, further communication needs will arise in the future with autonomous driving and the growing number of driver assistance systems, as well as for remote maintenance and software updates "over the air".

The numerous communication channels and the high number of ECUs in the vehicles, combined with several fieldbus systems, are widely open to hacker attacks. These follow a recurring pattern [Mil14]. Initially, intruders attempt to enter the vehicle IT via one of the established communication channels and then use malicious software to generate false data in the vehicle network, which are then read by safety-relevant or sensitive ECUs, depending on the objective of the attack. Their interpretation leads to malfunctions, such as the unintentional triggering of braking processes, malfunction of the engine electronics or blocking steering movements. The intrusion can also take place via other weak points such as, for example, remote control of the door lock, sensors for tire pressure control, Bluetooth mobile phone connection or downloaded Apps in the infotainment unit.

These examples of security issues and possible points of attack illustrate the need for special measures by manufacturers and their suppliers to secure the vehicle IT. They start with hardening the IT technology used in the vehicles. In the ECUs, identity management, encryption and active memory protection should be programmed for instance. Furthermore, safeguarding the applications and also the onboard wiring systems is needed, also with identity management, segmentation of the network areas and authentication [Bro16]. The basic principles of identification, authentication and authorisation with secure handling in a protected, tamper-proof environment have a very high protection function – and this holds true in the vehicle IT as well [Bon16]. One way to implement this is to create a security gateway, as shown in the solution overview in Fig. 8.14.

Within the security gateways, the identity management of each IT component is handled, as well as the secure storage namely of personal data and the entire communication with the vehicle. The integration of the back-end and thus the integration with the enterprise IT is encrypted and also done via the gateway. In the backend, identities, access keys, and authorisations are managed. Intelligent safety solutions detect anomalies and trigger preventive measures. These can, for example, be initiated and monitored in a so-called security operation centre, a service organisation at the manufacturer.

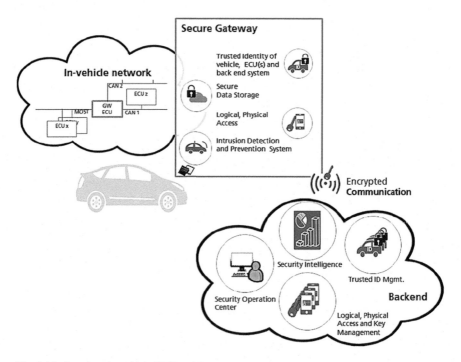

Fig. 8.14 Securing the vehicle IT [Bon16]

8.5 Case Studies on IT Transformations

This explanations round off the description of some important topics on the transformation of IT from a developed structure towards more innovation and agility. In the following, two case studies show how companies outside the automotive sector have met these challenges and achieved their goals.

8.5.1 Transformation Netflix

Netflix, although not being related to the automotive industry, is chosen as a transformation example here, because even though this company operates extremely customer-focused now, it had to adjust its business model massively three times within the short period of its existence. During this time, modern IT was a key enabler for this transformation and for implementing higher service and customer orientation. These experiences are also of interest to the automotive industry as the transformation is particularly needed on the sales, marketing and services side.

Netflix was founded in 1997 as a rental service for DVDs with mailing as a competitor to videotheques, ergo a rather traditional business model. In order to differentiate itself in the market, the company relied on good customer service and attractive prices. On this basis, the business grew steadily, and in 2007 with a portfolio of 35,000 films more than one million DVDs were shipped per day [Kee16]. During this time, the technological possibilities and bandwidth of the networks had also improved, so that precisely in this year a so-called "tipping point" was reached from which on the download costs of a film were cheaper than postal dispatch. The Netflix CEO recognised the threat at an early stage and urged his team to adapt the business model and become a highly efficient download provider.

This transformation was successful, but with streaming a new technology with disruptive potential and entry opportunities for competitors was soon establishing itself. Netflix also successfully adapted this swing and mutated from a download provider to the leading streaming provider.

But then another business model adjustment was required. In response to the incessant increase in the costs of screen rights for films, shows and other content, Netflix itself became a successful producer of films and series such as House of

Cards [Kee16]. At the same time, further adjustments were made, especially in the IT, in order to create the basis for growth and competitiveness. In the fiscal year 2016, Netflix achieved with approximately 3700 employees, sales of $8.8 billion. Earnings increased, and analysts see a good orientation of the company with growth potential through expansion into new markets [Fin16].

The key points of this successful transformation combined with massive adjustments to the business model are:

- Management with Leadership and Entrepreneurship
  - Early adaptation of disruptive technologies
  - Courage to adapt the business model
  - Consistent implementation of change
  - Corporate culture with a willingness to change
  - Screening of technology trends
- Uncompromising customer orientation
  - Focus on "experience" with a high level of service
  - Intensive evaluation of social media, feedback and market trends, early recognition of customer requirements by innovative IT solutions
  - Attractive offering which lives up to expectations: very comprehensive range of films, flexible use (rental, download, streaming), own content
  - Active "near-real-time" social media-based marketing (Facebook, Twitter, Instagram)
  - Attractive price structure: no shipping costs, no late fees, subscription model
- High-performance staff base [Kno16]
  - Hiring top performers ("A-Team")
  - Open performance assessments and result-orientation
  - High fixed salary (market benchmark); no bonus payments
  - Minimisation of internal regulations – e.g. no rules on holiday or travel expenses
- Use of IT as an enabler for innovation and transformation
  - Microservice-based application landscape; API opening
  - Complete Cloud orientation; no own IT infrastructure
  - Innovative analytics of Big Data to identify customer needs
  - Adaptation of new technologies
  - Crash tests to ensure availability

Further details on the aspects mentioned can be found in many contributions on the Internet, and it is certainly interesting to follow the further development.

Due to the thematic focus of this book chapter, the topic of IT is deepened in the following. Netflix has migrated the entire hosting of the applications into a Cloud and operates over 10,000 virtual instances spread over several time zones and regions [Tot16]. In the hosting structure, security concepts are active, whose performance is time and time again tested by specific scenarios. In doing so, virtual servers or even entire hosting regions are switched off for the exercise, while the responsiveness is checked in order to ensure a high availability of over 99.99% for the customers. The many terabyte data are therefore stored redundantly with a second Cloud company.

Netflix does not maintain any own servers for application operation. However, the company still operates in-house the network to the customer, its so-called Content Delivery Network (CDN). The aim is to maintain the know-how of a core technology for streaming and also to independently exploit analytically recognisable options to bundle products for commercial reasons. At peak times, Netflix uses one-third of the Internet bandwidth in the US.

Another core know-how of the company is Big Data Technologies and its own algorithms to forecast customer requirements. For example, the proactively communicated recommendations for shows and movies achieve very high hit rates. Analytics and prognostics are also used in marketing. By analysing social media data, one can recognise customer trends there and posts, for example in Facebook, regional information tailored to the region. In this way, IT supports agile and closely segmented appearances in the social media, which, in addition to up-to-date information, also show different images or extracts from new episodes. Decisions on which films and series will go into production and which content orientation they may take, are based on detailed analyses of customer expectations. The selection is not made by the Netflix executives, yet by the respective content managers.

The complete application landscape of the company is based on a microservices-based architecture and includes more than six hundred solution modules, for instance to handle the processes of registration, evaluations, recommendations and rental history. An overview on the architecture is shown in Fig. 8.15.

Customers can use any devices such as smartphones, web browsers or game consoles for access. In total, there are more than one billion hits per day, orchestrated by load balancers to distribute the loads, and managed via APIs published in the Open Source community. The APIs run via an intelligent buffer layer, which intercepts any errors and smooths them without failures. The administration of loosely coupled individually installable and upgradeable business services, as well as the system services, are provided by a processing system, a service registry. Data access to the distributed data storage is done via an access level. The technology base consists essentially of Open Source products, such as HTTP servers or Tomcat; Java, Ruby, Python and Go are used as programming languages, and for data management Casandra is employed [Tot16].

Also interesting is a look at the environment and the working conditions of the development. The services are created in parallel by many teams. These bear the complete responsibility for their solution modules, from development over deployment through to operation. There are no general guidelines for quality assurance for instance, release management or the defining of standards. Which technologies are used, is decided by the teams, and the optimal problem solution is in the foreground. The technology-fit stands above the demand for standardisation; innovative power and growth have priority over planning capability, and status statements and speed

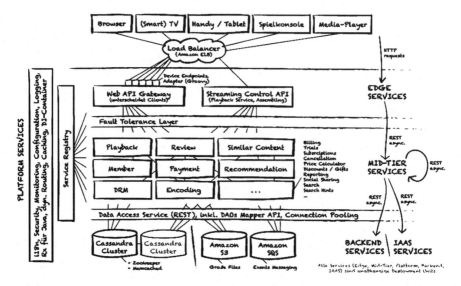

Fig. 8.15 Application architecture Netflix [Tot16]

of delivery are more important than being error-free. The work environment is aptly described with the heading of "freedom and responsibility". This approach and the microservice-based modular architecture founded on Cloud services provide many advantages. The entire application landscape is fault-tolerant due to the modularity and provides a very high availability. With the Cloud in the background, the scalability is secured, and innovations are available to customers very quickly. Currently, there are one hundred deployments on average per day.

The innovative environment is very attractive to young talents. The disadvantage is that the teams need to build up an extensive knowledge of the Netflix application landscape and the specific frameworks and tools used, and thus require a longer training period. Also, a heterogeneous technology portfolio is used. Furthermore, the independent services require comprehensive monitoring and logging. However, these disadvantages are compensated for by the described advantages of thorough customer orientation. The aspects of the Netflix approach are certainly worthy of note for manufacturers and can be implemented in the construction of new Connected Service and digital products.

8.5.2 Transformation General Motors

As a further case study, the IT transformation of General Motors (GM), a leading volume vehicle manufacturer, is being presented hereinafter. Of particular interest in this reference is the transformation of the IT from an operation which was characterised by third-party services into an orientation towards innovation, self-competence and agility. This type of reorientation is now a subject at almost all manufacturers and suppliers.

8.5 Case Studies on IT Transformations

First, some basic data of the corporation: General Motors is an internationally active US company present in more than 140 countries with over 170 production sites spread over ten brands with a total of 215,000 employees. With more than ten million vehicles sold per year, the turnover is well exceeding $150 billion with a 10% target return [GM16]. The main markets are the United States, China and Europe, with the focus for further growth being on the emerging countries.

Following the severe crisis in the automotive industry with its peak in 2009, the measures taken to increase efficiency and realign the company were effective, so that continuous increases in sales and earnings were achieved over the last few years, albeit with some backlog in certain regions and areas. The current strategy, summarised, is divided into four main areas:

- Earn customers for life

- Customer-relevant innovations such as connectivity and Connected Service at an early stage in the desired functionality in the market; strong focus on safety and quality; social communities
- Lead in technology and innovation
- Leader in 4G LTE (mobile radio standard of the 4th Generation); comprehensive mobility services (investments in Lyft, Maven); autonomous driving (investments in Cruise); e-vehicles
- Grow our brands
- Strengthening of Cadillac as an "iconic" luxury brand with focus on the USA and China; Chevrolet as a global volume brand; specific brand management
- Drive core efficiency

- Safe implementation of a program to increase efficiency with ambitious savings in administrative processes, production and development; lowering of break-even limits

In the implementation of this program, the IT is involved in all fields as an enabler with many projects and is also required to provide direct contributions, for example to the provision of innovations with simultaneous savings. To achieve this goal, IT has undergone a fundamental transformation. Key figures for this are shown in Fig. 8.16.

In line with many other companies, General Motors had almost completely outsourced its IT up to the year 2009. Essentially, there were only employees left in the company for supplier control, and the in-house competence in this thematic field faded. With the growing importance of IT as a driving force for innovation and digitisation, GM has since 2012 completely aligned the strategy in the direction of an insourcing concept, and hired 3000 IT employees of the previous outsourcing service provider [Sav12]. To further strengthen the in-house competence, a large number of additional experts and career starters from renowned universities were added, so that GM in 2016 employed a total of 11,500 own IT staff.

In order to keep the in-house knowledge up-to-date, comprehensive training programmes are available, and up to 500 university graduates enter the IT every year. With this consistent expansion of the own competences and "depth of IT production", the ratio of external service providers to own employees has become reversed, so that today 90% of all services are provided in-house. The work contents also

Metric	2009	2016
IT Staff	1.400	11.500
Performance Ratio External / Internal	90/10	10/90
Work Load, ratio Run / redevelopment	80/20	20/80
Number of main Data Centers	23	2
Number of core applications	4.000	3500 Target: 1500
Data Management	Distributed, heterogeneous	Central, EDWH
Governance	Decentralised Local interests	

Fig. 8.16 Key figures of the General Motors IT transformation (Source: author, following [Pre16])

fundamentally changed towards process improvements, global operation teams and focus on automation. In 2016, 80% of the capacity was used for new developments and innovation, while the operation of the existing IT landscape needed just 20%. Global governance with corporate-wide responsibilities and synergy in the service teams across brands is certainly also a basis for this change. This avoids duplicated work as not every brand maintains a server operation team for instance, rather there is only one team in the world.

Along with the establishment of the GM IT team, the computing centres, formerly 23, were consolidated into now two main computing centres in mutual backup. The number of central applications shrank from 4000 to 3500, always moving towards the goal of 1500 applications. At the same time, a considerable number of applications from the "shadow IT", i.e. IT solutions operated in the specialist departments, were transferred to IT responsibility and thus to safe operation. Data management is now fully centralised, and in the central computing centre a global data warehouse (EDWH Enterprise Data Warehouse) is located first for North America, where all structured and unstructured data are stored and are available for evaluations. As an example, on this basis it is possible for the first time to carry out detailed cost analyses on specific vehicle models in individual markets in order to determine contribution to margins in advance or to simulate measures that increase profitability. A few further summarised innovations go here:

- Consolidation of social media applications; central service centre for centralising 30 separate applications and providing an application in the service centre; a

customer view for enquiries or also sales activities → Increase customer satisfaction
- Expansion of high-performance computing to increase the use of simulations in the development, for example, to optimise the consumption and material use of the vehicles → Shortening the development time
- Use of predictive analytics in paint shops and robotics → Increase output and availability
- Real-time inventory in the central spare parts warehouse in Brazil → Increase service level
- Innovation Centre for Connected Services; for example remote access to vehicles via an App for tire pressure control or operation of the air conditioning system or heating; onboard diagnostic solutions → Increase competitiveness

These examples demonstrate how the IT and its projects contribute directly to the implementation of the company strategy. Overall, GM has succeeded in aligning the ability to act and the possibilities of the IT with the expectations of the specialist departments and customers. Requirements are taken up competently and solutions are developed within a short time in an agile approach in close cooperation with the specialist departments. The IT provides up-to-date solutions that can be used on mobile end devices via graphical user interfaces. The benefits of IT and the role of enabler and moderator for innovation and digitisation are more and more recognised. The precondition for this success was the establishment of a change culture among the employees and on this basis the transformation along a clearly communicated roadmap.

8.6 Conclusion

Overall, IT plays an important role in digital transformation. It provides the platforms for new digital business models and the use of new technologies such as 3D printing and augmented reality as the basis for new learning methods. It also becomes a consultant to the specialist departments on the possibilities of the latest IT solutions, for example in order to achieve savings through the automation of business processes and the use of Apps.

At the same time, IT has to renew itself technologically and, for example, switch to Microservices, Data Lakes and Cloud, all of this financed by consolidations and optimisations in the legacy system. In this transformation, the matter of security is of new importance, not only in the company IT, yet also in the factory and vehicle IT. Successful case studies show that it is crucial to promote entrepreneurship and leadership, to establish a change culture, and then to move forward together with the specialist departments in both the existing and new fields, employing agile project methods [Gen 14]. The following chapter explains some successful digitisation projects that have been made in the light of these success criteria.

References

[Ara04] Arasu, A., Babcock, Babu, S., et.al.: STREAM: the Stanford data stream management system, White Paper Department of Computer Science, Stanford University (2004). http://ilpubs.stanford.edu:8090/641/1/2004-20.pdf. Drawn: 15.01.2017

[Bac16] Bachlechner, D., Behling, T., Bollhöfer, E., et.al.: IT-Sicherheit für die Industrie 4.0 – Produktion, Produkte, Dienste von morgen im Zeichen globalisierter Wertschöpfungsketten (IT security for Industry 4.0 production, products, services of tomorrow under the banner of globalised value chains), final report, Berlin 2016, Hrsg: Federal Ministry of Economics and Energy. http://www.bmwi.de/BMWi/Redaktion/PDF/Publikationen/Studien/it-sicherheit-fuer-industrie-4-0-langfassung,property=pdf,bereich=bmwi2012,sprache=de,rwb=true.pdf. Drawn: 22.01.2017

[Bec16] Beckereit, F., Wittmann, I., Keller, L., et al.: Überblick Software Defined "X" – Grundlage und Status Quo, Bitkom 2016. https://www.bitkom.org/noindex/Publikationen/2016/Leitfaden/Software-Defined-X/160209-LF-SDX.pdf. Drawn: 15.01.2015

[Bon16] Bongartz, M., Chen, H., Fricke, V., et al.: IT security for the connected car. White Paper Giesecke & Devrient, IBM, München (2016). https://www.gide.com/gd_media/media/documents/brochures/mobile_secu</g>rity_2/IT_Security_for_the_Connected_Car.pdf. Drawn: 23.01.2017

[Bos15] Bossert, O.: DevOps in einer Two-Speed-Architektur (DevOps in a two- speed architecture), CIO by IDG, 22.12.2015. http://www.cio.de/a/devops-in-einer-two-speed-architektur,3251208. Drawn: 13.01.2017

[Buc16] Book, M.: OpenStack and Co. – Veni, vidi, vici, Crisp Research, 26.09.2016. https://www.crisp-research.com/openstack-und-veni-vedi-vici/#. Drawn: 18.01.2017

[Bro16] Brown, D., Cooper, G., Gilvarry, I., et al.: Automotive security best practices, White Paper McAfee, 2016. http://www.mcafee.com/de/resources/white-papers/wp-automotive-security.pdf. Drawn: 23.01.2017

[BSI09] Bundesamt für Sicherheit in der Informationstechnik: Informationssicherheit: Ein Vergleich von Standards und Rahmenwerken (German Federal Office for Information Security: Information security: A Comparison of Standards and Frameworks), ed.: Bundesamt für Sicherheit in der Informationstechnik, Bonn, 2009. <gid="2">https://www.bsi.bund.de/SharedDocs/Downloads/DE/BS I/Grundschutz/Hilfsmittel/Doku/studie_ueberblick-standards.pdf?_blob=publicationFile. Drawn: 20.01.2017

[BSI16] Bundesamt für Sicherheit in der Informationstechnik: Anforderungskatalog Cloud Computing – Kriterien zur Beurteilung der Informationssicherheit von Cloud-Diensten (Requirements catalog Cloud Computing – Criteria for assessing the information security of Cloud services), ed.: Bundesamt für Sicherheit in der Informationstechnik, Bonn, 2016. https://www.bsi.bund.de/SharedDocs/Downloads/DE/BSI/CloudComputing/Anforderungskatalog/Anforderungskatalog.pdf?_blob=publicationFile&v=6. Drawn: 20.01.2017

[Büs15] Büst, R.: Microservice: Cloud und IoT-Applikationen zwingen den CIO zu neuartigen Architekturkonzepten (Microservice: Cloud and IoT applications force the CIO to develop innovative architecture concepts), Crisp Research AG, 30.04.2015. https://www.crisp-research.com/microservice-cloud-und-iot-applikationen-zwingen-den-cio-zu-neuartigen-architekturkonzepten/#. Drawn: 13.01.2017

[Cox16] Cox, I.: Developing the right IT strategy – how to support business strategy with technology, CIO Uk from IDG, 12.05.2016. http://www.cio.co.uk/it-strategy/developing-right-it-strategy-how-support- business-strategy-with-technology-3430400/. Drawn: 11.01.2017

[DHS16] U.S. Department of Homeland Security: Mobile Application Playbook (MAP), U.S. Department of Homeland Security (DHS), Office of the CTO; 2016. http://www.atarc.org/wp-content/uploads/2016/04/DHS-Mobile-Application-Playbook.pdf. Drawn: 15.01.2017

[Dit16] Dittmar, C., Felden, C., Finger, R., et al.: Big Data – Ein Überblick, dpunkt. verlag GmbH, Heidelberg (2016). https://emea.nttdata.com/uploads/tx_datamintsnodes/1606_DE_WHITEPAPER_BIGDATA_UEBERBLICK_TDWI.pdf. Drawn: 15.01.2017

References

[Fin16] Finanzen.net: Netflix Aktie – Unternemensübersicht, GuV (Netflix share – company overview, P & L). http://www.finanzen.net/bilanz_guv/Netflix. Drawn: 24.01.2016

[Fow15] Fowler, M., Lewis, J.: Microservices: Nur ein weiteres Konzept in der Softwarearchitektur oder mehr? (Microservices: Just another concept in software architecture or more?), OBJEKTspektrum 01/2015. http://www.sigs-datacom.de/uploads/tx_mwjournals/pdf/fowler_lewis_OTS_Architekturen_15.pdf. Drawn: 13.01.2017

[Fre16] Freitag, A., Helbig, R.: Finanzplanung und -steuerung von Unterneh- mensarchitekturen, CONTROLLING-Portal.de, 25.02.2016. http://www.controllingportal.de/upload/old/pdf/fachartikel/software/Finanzplanung_und_-steuerung_von_Unternehmensarchitekturen.pdf. Drawn: 11.01.2017

[Gad16] Gadatsch, A.: IT-Controling für Einsteiger – Praxiserprobte Methoden und Werkzeuge (IT-controling for beginners – practical methods and tools)

[GadA16] Gadatsch, A., Landrock, H.: Big data vendor benchmark 2017, Investigation of Experton Group AG, Munich, December 2016. http://www.experton-group.de/fileadmin/experton/consulting/bigdata/BDVB17/Table_of_Content.pdf. Drawn: 15.01.2017

[Gen 14] Gene, K., Behr, K., Spafford, G.: The Phoenix project – a novel about IT, DevOps and helping your business win, rev. edn. IT Revolution Press, Portland (2014)

[GM16] General Motors: General Motors – strategic and operational overview, Detroit, 28.10.2016. https://www.gm.com/content/dam/gm/events/docs/GM%20Strategic%20and%20Operational%20Overview%2010-28-16.pdf. Drawn: 24.01.2017

[GSM16] GSM Association: IoT big data framework architecture Vers.1.0, GSM Association, 20.10.2016. http://www.gsma.com/connectedliving/wp-content/uploads/2016/11/CLP.25-v1.0.pdf. Drawn: 08.03.2017

[Gue16] Guevara, J.: IT budget: enterprise comparison tool, Gartner Sample Report, 12.02.2016. http://www.gartner.com/downloads/public/explore/metricsAndTools/ITBudget_Sample_2012.pdf. Drawn: 11.01.2017

[ITG08] IT Governance Institute: Align COBIT, ITIL and ISO/IEC for Business Benefit, IT Governance Institute, 2008. https://www.isaca.org/Knowledge-Center/Research/Documents/Aligning-COBIT-ITIL-V3-ISO27002-for-Business-Benefit_res_Eng_1108.pdf. Drawn: 11.01.2017

[Joh14] Johanning, V: IT-Strategie – Optimale Ausrichtung der IT an das Business in 7 Schritten (IT strategy – optimal IT orientation to business in 7 steps). Springer Vieweg Verlag (2014)

[Kee16] Keese, C.: Silicon Germany – Wie wir die digitale Transformation schaffen (Silicon Germany – how we create the digital transformation, 3rd edn). Albrecht Knaus Verlag, Munich (2016)

[Kni14] Knittl, S., Uhe, C.: SABSA-TOGAF-Integration: Sicherheitsanforderun- gen für Unternehmensarchitekturen aus Risiko- und Business-Sicht (SABSA-TOGAF integration: security requirements for corporate architectures from a risk and business perspective), OBJEKTspektrum 03/2014. Drawn: 20.01.2017

[Kno16] Knoblauch, J., Kuttler, B.: Das Geheimnis der Champions: Wie exzellente Unternehmen die besten Mitarbeiter finden und binden (The secret of the champions: how exceptional companies find and tie the best employees). Campus Verlag (2016)

[Kri16] Krill, P.: Docker, machine learning are top tech trends for 2017, InfoWorld from IDG, 07.11.2016. http://www.infoworld.com/article/3138966/application- development/docker-machine-learning-are-top-tech-trends-for- 2017.html. Drawn: 13.01.2017

[KriS16] Krishnapura, S., Achuthan, S., Jahagirdar, P., et.al.: Data center strategy leading Intel's business transformation, Intel White Paper, Mai 2016. http://www.intel.de/content/www/de/de/it-management/intel-it-best-practices/data-center-strategy-paper.html. Drawn: 20.01.2017|

[KRIT16] UP Kritis: Best-Practice-Empfehlungen für Anforderungen an Lieferan ten zur Gewährleistung der Informationssicherheit (Best practice recommendations for requirements on suppliers to ensure Information Security), UP KRITIS, 05.07.2016. http://www.kritis.bund.de/SharedDocs/Downloads/Kritis/DE/Anforderungen_an_Lieferanten.pdf?_blob=publicationFile. Drawn: 20.01.2017

[Kur16]　Kurzlechner, W.: Analysten-Prognosen für IT-Budgets 2017 (Analyst for IT Budgets 2017), CIO of IDG, 19.12.2016. http://www.cio.de/a/analysten-prognosen-fuer-it-budgets-2017,3260930,2. Drawn: 11.01.2017

[Low16]　Lowe, S., Green, J., Davis, D.: Building a modern data center – principles and strategies of design. Atlantis Computing, 2016. http://www.actualtechmedia.com/wp-content/uploads/2016/05/Building- a-Modern-Data-Center-ebook.pdf. Drawn: 15.01.2016

[Mar15]　Marz, N., Warren, J.: Big data: principles and best practices of scalable realtime data systems. Manning Publications Co. (2015)

[Men16]　Menzel, G.: Microservices in cloud-based infrastructure – paving the way to the digital future, White Paper Capgemini, 7.6.2016. https://www.capgemini.com/resource-file-access/resource/pdf/microservices_in_cloud-based_infrastructure_0.pdf. Drawn: 18.01.2017

[Mil14]　Miller, C.; Valasek, C.: A survey of remote automotive attack surfaces, White Paper, 2014. http://illmatics.com/remote%20attack%20surfaces.pdf. Drawn: 23.01.2017

[MSV16]　MSV, J.: Managing persistence of docker containers, White Paper, 24.09.2016. https://www.janakiram.com/posts/blog/managing-persistence-for-docker-containers. Drawn: 13.01.2017

[New15]　Newman, S.: Microservices – Konzeption und Design (Microservices – conception and design). MITP Verlag Frechen (2015)

[NIS17]　National Institute of Standards and Technology: Framework for improving critical infrastructure cybersecurity; National Institute of Standards and Technology NIST, 10.01.2017. https://www.nist.gov/sites/default/files/documents/2017/01/17/draft-cybersecurity-framework-v1.1.pdf. Drawn: 20.01.2017

[Old15]　Oldag, G.: 5 Gründe warum Legacy-Systeme keine Zukunft haben (5 Reasons why legacy systems do not have a future), IT Management Blog, 10.06.2015. http://www.it-management-blog.de/it-strategie/5-gruende-warum-legacy-systeme-keine-zukunft-haben/. Drawn: 11.01.2017

[Pla16]　Plattner, H.: In memory data management, 22.11.2016, in online Lexicon: Enzyklopädie der Wirtschaft (Encyclopedia of the economy). http://www.enzyklopaedie-der-wirtschaftsinformatik.de/lexikon/daten-wissen/Datenmanagement/Datenbanksystem/In-Memory-Data-Management. Drawn: 15.01.2017

[Pre15]　Preissler, J.; Tigges, O.: Docker – perfekte Verpackung von Microseries (Perfect packaging of microseries), Online-Special Architektur 2015; OBJEKTspektrum. https://www.sigs-data-com.de/uploads/tx_dmjournals/preissler_tigges_OTS_Architekturen_15.pdf. Drawn: 13.01.2017

[Pre16]　Preston, R.: General Motors' IT transformation: building downturn – resistant profitability, ForbesBrandVoice;14.04.2016. http://www.forbes.com/sites/oracle/2016/04/14/general-motors-it-transformation-building-downturn-resistant-profitability/#f7382ea63ad3. Drawn: 24.01.2017

[Qui15]　Quintero, D., Genovese, W., Kim, K., et al.: IBM Software defined environment. IBM Redbook 08/2015. http://www.redbooks.ibm.com/abstracts/sg248238.html?Open. Drawn: 18.01.2017

[San15]　Sandmann, D.: Big data im banking: data lake statt data warehouse? (Big data in banking: data lake instead of data warehouse?) Banking Hub by zeb, 01.03.2015. https://bankinghub.de/banking/technology/big-data-im-banking-data-lake-statt-data-warehouse. Drawn: 15.01.2017

[Sav12]　Savitz, E.: Outsourced reversed: GM Hiring Back 3000 people From HP, Forbes/CIO Next; 18.10.2012. http://www.forbes.com/sites/ericsavitz/2012/10/18/outsourcing-reversed-gm-hiring-back-3000-people-from-hp/#744cb87d1377. Drawn: 24.01.2017

[Sch16]　Schlosser, H.: CloudFoundry: Auf dem Weg in die Cloud-Avantgarde (On the way to the Cloud Avant-Garde), S & S Media, JAXenter 05.10.2016. https://jaxenter.de/cloud-foundry-summit-47668. Drawn: 18.01.2017

[Ste15] Steinacker, G.: Von Monolithen und Microservices (About monoliths and microservices), Informatik Aktuell, 02.06.2015. https://www.informatik-aktuell.de/entwicklung/methoden/von-monolithen-und-microservices.html. Drawn: 13.01.2017
[Tie11] Tiemeyer, E.: Handbuch IT-Management: Konzepte, Methoden, Lösungen und Arbeitshilfen für die Praxis (Handbook on IT management: concepts, methods, solutions and work aids for practice, 4th edn). Carl Hanser Verlag (2011)
[Tot16] Toth, S.: Netflix durch die Architektenbrille – Die umgekehrte Architek- turbewertung eines Internet-Giganten (Netflix through architect's eyes – the reverse architectural evaluation of an Internet giant), EMBARC JUG Darmstadt, 09.06. 2016. http://www.embarc.de/wp-content/uploads/2016/06/JUG_DA_2016_stoth.pdf. Drawn: 24.01.2016

Chapter 9
Examples of Innovative Digitisation Projects

9.1 Framework

The author's numerous discussions in the automotive industry have revealed that there is a high degree of uncertainty about how the issue of digitisation is to be approached. It is evident to all the responsible persons in this industry and its supplier sector that something needs to be done, but how to start? In many cases, one waits for guidelines from management or starts without an overarching planning just with some minor beacon projects.

The intention of this book is to provide assistance in this situation. The previous chapters have laid the foundations for this and structured the topic. First, the background was highlighted, such as IT drivers and the transformation of the industry towards mobility services. This was followed by an explanation of the relevant technologies available for digitisation, Such as IoT, 3D printing, and Cloud computing. In the following, a forecast was developed as to how the industry might evolve by the year 2030, and based on an assessment of the current digitisation status of some manufacturers, proposals were presented for the development of a roadmap to drive digital transformation holistically. Finally, change management and corporate culture are important success criteria as well as IT transformation which as an enabler and pioneer supports the specialist departments in a targeted manner.

As quintessence of the book, Fig. 9.1 summarises the essential steps to establish a digitisation programme for specialist departments or plants.

In the first step, the framework and the vision of digital transformation are defined, based on the fundamental decisions on the company strategy and the business objectives. This vision determines the orientation and successive steps of the implementation. At the initiation stage, it is about identifying possible disruptions in the previous business processes and to assess competitors, neighbouring companies and new technologies. Chapters 2 and 4 give suggestions on this.

In step 2, the vision for the digitisation of the area has to be defined, and the existing processes have to be checked for efficiency potentials, with the relevant

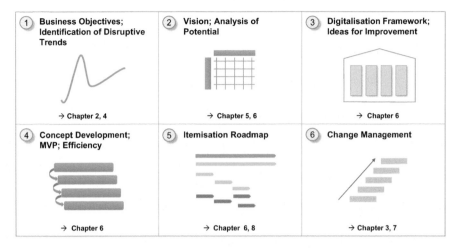

Fig. 9.1 Development steps of a digitisation road map (Source: author)

methods presented in Chaps. 5 and 6. In step 3, the direction of the organisation's digitisation must be described, and first ideas for the implementation have to be developed. Subsequently, further workshops concretise the prioritised ideas in step 4, carry out first cost-effectiveness assessments and create functional models (MVPs) of the prioritised approaches in order to ensure feasibility. Together with the IT, a detailed roadmap is created in step 5 which is integrated into the communication of change management in step 6. There, with the help of suggestions from Chap. 7, all employees should be inspired and motivated to join in order to achieve the vision and the goals on the basis of the roadmap together.

The foregoing explanations contained numerous reference examples on the respective context. In order to further enhance the practical relevance of this book, below more successful examples of innovative digitisation projects as well as additional ideas and impulses for work in the area of digitisation are presented. The chapter follows the four "support pillars" of the proposed digitisation framework (cf. Fig. 6.11), which is shown in Fig. 9.2 as a reminder, with the pillars being highlighted.

Each of the pillars is described in the following by means of examples. Content details, approaches and background information can be found in Chap. 6.

9.2 Connected Services/Digital Products

All car manufacturers are placing high importance on Connected Services and are working intensively to develop offerings for this important growth market. Innovative solutions are seen as an opportunity to be the first provider to differentiate oneself from the competitors and to demonstrate innovative power, especially

9.2 Connected Services/Digital Products

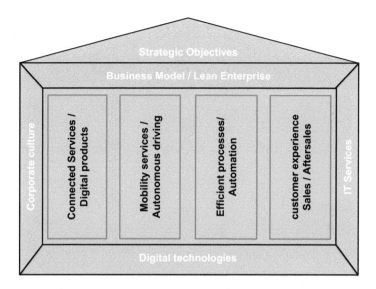

Fig. 9.2 Pillars of digitisation in the automotive industry (Source: author)

for younger buyers. In addition, the subject provides the foundations for new business segments as well as for mobility services and autonomous driving. The opportunities and chances in a market with around 80 million new vehicle registrations every year are also detected by non-industry providers, especially innovative IT companies with their Apps. It is therefore not only important to exploit sales and profit potentials with Connected Services and new business models, but also to improve access to customers for complementary offers and marketing, and to consolidate the own position with customers in respect of mobility services and new offerings, such as intermodal transport services.

In this new competition between manufacturers and newcomers from the IT environment, the infotainment unit of the vehicles emerges as a strategic control point. These units have since long been not just control panels for radio, navigation and telephone as well as display units for vehicle systems, but also the control centre for the use and operation of Apps. This is where Google and Apple get into the car and to the driver via mirroring solutions. It is the "rendezvous" of vehicle electronics and the mobile App-world, as illustrated in Fig. 9.3.

Here the value chain of the infotainment units can be seen. On the hardware of these units with the associated operating system (OS), a middleware level is installed. It comprises a software layer with core services, for example, for operation and communication. Here is also the so-called mirroring positioned which Apple uses with its CarPlay solution and Google with Android. This feature transfers Apps from the smartphone to the infotainment unit, displays them there and makes them usable instantly. The next step in the chain is the integration of services for system components such as radio and media. Then the vehicle IT is integrated, followed by the application services.

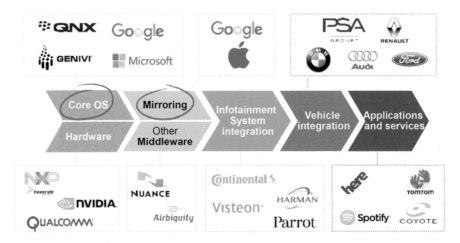

Fig. 9.3 Value chain infotainment unit [Cou16]

Various providers offer systems to implement the value chain. Important control points in order to control the solution area for the customers are the operating system and the mentioned mirroring, as marked in red colour in Fig. 9.3. In the operating systems, the QNX software has a high market share of approx. 50%, while with GENIVI and Google Android, two competitors on Open Source basis, follow [Cou16]. Due to their market leadership in smartphones, Google and Apple have by far the highest market shares in the field of mirroring. Other challengers are open manufacturer consortia such as Ford and Toyota with the SynchAppLink solution.

In the following sections of the value chain, other vendors such as Nvidia and Qualcom are positioning themselves in the hardware section; Continental and Visteon as integrators, and in the area of solutions the company 'here' with map information as well as Spotify for music streaming. It is important that the mirroring services determine the content that runs through this interface. For example, Google will surely favour its own Maps rather than 'here'. There is a conflict of interests between manufacturers and IT providers.

At present, this field is not dominated by the manufacturers and they have to weigh up how they can strengthen themselves in the environment of smartphone connection. The mirroring should therefore become part of the proposed integration platform with a device management function as shown in Fig. 9.4. The details, such as the function and modules of the integration platform, were explained in Sect. 6.2.1.

Two vehicles are sketched below the integration platform. On the left is a vehicle with a very heterogeneous embedded IT landscape with many individual control devices and complex linkage, and on the right is as a vision – explained in detail in Sect. 5.3.5 – a vehicle with central and backup computers with simple digital linkage. The integration platform provides the integration of the vehicles into different services, which is complex in the heterogeneous world yet considerably easier in the centralised architecture. The services are available through programming interfaces (APIs) for application development. The applications use this interface

9.2 Connected Services/Digital Products

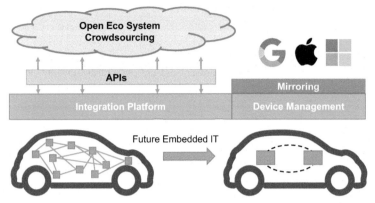

Fig. 9.4 APIs as basis for an open ecosystem and for crowdsourcing (Source: author)

for installation on the platform, are displayed on the infotainment unit and operated from there. Alternatively, the device management services enable the integration of different smartphones as the basis for the mirroring of Apps. Through this use of the integration platform, the manufacturers have opened up the alternative, in addition to the mirrored Apps of the IT manufacturers to also bring further applications into the vehicle.

Basically, manufacturers have three options available to position themselves in the competition against IT providers. They could entirely rely on the mirroring or the smartphone-based way, thus leaving the management of customers to the IT providers. As a second option, they could independently develop Apps and establish a manufacturer's App store, including "customer ID", parallel to the existing stores of IT competitors. Finally, a third option would be to rely on the integration described above and to provide an open, attractive ecosystem for developers with APIs, training, and support, similar to Apple and Google, to promote crowdsourcing and open innovation in such a manner. Namely the third model offers high scalability, and a large number of innovative solutions can be created within a short time. Thus, there is a chance to keep up with the many App offers in the smartphone environment and to attract the attention of the customers. This is the reason why some manufacturers like Ford, Toyota and PSA are using this promising concept [Gra17], [Ber13].

The French manufacturer PSA is discussed below as a case example of Connected Services. The company is the second largest European manufacturer [PSA16], with a turnover of over 50 billion Euros annually and more than three million sold vehicles. In its focusing on innovation and the acceleration of the digital transformation, the company is relying on the "Customer Connected Company" and "Smart Company" initiatives. In their implementation, Connected Services are of great importance. PSA's goal is to create a new ecosystem around the fields of data services, smart services and mobility, which is attractive for customers and creates new purchasing incentives but also generates additional sales, as displayed in Fig. 9.5.

Category	Service	Value for customer	Business Model	Revenue for PSA Group
Data services	Aftersales leads	Time saving	Activated by default on new cars	Aftersales business
	Enhanced leads	Satisfaction improvement	Leads bought by dealers	Sale of leads Aftersales business
	Smarter Cities	Enhanced infrastructures	Partnering with IBM to supply anonymized traffic data	Revenue sharing
Smart services	Car Locator Stolen vehicle tracking	Convenience & Security	Subscription	Additionnal turnover
	Live traffic, speedcam info	Time saving & Security	Subscription (1st 3 years included)	Additionnal turnover
Mobility	Fleet Management	Reducing Total Cost of Ownership	Additional contract to fleet sales	Competitive edge Additionnal turnover

Fig. 9.5 Connected services at PSA [Col16]

The picture shows solutions from the area of Connected Service in three target areas. For each solution, customer benefit, the business principle and the affected sales area are listed. For example, an App is offered to locate stolen vehicles. This App is used by customers on the basis of a subscription, which leads directly to sales at PSA. The 'Smarter Cities' App is about an alliance manufacturer/city to offer parking facilities in advance and to avoid traffic jams by means of a proactive intelligent traffic management. For this purpose, an IT partner company compiles movement and infrastructure data, evaluates these and derives relevant traffic projections. The business principle is based on a sharing model for costs and sales.

When it comes to creating solutions, PSA relies heavily on crowdsourcing. Digital Natives are offered an interesting technological environment with exciting questions and challenging problems, and they are inspired to meet there with many other developers and to further generate topics together as well as to create new ideas. As the basis for this model, a high number of APIs from the development environment for Connected Services was published. These provide, for example, vehicle signals such as oil temperature, tyre pressure or movement data. To distribute the APIs, PSA has created a platform which also provides documentations, blogs and support functions. To encourage developments, PSA is organising 4-week competitions, so-called accelerator events [PSA17]. These take place under preset key topics, and all interested developers or developer groups can participate. The winners will be awarded a price money and also a coaching for the professionalization of their App, leading up to deployment. Through this crowdsourcing, many new ideas can be tested simultaneously with first Apps and implemented step by step through customer proximity.

Similar approaches are also pursued by other manufacturers. As a matter of fact, Ford has even arranged a comparable competition on a global basis as part of the implementation of its mobility strategy. As the overview in Fig. 9.6 shows, twenty-five initiatives were launched.

The picture shows the initiatives with their respective themes. A distinction was made between experiments with different innovation partners and an open competition between developers. The tasks are distributed throughout the world, so that all

9.2 Connected Services/Digital Products

Fig. 9.6 Development of solutions in the connected services environment [FOR16]

relevant regions with "start-up and IT spirit" as well as important Ford markets were involved. Experiments were conducted in the USA both in the Silicon Valley and in Detroit; the UK, Germany, China and India were also included. A broad spectrum of topics has been covered, focusing on car- and shuttle sharing in large cities. New business fields such as health and insurance and well-known topics such as parking and fleet management were the focus.

The program has several advantages for Ford. In a short time many solutions were developed in different markets to cover different customer expectations, but also many ideas for future projects. The open development environment was subject to a broad field test and was further enhanced by extensions and adaptations. The implementation of the projects resulted in high public visibility, which strengthened the manufacturer's image in terms of innovation, mobility and openness. Also, a wide developer community was formed which is also available for future tasks. Therefore, this approach is certainly worth considering for other manufacturers as well.

In summary, it is to be noted that almost all manufacturers are active in the field of Connected Services, however with very different degrees of maturity. The update of the embedded software of the vehicles "over the air" is established at Tesla Motors for instance, while most other manufacturers still have to make up ground here [McK16], [Bul14]. This so-called OTA function (over the air) provides the customer with many benefits and is expected for the other vehicles also. As an example, Fig. 9.7 shows the functionality of the current software updates of the Tesla vehicles.

The listed functions were available free of charge in 2017 with a software update per download, similar to the update of a smartphone App, even for older Tesla vehicles. These are not just simple functions, yet rather significant improvements, such as the more precise temperature control in the vehicle interior or the complex navigation via motorway junction points. Also the increased responsiveness from the monitoring of two preceding vehicles as well as the correction of the vehicle posi-

Functions Tesla Software Update Release 8.0	
➢ Avoidance of Overheating of Vehicle Interior ➢ Improvement of Autopilot • Adjustment of cornering speed based on fleet data • Automated Correction of Position in Lane during Overtaking Manoeuvre • Improvement of Reactivity, Monitoring of the twoVehicles in Front ➢ Map Updates; improved Trip Planner ➢ Extention of Voice Control	

Fig. 9.7 Functional scope of the software update "over the air" for Tesla vehicles [Gri16]

tion after overtaking in relation to the overtaken car are part of the update. With the offer, Tesla increases the comfort and safety of its customers by problem-free software updates and can be considered as the benchmark for many manufacturers. In other topic areas which were subject in Sects. 5.3 and 6.2.1, the established manufacturers should, in the author's opinion, also strive for progress in order not to be overtaken by the "newcomers" of the industry. Particular mention should be made of the following:

- Centralised architecture for embedded IT (cf. Sect. 5.3.5)

- Transformation of the heterogeneous infrastructure, which is difficult to operate safely with over one hundred control units and several bus systems, into a centralised approach (simplified according to Fig. 9.4), in order to facilitate operational safety and integrability, and to secure future viability
- Strengthening the integration platform
- Development of an open, cross-brand platform; integration of the mirroring, so that upon the user entering the vehicle, a complete synchronisation of Apps in the vehicle with the user's smartphone is made automatically. The success criterion is the omission of the mobile phone holder, as well as the automatic driver recognition when the user gets into the vehicle, and the mirroring of the personal Apps
- APIs with appropriate system environment
- Define API strategy, build developer platform with APIs, social media and support environment for developer communities and interested commercial users, such as insurance companies or retail companies
- Establishing a business model for digital new business

- Development of a business model for Apps, API and data including the necessary processes such as payment processing, distribution and regulation of data use with drivers

9.2 Connected Services/Digital Products

Fig. 9.8 Infotainment unit display with different offers [Bur16]

Connected services will be increasingly integrated into the environment of the vehicles with an increasing number of functionalities and thus become an important purchase criterion and differentiating characteristic [Kni15]. This trend is reinforced by announcements from General Motors and BMW for instance. Based on its long-standing proven OnStar connectivity platform with basic functionality, General Motors is partnering with IBM to bring an innovative application with cognitive capabilities into the vehicles [Bur16]. On the one hand, this solution is to be understood as a learning assistant who in the background proactively recognises technical problems and proposes actions to the driver or independently provides navigation information for the next appointment from the driver's calendar and address book. On the other hand, the software is a marketing and sales platform on which service companies, as shown in Fig. 9.8, present themselves with their offerings.

The logos of various companies appear as icons on the initial screen of the infotainment unit of a GM vehicle. Here the user can scroll through and make its selection. If, for example, the driver clicks the ExxonMobil icon, stations and individual fuel pumps appear in the vehicle neighbourhood. Payment after refuelling is made from inside the car. Mastercard facilitates the payment procedures, and Parkopedia guides to free parking spaces. The platform is open, so other interested companies can contribute. When these services are used, the manufacturer receives a commission fee.

BMW is marketing the new 5-series model exclusively by highlighting the functionality of Connected Services, pointing at the conquest of the digital world, and no longer refers to motorisation or fuel consumption values. In radio advertising and on YouTube it says:

> In a digital world, my car tells me when it's time to drive off. It understands my words, my gestures. And at the destination it is a search engine for free parking spaces ... my car knows my name, my destinations, learns my routes – and supports me when I want it. Some call it progress, to me it means freedom. [NDR17]

The examples illustrate how Connected Services can help companies to differentiate themselves in the market and address potential customers. They underline the need to vigorously attend to this field in order to be close to the customer.

9.3 Mobility Services and Autonomous Driving

The second pillar of the digitisation framework includes mobility services and autonomous driving. Before a reference is presented, a brief overview on the current situation on the market and the developments is next, which also serve as a positioning aid for many new project ideas.

The market for mobility services continues to develop at high growth rates. Lyft, the challenger of Uber, was able to increase the number of handled journeys in the US to 162.6 million in 2016, thus tripling the 2015 results [Sol17]. Despite the considerable growth of its competitor, Uber remains the undisputed market leader with its presence in more than 70 countries. In January and February 2016 for instance, Uber completed more trips than Lyft throughout the whole year. However, the impressive growth goes in both Lyft and Uber along with significant losses [Haw17]. This does not prevent investors though from further infusions of capital, nor does it stop manufacturers from seeking participations. For example, General Motors has its own mobility service organisation named Maven and also holds a stake in Lyft. BMW operates DriveNow, Volkswagen holds a majority interest in the Israeli company Gett and Daimler has been involved with Car2Go for many years. After Toyota has significantly invested in Uber, Daimler is also entering into a strategic partnership with Uber to push autonomous driving as the basis for mobility services [Ger17].

In a further strategically important field, autonomous driving is promoted by all manufacturers, and a fierce competition has emerged for who will be the first to present this technology as ready for series production. The path to autonomous driving is divided into five technological steps or degrees of maturity (cf. Sect. 5.3.3). Up to Level Three, vehicles are already established in the market. With technology Levels Four and Five, first pilot tests are running since the middle of the 2010s, and manufacturers are exhibiting autonomously-driving vehicles at all relevant trade fairs [WELT17]. A summary of the market introductions planned by the manufacturers is shown in Fig. 9.9.

The introduction dates are shown over the time axis, classified according to demonstrators, mobility service providers and manufacturers. First pilots in limited road areas already run in Singapore, Greenwich and Pittsburgh. At Level Four with high automation, GM, Volvo and Audi are planning to be the first on the market, closely followed by Tesla. Daimler and Nissan announced the release of fully autonomous vehicles for the year 2020, followed by BMW for 2021. Uber foresees fully autonomous mobility services, so-called robotaxis, to be in use in the year 2030.

The availability of autonomously driving vehicles will trigger a further significant growth momentum for mobility services and car sharing models as shared use also promises cost advantages. Figure 9.10 shows a comparison of the costs to drive electric cars in different usage models in the years 2016 and 2025.

Driving costs in Dollar per mile for conventional taxis are between $ 2.85 in 2016 and $ 2.76 in 2025. In comparison, costs in sharing models are significantly lower, ranging from $ 1.36 to $ 1.32. The cheapest are private vehicles with driving costs of $ 0.56 and thus well below those of the mobility services.

9.3 Mobility Services and Autonomous Driving

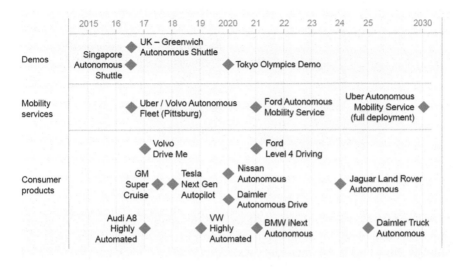

Fig. 9.9 Planned introduction dates of autonomous vehicles [Han16]

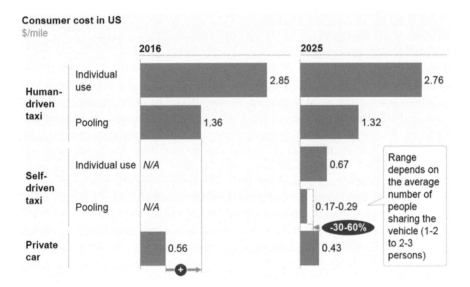

Fig. 9.10 Trend of costs to drive electric vehicles [Han17]

The situation is changing fundamentally with autonomously driving vehicles. In this way, in 2025 cost in the sharing model will drop to $ 0.17 when using the vehicles with two to three people, and to $ 0.29 in lower usage with one to two people. The cost advantage of 30%–60% versus private vehicles with $ 0.43 will bring a significant increase in the use of mobility services.

Fig. 9.11 Electrically powered, autonomously driving shuttle bus (according to [Mol16])

Autonomously driving electric vehicles will thus push back private car ownership in the long term. From this point of view it is interesting to get to know the Minibus Olli which is shown in Fig. 9.11.

This shuttle bus for up to twelve persons was developed in 2015 under the leadership of Local Motors by voluntary co-developers in a public competition over 6 weeks. After the selection of the winning design, which gave the winner $ 28,000 of price money plus future royalties from the vehicle sale, it took only 3 months until the SOP (Start of Production) [IDE17]. Other than the windows and the aluminium chassis, the vehicle components are created using 3D printing techniques. In the relatively small number of pieces, this production process is economical. The vehicle does not have a steering wheel, is equipped with about thirty sensors and runs fully autonomously. At a speed of 20 km/h the range is 58 km, which is sufficient in shuttle operation. The cognitive platform is being taught, also via crowdsourcing, and is available to the passengers for dialogue. The ability to communicate is trained on the operating environment and further develops in self-learning. Several Ollis can also communicate with each other and at peak times organise themselves independently as a group. First vehicles are used for shuttle services in Washington DC, and there is great interest in the use of additional vehicles in the USA and Europe, initially as pilots on non-public grounds.

Both the cognitive platform as well as the structure and interior of the vehicles can be adapted to suit the requirements. In this way, Olli could mutate into a driving café or gym. Local Motors and its project partners also see the vehicle as a learning platform for further projects of this kind. The goal is to set up micro factories in all relevant markets in order to quickly identify and implement the specific customer requirements as well as to minimise the logistics costs for components and vehicle deliveries. As a vision, Olli is supposed to make its way to "his customer" on its own. Later he could wait outside the front door if the integrated cognitive platform sends the next available vehicle to the customer after a comparison of calendar functions, user behaviour and weather. Sure, this is still a vision, but it is in fact to be considered by manufacturers [Jun16].

The idea of grouping vehicles is also studied as "platooning" in the truck sector. This term describes the driving of several trucks as a convoy, following at close distance between each other a guidance vehicle. In the process, the trucks communicate with each other in near-real-time, so that braking manoeuvres of the guidance vehicle are directly carried over to the following vehicles for instance. The trucks are quasi connected by an electronic drawbar. Thereby, the following vehicles are intended to drive autonomously [Vol16]. The advantages of this concept are significantly lower fuel consumption and thus less pollution. Further benefits result from lower space requirement on the road due to the narrower between the following trucks, relieving the drivers by the even speed, and by fewer traffic jams due to the superordinate coordination. Currently, the process is being piloted in Singapore with the participation of Scania and Toyota [Eck17]. As far as potential readiness is concerned, similar as with autonomous driving, the statutory regulations for admission to public roads are to be enacted first.

The promising platooning could also be an option in the area of mobility services in passenger transport. If autonomously driving cars are on the same route, they could automatically join together, at least temporarily on longer stages, in order to use the aforementioned advantages of the concept. Further visionary aspects in this field extend towards the creation of offerings to make meaningful use of the passengers' time in the vehicle. If personal driving and the associated seating arrangement and orientation are no longer necessary, the interior of the vehicles can also look completely different. The Olli-café or the Olli-fitness-studio have already been mentioned. There could be a conference room or restaurant as well during the trip – there is no limit to the imagination. It is important that the manufacturers adapt themselves to these developments and build up a sufficient degree of expertise in order to be prepared for these new issues and be able to offer competitive solutions.

9.4 Efficient Processes and Automation

The third pillar of the digitisation framework is about the efficiency improvement of processes up to the complete automation through digitisation. The thesis of the "Digital Darwinist" Karl-Heinz Land applies to this context: "Everything that can be digitised, will be digitised. What can connected, will in fact form a network. And what can be automated, will in fact become automated. This applies to every process in the world "[Lan16].

Section 6.2.3 explained a general method on how to approach the digitisation of processes and developed specific initiatives for three business areas. The following innovative projects in this field complement the comments made there.

As a result of the German Federal Government's long-standing and broad-based initiative, the thematic area of 'Industrie 4.0' has become a focus of all manufacturing companies there since the mid-2010s, and many initiatives and projects have been launched. All in all, the goal is to achieve a horizontal and vertical process integration through digitisation, thus making companies more flexible and efficient. In addition to the increase in responsiveness and thus customer orientation, according to a Fraunhofer Study, direct savings of on average 10–20% can be achieved in the

Fig. 9.12 Saving potentials Industrie 4.0 [Win16]

automotive industry. The breakdown into the organisational areas is shown in Fig. 9.12.

The reduction of complexity through modular and standardised products and the simplification of processes and interfaces allow the highest savings potential of up to 70%, while inventory reductions due to harmonised production processes and foresighted, tightly synchronised retrieval of supply materials reduce costs by up to 50%. In the other areas such as production, quality and maintenance, savings of up to 20% can be expected.

Manufacturers should apply these parameters in their projects. According to the author's experience, the highest potentials are at the intersection between organisational boundaries and process interfaces. However, the challenge to realise these is often not in technological problems, yet rather in cultural issues when motivation for cross-interface cooperation needs to be created. A technological platform which extends across these boundaries can be helpful. Figure 9.13 shows a simplified Industry 4.0 scenario with such a level, called Shop Floor Integration Layer here.

Three stations of a painting line are shown as an example. In the first processing step, the body is painted by robots, followed by manual-assembly and then a final inspection for quality control. On the line are various robots with their respective controls, sensors, cameras as well as tablets for the workers. The various IT components are integrated into the business IT via the Shop Floor Integration Layer. On the software level, basic services are available, for example, for communication and data handling (cf. Sect. 6.2.3.3). The integration layer can connect the entire painting line and also other production areas so that access to sensors and controls of other areas is possible to perform overarching data analyses. The layer provides standardised APIs for use by application solutions that are successfully implemented at different manufacturers. The following is an overview on the solutions mentioned in the picture:

9.4 Efficient Processes and Automation

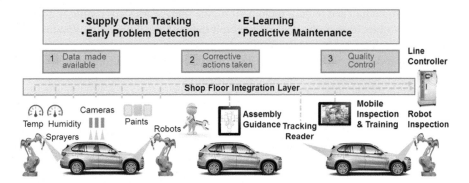

Fig. 9.13 Shop Floor Integration Layer example (Source: author)

- Supply chain tracking
- This application monitors the logistics chain in order to identify supply shortfalls at an early stage and initiate countermeasures. For critical parts, monitoring goes beyond the direct supplier, right up to their provider.
- Early problem detection
- This analytical solution combines a large number of sensors, plant- and order data, analyses them and, with the aid of forecasts, identifies production stations where problems may arise, for example due to missing parts.
- E-learning
- For the operation of complex machines or also for complex service work on machines, QR code labels can be read out there using a tablet and thus the special error situation can be localised. From a learning management system, the worker then loads the learning modules onto his tablet, which he can work through and thus obtain information on how to carry out the work.
- Predictive maintenance
- The application for preventive maintenance is about avoiding machine failures through proactive service measures and thus achieving a consistently high output performance. For this purpose, machine data are continuously being recorded, trends and deviations from specifications are identified, forecasts are made using models, and actions are proposed.

Such solutions can be effectively developed on the basis of an integration layer, since this simplifies the application development by using the platform basic services. More importantly, the replacement of machines can be easily depicted in the integration layer, and the application does not require any adjustments so that it can continue to grow step by step, even across organisational areas. Similarly, the rollout of the applications to other production locations is facilitated and quasi becomes a download, even if there the integration layer is installed as the basis.

Another subject area which is being driven forward within the context of Industry 4.0 initiatives, is the direct collaboration of workers and robots, as already described in Sect. 4.8. This is not like in the 1980s, when it was about integrating robots into the flow production and to have them perform the same work steps routinely as reliably as possible. Rather, the goal is to exploit the high flexibility from advances in sensor technology, kinematics and software.

Robots today are much more sensitive, more responsive and more mobile than the first pick and place- or welding robots. With these capabilities, robots can also be useful in a wide range of other areas. Uses in the household, nursing and also in the operating room of a hospital are just as conceivable as teaming with workers in automobile manufacturing [Buc17]. Here the robot is used according to its strengths in a targeted way to support humans. A comparison of the abilities of humans and robots and an application example is shown in Fig. 9.14.

In the assembly of complex components, flexible decision making in the manufacturing process and also in the balancing of tolerances and insertion movements, humans certainly have an advantage over their iron colleagues. Robots play their advantages when it comes to handling heavy and sharp-edged loads, repeatability and endurance. Stoically they carry out the same work steps over and over with the same quality. As a team, both partners are even stronger, as the example of the hardtop assembly in Fig. 9.14 demonstrates. The robot brings the bulky and heavy component precisely to the nearest assembly position above the body. The worker then undertakes the final orientation and insertion of the roof.

The utilisation potential of robots continues to grow with the increasing performance of the software. Today, programming is relatively simple through teach-in procedures or graphical configuration on a tablet. Due to that, IT specialists are no longer needed for programming which rather is made by employees from the production department. It is foreseeable that production robots will have cognitive abilities in the future which will open up even more flexible application possibilities. Thus, the stand-in on the final assembly line, as well as the scheduler for the fine-tuning of the occupancy planning, could soon be a robot in order to absorb the fears of a looming skilled workers shortage.

Not only in manufacturing is under the heading Industry 4.0 the subject of process optimisation through digitisation a hot topic, but also in all other divisions. In

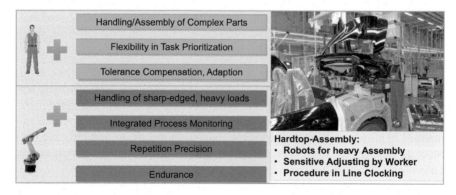

Fig. 9.14 Comparison of the skills of workers versus robots in hardtop assembly (According to [Kos14])

9.4 Efficient Processes and Automation

development for instance, to increase the reuse of parts by means of intelligent search systems or the prototype-free testing of new products up to the feasibility test with Augmented Reality is imaginable. In human resources, machines run the applicant screening, and employees use Apps to maintain personal master data more easily.

In the future though, cognitive solutions are expected in totally different areas as well by which robots can independently further expand their capabilities. The example of Pepper is shown in Fig. 9.15.

Pepper is a humanoid robot that, with the help of the cognitive IBM platform Watson, is able to find its way around and communicate with people. The system has to learn the surrounding details and also the subject area for the dialogues in a first training phase. Based on this basic knowledge, the system then learns from the feedback in the interaction. At present, Pepper is active in shopping malls for customer reception and guidance [Ada17]. Further uses are also thinkable at motor shows or at the dealership in order to answer questions concerning vehicle models.

These uses, which have already been implemented several times, are not further deepened here. Instead Blockchain is presented as a means to increase efficiency with possible application examples in the automotive industry. Background and function of the Blockchain technology were explained in Sect. 4.7. It is often referred to in the Internet as "the next big thing" [Bre16]. Simplified and summarised, these are encrypted data records for the documentation of transactions, which are continually updated, including verifications of correctness, through a transaction chain called "blockchain" and are stored in a distributed database. The process ensures the worthiness and flow of Bitcoins and forms the heart of this Internet means of payment.

Although the Blockchain technology is closely linked to the Bitcoin currency, it is irrespective of that also transferable to many other applications independently, and has a disruptive potential as the application processes and structures change entirely. For example, no bank is needed to process a payment between two parties using Blockchain technology. The transfer takes place directly, quickly and cost-effectively through a web-based secure process running in the background. Personal data protection is also not an obstacle, since no open names appear, rather each user

Fig. 9.15 Robot Pepper in conversation [Ada17]

is assigned a separate code made up from numbers and letters. Further protection is provided by the use of distributed databases on which many copies of the transaction chains are stored. Manipulations become almost impossible.

The main advantages of the procedure are the high security, the simplicity in the verification of the transaction and the potential to simplify processes, since additional control functions, for example for invoicing or checking contracts can be dropped. The disadvantages are the extent of the growing blockchains, limitations in performance, in throughput, and the effort needed for authorisation management.

As the advantages are predominant, there is a great interest in the topic, and many manufacturers are starting their first projects, often in the area of so-called Smart Contracts. These are computer protocols and software-based algorithms which map contract contents. The processing and fulfilment of the contracts is automatically monitored and documented so that the paper form is no longer required, and lawyers are unnecessary in the drafting of contracts and supervision of execution [Kal16]. Blockchain is also the basis for the use of technology in the usage processes of vehicles, Fig. 9.16.

The picture shows the life cycle of a vehicle from the transfer of the car from the manufacturer to the dealer, which involves a lessor to finance the purchase, so that the car is handed over to the customer (Lessee) on the basis of a leasing contract. At the end of the usage, a used car dealer takes over the car, and the lease ends on the basis of an expert appraisal. The entire process is documented in a Blockchain, stored step by step in distributed databases (nodes). In addition to the vehicle transfer, the contractual vehicle status is also directly checked in the Smart Contract

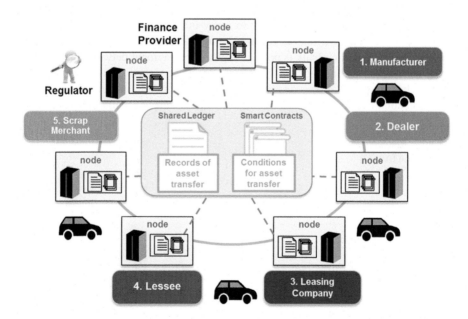

Fig. 9.16 Blockchain technology in the life cycle of a vehicle [Hon16]

method and stored in the information blocks of the chain. The simple sequential procedure results in a complete documentation of the vehicle condition, which can also be used when changing users or exchanging parts, in order to secure the use of original parts, for example.

Another example of the use of Blockchain technology in the automotive domain comes from electric vehicles. The refuelling is carried out at charging points which are often supplied by different electricity suppliers and mostly also use different methods of payment. These range from coin-operated machines, customer and credit cards to payment via smartphone App. Partially, electricity is also only sold to contractual customers of the supplier. In order to simplify procedures, the Blockchain method based on Smart Contracts is also suitable here. During each charging process, the customers conclude a contract with the respective supplier by identifying at the station, whereupon the transaction including the payment is processed in the background [Roe16].

In a similar way, a wide range of possible applications are available, such as the processing of:

- Payment transactions, orders, invoice processing
- Supply chains
- Rental agreements incl. issuing of smartkeys for opening the vehicle
- Services
- Repairs
- Solar power (feeding back and supply)
- Mobile phone usage at different providers (roaming).

These examples illustrate the potential of the blockchain technology, especially since the use often also brings a process simplification by the elimination of control functions or intermediaries and thus an increase in efficiency.

Another technology to simplify procedures are so-called Chatbots, derived from "chat" and "bot" (second syllable of 'robot'). They include automated communication, for example in software applications as a service offer to answer queries regarding usage or as an information system in public transport. Communication can be in writing via dialogue boxes or via language. The solutions work according to the principle of pattern recognition. The requested characteristics are used to search for suitable matches in databases and to derive an appropriate response from the information found. The demand for these solutions is growing with increasing performance and simplicity of use, which are known from using the chatbots of smartphones such as Google or Apple [Frü16]. Enriched by cognitive possibilities, the performance of the solutions will continue to increase and the utilisation extend to further areas.

Another application area of chatbots is as personal digital assistants in the office or in the private sphere, realised for instance in Amazon's Echo or by Microsoft with the Cortana Platform [Jun17]. These systems are capable of learning and answer user queries by accessing Internet data, remind their owners of their respective appointments as per the diary, or initiate the playback of desired music.

In a next development stage, it is expected that chatbots will be used to control different applications via voice control and then learn to perform certain work procedures automatically. A possible usage scenario is shown in Fig. 9.17.

The Chatbot solution works with access to databases of different applications and is in direct dialogue with applications of a specific business area, such as the finance department. However, a chatbot could also be addressed directly in order to perform certain activities in interaction with the connected IT systems. For example, it could display outstanding delivery items or settle a specific invoice under deduction of a cash discount. The working steps are listed on the right in the picture. After input via voice or via dialogue window, the chatbot identifies the user and checks authorisations, interprets and understands the task and subsequently performs the necessary steps to display the desired results or to report the execution of the task. To perform this transaction manually, the user might have had to approach three different systems to find the invoice, check the goods delivery and settle the bill.

This little example already illustrates the potential for increasing efficiency in process execution. It is particularly useful in the handling of financial transactions, travel bookings, help desk services and also in the area of purchasing and personnel. Solution examples are already known for mobility services to book trips or to search for intermodal transport connections [Jun17]. Since the integration of chatbots into existing solution fields is relatively easy, operation is intuitive and the added value is high, this technology is suitable for many digitisation projects, such as in the next major topic area.

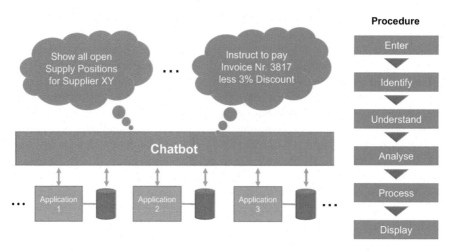

Fig. 9.17 Example of use and work steps of a chatbot (Source: author)

9.5 Customer Experience – Marketing, Sales, After-Sales

Digitisation and the change in the automotive business lead to a shift in the existing sales structures towards an Internet-based multichannel distribution. At the same time, the direct contact points with the customers in the areas of marketing, sales and service change through the use of innovative digitisation technologies. Drivers for this are the customer expectations, stemming from the familiar handling of Apps on mobile end devices such as smartphones and tablets. For the details of the transformation and a possible road map, see the comments in Sects. 5.3.7 and 6.2.4. The current situation and the future changes in the sales structure are summarised in Fig. 9.18.

As can be seen in the illustration on the left, the manufacturers in sales and service do not have a direct connection to the end customer so far. Importers and dealers handle the sale and the services of the vehicles and shield all direct customer contacts like a "trade secret" from the manufacturer. They provide marketing materials, market information, technical support for the service facilities and more and more also customer relationship management (CRM) functions such as campaign- and lead management.

With the future digital services, direct relationships between manufacturers and customers will develop. These also include Connected Services or the reading and analysing of vehicle data in order to give the customer driving instructions or to diagnose and communicate preventive service requirements via diagnostics. For the upcoming online sales of vehicles, a direct interaction between the manufacturers and the end user takes place. In the course of the following business transactions, the established manufacturers at present usually involve the existing sales structure [Tan12]. It remains to be seen if this will continue or whether the intermediary lev-

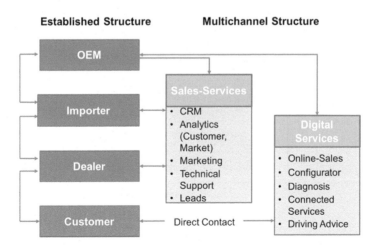

Fig. 9.18 Future change in the car sales structure (Source: author)

els will cease in the long run, or at least change considerably. Tesla, for example, operates outright directly online in sales without any levels of trading.

In the future, a direct communication and business relationship between the customer and the manufacturer will be established in order to transact business via this additional distribution channel and to gain more customer information. The case of General Motors which was presented by Fig. 9.8, again serves as a practical example for such solution. In this way, the drivers receive via the infotainment unit service offerings from ExxonMobil, Parkopedia or Mastercard for instance. To handle these transactions, an integrated service platform is implemented, which is shown in a simplified depiction in Fig. 9.19.

The vehicles are connected to the different services via a public API gateway. As an alternative to the infotainment unit, customers can also use a smartphone App or a web portal solution for the processing. A data management level prepares, for instance, the user, vehicle and motion data, continuously analyses these and identifies suitable offerings for the drivers from the portfolio of the connected providers (merchants), stored in the target marketing Cloud. The results are recorded by the service department which generates specific offers.

Thus the system could detect that the remaining tank content of the vehicle only has small reserves. The information is provided along with the current localisation data of the vehicle to the service platform. This processes the information and, together with the partner Exxon, creates an offer with price and a potential discount as an incentive for a nearby petrol station and displays it on the infotainment unit. The driver can accept the offer by clicking and then refuel at a pre-reserved pump. The payment is made via the further partner Mastercard, without the driver having to enter the cash till area of the service station.

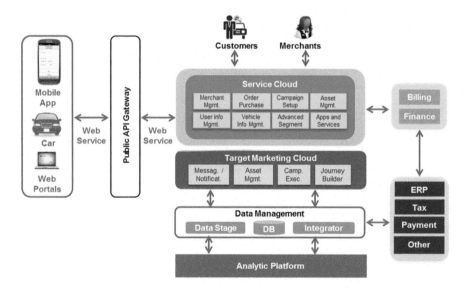

Fig. 9.19 General Motors' platform for handling of third-party business (According to [Sat16])

9.5 Customer Experience – Marketing, Sales, After-Sales

The service Cloud provides the services required to process these transactions and also manages customer, vehicle, and vendor information. Another component of the platform, the Target Marketing Cloud, runs customer-specific marketing campaigns. Within an as narrow customer segments as possible, the system analyses data from various sources and converts them into customised offers in order to achieve a high acceptance rate. The entire solution is integrated into the manufacturer's back-end systems, so that, in addition to the master data management, the accounting and payment of taxes is also done there. The solution is operated by a service partner. The business model is based on the risk sharing approach in which manufacturers, IT service providers and vendors participate in a proportionate share of the sales revenue of the transaction. The platform is open in order to include further partners in the future or to expand the solution into further markets and to involve local partners there.

All in all, GM has built up a direct sales and communication channel between the manufacturer and the customer, which means additional convenience for the customers and at the same time provides additional business for the platform parties.

Another important subject in this sales transformation is the development of an internet-based sales channel. This is also confirmed by studies which assume that in 2020 every third vehicle is traded online [Kal16]. Leading in this field is Tesla, which exclusively rely on online sales. However, other manufacturers are also working hard to offer similar solutions [Kai15, McN15]. BMW for example has set up a dialogue-oriented option on its Internet site for online purchase in the UK, which is summarised in Fig. 9.20.

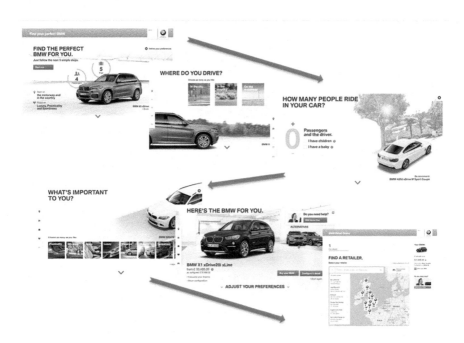

Fig. 9.20 Process of BMW online purchase in the UK (According to [BMW17])

After the start of the process, the customer is guided by an intuitive configurator, which is characterised rather by a recording of the planned vehicle use and personal preferences than by often deterrent technical matters in traditional solutions. First, the customer selects his driving environment from the options city, country, long distance, then the future passenger number and the typical luggage type. Finally, the system asks for the lifestyle that the vehicle should meet, ranging from luxury over pragmatic through to sporty. Subsequently, under evaluation of previous configuration sequences and learning from these, a vehicle with the best match, including the price and delivery date, is proposed. In addition, a second configuration appears, in which in 80% of the cases the initial proposal is accepted [BMW17]. Finally, the customer can select a dealer. The final commercial steps, if desired, including a leasing option and, if necessary, the takeover of any existing vehicle, are then carried out also online in the system environment of the selected dealer without a system break for the customer.

This solution provides customers with a comfortable environment for the online purchase of a car. The processing is carried out in a network of manufacturer and dealers. As dealers in this case do not lose any turnover, and there is no competition between the parties, the acceptance on part of dealers is very high. Almost all BMW dealers in the UK are using this solution [McN15].

A further example of Audi is presented below. In addition to online sales, Audi has developed an innovative showroom solution for its car dealerships. An impression is provided in Fig. 9.21.

The "Audi City" solution offers a virtual customer experience. All Audi models appear on high-resolution large screens in almost real size. The vehicle can be displayed in different driving situations and one can enter the car virtually. Operation is done via gesture control or multitouch tables. The customer can, for example, call up a previously selected configuration via his online ID by smartphone, view "his" vehicle in different situations, make changes to colour, equipment and motorisation and watch these changes directly online in the virtual world. Complementary, the systems are equipped with augmented reality devices as well as sound machines to make the experience even more realistic.

Fig. 9.21 Virtual AUDI showroom [AUDI16]

The comprehensive system technology necessary for this purpose has been installed by Audi in its flagship stores in the inner cities of large metropolises such as London, Berlin and Beijing. At these locations, there would not be enough space anyway for the physical presentation of several models. The challenge of finding a balance between the available showroom space and the scope of exhibits is on all dealers. Due to the growing model range and the increasing variant diversity, the "virtual showroom" is an alternative. Parts of the solution, such as virtual vehicle models, are also available to smaller dealers. The next version of "Audi City" includes the integration of learning configurators who already know a customer upon entering the showroom from the social media and their history with the manufacturer, and then pro-actively load configuration suggestions on the screens.

Both solutions presented as well as in many other situations there are different business contacts with the customer. The customer could be the owner of several vehicles of different brands of the same manufacturer or a customer of the finance area, have already completed different test drives and have had a guarantee claim as a customer of the service organisation. All of these situations generate customer data, which from different systems go beyond the sales structure, without being put together in an integrated view at least within the manufacturer. In addition, the same customer may express himself in the social media about his experiences with the vehicles or discuss his future vehicle interests with friends or in public forums. This information also provides valuable information to the manufacturers and should therefore be combined with the integrated manufacturer-internal information.

The need for action is evident to many manufacturers, and corresponding data consolidation projects are in progress. Typical technologies used in the course of this are master data management, Data Lake and Hadoop database technologies. For a deepening of the technological topic using tried and tested architectures, reference is made to the relevant literature, e.g. [LaP16]. It is important to actively address this issue in order to be able to respond to customers consistently and develop them. Typical further digitisation projects in the field of marketing, sales and after-sales are:

- Cognitive solutions in customer service, for example digital assistants at the repair workshop reception or the customer help desk
- Online workshop booking and repair tracking
- Control of digital marketing; reporting of customer feedback
- Social media monitoring; customer segmentation; next best action in marketing
- 3D printing for local customised spare parts production [Lec16]
- Big Data-based long-term forecasting for parts stockage in spare parts management
- Demand-oriented pricing based on market trends

To sum up, the digital transformation at the customer interface is at all manufacturers in progress with much dedication. However, successes in individual projects should be secured with an overall strategy and the integrated roadmap derived from it.

9.6 Corporate Culture and Change Management

In addition to structured holistic planning, as was already detailed in Chap. 7, a fresh agile corporate culture with the appetite for change and digitisation as well as an efficient change management are important preconditions for a successful transformation. Therefore, some examples and experiences from this area are given below.

The digital transformation must be accompanied by clear communication from part of the company management about its necessity and goals in order to encourage the entire work force to participate with full commitment. They are the ones who have to implement the individual projects and thus directly determine the success of the journey. Communication starts at the top of the company through authentic behaviour and is an important success criterion [Stö16]. The executive team is at times of change under particularly close observation by the employees. The leadership must radiate the willingness for digitisation, the motivation for innovation, and the determination to implement it, for instance by using new communication channels, exemplifying an open dialogue and by proving new partnerships. Other aspects to consider in the communication are:

- Mobilise the leadership team and involve opinion-formers
- Formulate clear objectives as to what is to be achieved in the individual fields with the digital transformation, why it is important and what it means to the employees
- Define performance figures for the targets
- Create roadmap for implementation with milestones
- New values and behavioural patterns to become an integral part of the corporate culture
- Support multi-dimensional yet consistent communication using multiple channels
- Stimulate dialogue and feedback and embrace it in actions
- Aim for and highlight swift successes; also accept and name failure
- Maintain continuous communication with no break-off

It is essential that innovative solutions and ways are used to communicate in accordance with these rules. In addition to established tools, in-house videos, wikis, forums, chats and collaboration tools are expedient. The rule however is: Not too much at once, but rather take the chosen path in a consistent manner to a breakthrough. It is crucial that the management is actively involved – in fact, personally and the matter not delegated to any assistants.

The introduction of a collaboration platform, for example at Coca Cola or Bayer AG, became a success only after the executives were actively involved in the dialogue. Further examples of how to convincingly exemplify the role model in communication are: [Wes14]:

- Societe Generale

- Mobilising sixteen thousand employees in nineteen countries in an exchange of views on an internal social media platform to underpin the digitisation roadmap with suggestions for initiatives and to specify the elements of IT equipment

9.6 Corporate Culture and Change Management

- Pernod Ricard
- Development of the digitisation roadmap in internal crowdsourcing
- IBM
- Innovation jam for the adjustment of the company's basic values (cf. Chap. 7)
- Virgin Group

- Exemplifying the customer orientation; CEO Richard Branson invites customers via Twitter under the hashtag #AskRichard to enter into direct dialogue

The examples illustrate that creative ideas, authentic input and dialogue are equally important to success. It is important to avoid the very common mistake of putting the focus of the communication on new IT tools. When implementing collaboration tools, the effort and costs for training and change management must be taken into account, and the project does not end with the installation. The continuous use of the tools, including the involvement of executives and opinion leaders, as well as the transparent change in working behaviour through the use of the new tools, ensures sustainability and thus influences the corporate culture.

The culture is also significantly changed by interdisciplinary cooperation among employees, using agile methods such as Scrum in development projects and Design Thinking in innovation workshops. The methods were explained in detail in Sect. 7.2. There are many positive references namely with respect to Design Thinking [deS16]. Bosch, Telekom, SAP, Volkswagen, Bayer and Lufthansa, among others, apply this method. The success is particularly due to a strong customer orientation and interdisciplinary cross-divisional cooperation. Practical experiences of the author may encourage the application of this approach illustrated by 6 steps in Fig. 9.22.

In this case, three groups with a total of fifteen participants from the areas of purchasing, the organisation of the CDO and IT worked on the situation in the purchasing area of a manufacturer. The goal was to find new ideas for digital transformation. In preparation for the workshop, the workflows were recorded in the department, and in a "day in a live" format – which meant to identify a purchaser's daily work with his tools – presented at the beginning of the group work, along with the case study of another company. In the first step of the process, the situation of

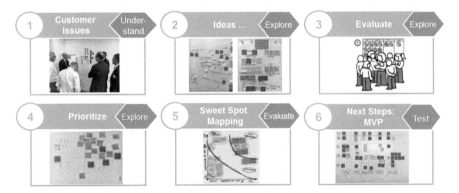

Fig. 9.22 Adaptation of the Design Thinking method for a one-day innovation workshop (Source: author)

the purchasing department was analysed, followed by the identification of the deficits, requirements and needs for the improvement of workflows.

Based on this, innovation ideas were developed in further steps using the moderated metaplan method to simplify workflows. After examining the concrete benefits and the feasibility (duration, effort), a list of priorities with a little more than ten ideas was achieved. For the two best ideas, the team decided to present them in a simple prototype as an App for the purchasers and to be used as a basis for joint follow-up workshops. All in all, the day was considered very successful and, in addition to the ideas developed, especially the overarching cooperation was seen positively.

From experience, as a result of the method and the pragmatic approach, a new behaviour and an open exchange and dialogue will establish and can in future live on in collaborative tools in the community. Namely the use of innovative tools stimulates the motivation to participate and to promote new ideas off the beaten track as well. For example, the creative process of graphic recording can also be used to document workshops and meetings instead of the text form or the traditional PowerPoint images, Fig. 9.23.

There are two graphs which summarise the results of two workshops. Both depictions were made during the event and serve as a basis for the final documentation. The impressively designed pictures remain a lasting memory and give the events an innovative character. The creation is done by designers or can after some training be produced by the "keeper of the minutes" as a personal contribution. Various easy-to-use Apps are available for this task.

A practical example from the field of workplace of the future with the focus on office equipment may serve as a further element for influencing corporate culture. From the point of view of the "war for talents", i.e. the competing for well-trained IT staff, and to initiate the motivation of the employees, the offices have to be designed with new requirements in mind. Future work structures are much more open and flexible than previously. Fixed workplaces and inflexible, regulated working hours are a matter of the past. An impression of such solutions is given in Fig. 9.24.

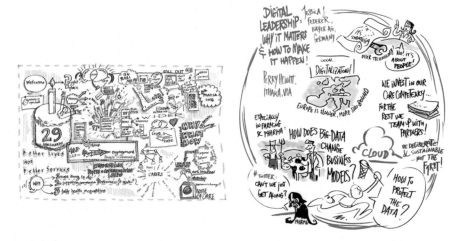

Fig. 9.23 Usage examples of graphic recording [DD17]

9.6 Corporate Culture and Change Management

Fig. 9.24 Innovative office environment at Google in Zurich [BW17]

The photos show office areas of Google at its Zurich location. There are shared-desk concepts with large displays next to team rooms with can be flexibly partitioned. Retreat zones in planted areas and the use of hanging chairs with a view of the mountains offer varied environments for undisturbed and creative work. All workstations are connected to high-performance networks, thus securing access to data and modern IT solutions for work support. In the future, digital assistants are supposed to take on routine tasks such as travel bookings and coordination of appointments in the background.

The design of the workplace of the future is a central theme of the upcoming transformation in the enterprises. What does "Work 4.0" look like tomorrow, and what are the framework conditions which must be created for this? How can IT solutions help to make processes in human resources more efficiently and support them, and how take successful initiatives in education and recruitment to attract the right employees for successful digital transformation? It is important to find convincing answers to these questions, make them an integral part of the digital road map and thus contribute to a change in the corporate culture.

In order to make progress in the field of innovativeness and the ability to innovate, many companies have established so-called "labs", the name often being combined with additions such as innovation, digitisation or mobility, or other sector-specific terms. In Germany, there were more than 60 of such facilities in 2017 [Kel16]. "Labs" are separate organisations located outside the enterprise location, usually in attractive IT-affluent cities. Hip office equipment as well as many free spaces give the teams a start-up feeling and should increase creativity and innovation speed. Reports on the "Labs" of almost all manufacturers are available on the Internet as a report and also on YouTube.

Moreover, many manufacturers have scouting units installed at the international "IT melting points", such as Silicon Valley, Tel Aviv, London, and Bangalore, as well as incubator units to promote collaboration with new partners, relevant universities and research facilities. At the same time, temporary assignments of employees are sent to these organisations, or in-house fairs are held for the presentation and dissemination of the ideas.

The very challenge for the Labs is in many cases the transfer of the work and findings out of the Labs, and especially the start-up culture, into the existing organisation. It is vital to successfully bring a digital spirit into the business and to make it part of the culture. The organisation should be motivated and appropriately prepared and equipped to quickly and sustainably benefit in business from the new digital possibilities, it should acquire a "digital dexterity" [BW17], [Sou16].

This property can be characterised by four thematic fields, which are shown in Fig. 9.25. Digital solutions are always preferable when changes are pending or new technical possibilities are emerging. The goal should always be a full process automation. Wherever possible, data from a variety of sources must be used to derive improved decisions or new initiatives. Comprehensive collaborations based on innovative tools should be common practice. Building knowledge on digitisation should be of high priority to every employee, and the commitment to digital projects must be considered and advanced in the sense of new business models, cross-functionally as well as across company boundaries.

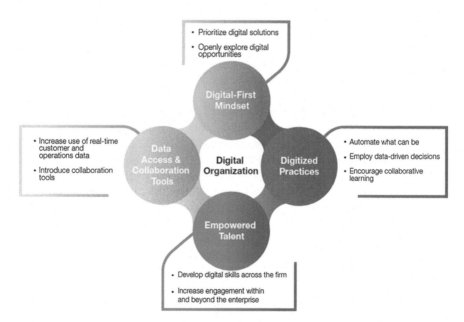

Fig. 9.25 Core competencies of a digital organisation [Bon17]

In conclusion, the innovative projects presented in this chapter from the individual fields of digitisation show that there are already many successful projects and references. From the author's point of view, the potential for improvement is in the speed of implementation and the width of the projects.

References

[Ada17]   Adams, D.: The machines ham it up again: we found the best robots at CES, Digital Trends, January 6, 2017. http://www.digitaltrends.com/cool-tech/machines-ham-best-robotsces-2017/. Drawn: 09.02.2017
[AUDI16] AUDI: Audi City, Audi Media Center, 30.062016. https://www.audi-mediacenter.com/en/press-releases/audi-city-6195. Drawn: 15.02.2017
[Ber13]   Bergert, D.: Ford setzt bei Auto-Apps auf Open Source (Ford is relying on Open Source), CIO of IDG, 19.08.2013. http://www.cio.de/a/ford-setzt-bei-auto-apps-auf-opensource,2926838. Drawn: 02.02.2017
[Bon17]   Bonnet, D., Puram, A., Buvat, J.: Organizing for digital: why digital dexterity matters. Capgemini Consulting 2015. https://www.capgemini-consulting.com/resource-file-access/resource/pdf/digital_orgns_cover_08-12.pdf. Drawn: 15.02.2017
[BMW17]BMW: BMW UK Website – New Cars/Buyers Tools. https://findyour.bmw.co.uk/ZXX5A.html?share=6nxwnb. Drawn: 15.02.2017
[Bre16]   Brennan, C., Lunn, W.: Blockchain – The Trust Disruptor, Equity Research Credit Suisse, 03.08.2016. https://www.finextra.com/finextra-downloads/newsdocs/document-1063851711.pdf. Drawn: 09.02.2017
[Buc17]   Buchter, H.: Neue Produktionsroboter wie Baxter arbeiten Hand in Hand mit Menschen zusammen (New production robots like Baxter work hand in hand with people), Zeit online, 12.01.2017. http://www.zeit.de/2017/01/industrieroboter-jobs-baxter-produktion-arbeitsplaetze. Drawn: 09.02.2017
[Bul14]   Bullis, K.: Tesla Motors' over-the-air repairs are the way forward, MIT Technology Review, 14.01.2014. https://www.technologyreview.com/s/523621/tesla-motors-over-the-air-repairs-are-the-way-forward/. Drawn: 12.02.2017
[Bur16]   Burden, M.: GM OnStar Go taps IBM Watson for in-car marketing.The Detroit News, 25.10.2016. http://www.detroitnews.com/story/business/autos/general-motors/2016/10/25/gm-onstar-go-taps-ibm-watson-car-marketing/92757178/. Drawn: 03.02.2017
[BW17]   BÜROWISSEN: bürowissen – Innovative Bürokonzepte (office know-how – Innovative office concepts), photo gallery; Google Zurich. http://www.buerowissen.ch/Zukunftsvisionen/Innovative-Burokonzepte-/#!prettyPhoto. Drawn: 17.02.2017
[Col16]   Colin, G.: The connected car: at the heart of PSA Groups's strategy – Creating value for both customers and carmaker; PSA Innovation Day 2016; 24.05.2016. https://www.groupe-psa.com/en/finance/analysts-and-investors/investor-day/. Drawn: 02.02.2017
[Cou16]   Coutris, J., Grouvel, A.: Automotive infotainment: how the OEMs contain the digital giants surge into the cockpit, Emerton Market Insights, October 2016. http://www.emerton.co/app/uploads/2016/10/Emerton-Connected-Mobility-Market-insights-Oct-2016.pdf. Drawn: 02.02.2017
[DD17]   DesignDoppel: Examples in picture gallery company designdoppel. http://www.designdoppel.de/graphicrecording/. Drawn: 17.02.2017
[deS16]   de Souza Soares, P.: Design thinking – eine neue Denkschule erobert Deutschlands Strategie-Abteilungen (Design Thinking – a new school of thought conquers Germany's strategy departments, departments, Manager Magazine, 22.01.2016. http://www.manager-magazin.de/magazin/artikel/design-thinking-eine-kreativitaetstechnik-erobert-konzernzentralen-a-1086472.html. Drawn: 17.02.2017

[Eck17] Eckardt, S.: Scania setzt Platooning-System in Singapur ein (Scania is using platooning systems in Singapore), Elektronic automotive, 16.01.2017. http://www.elektroniknet.de/elektronik-automotive/assistenzsysteme/scania-setzt-platooning-system-in-singapur-ein-137605.html. Drawn: 06.02.2017

[FOR16] FORD: Ford smart mobility map. https://media.ford.com/content/fordmedia/fna/us/en/news/2016/02/19/ford-smart-mobility.html. Drawn: 03.02.2017

[Frü16] Früh, F.: Chatbots und die Automatisierung von Kommunikationsprozessen, Bitcom online Artikel. https://www.bitkom.org/Themen/Technologien-Software/Digital-Office/Chatbots.html (2016). Drawn 11.02.2017

[Ger17] Gerster, M.: Nutzung der Plattform: Daimler und Uber kooperieren (Use of the platform: Daimler and Uber cooperate), Automobilwoche, 31.01.2017. http://www.automobilwoche.de/article/20170131/NACHRICHTEN/170139981/nutzung-der-plattform-daimler-und-uber-kooperieren. Drawn: 05.02.2017

[Gra17] Graser, F.: Ford und Toyota streben Open-Source-Standard für Auto-Apps an (Ford and Toyota aim for Open-Source standard for car apps), Elektronikpraxis, 04.01.2017. http://www.elektronikpraxis.vogel.de/themen/embeddedsoftwareengineering/softwarekomponenten/articles/570409/. Drawn: 02.02.2017

[Gri16] Gritz, C.: Tesla bringt neues kostenloses over-the-air Softwareupdate (Tesla with new free over-the-air software update), Windkraft-Journal, 23.09.2016. http://www.windkraft-journal.de/2016/09/23/tesla-bringt-neues-kostenloses-over-the-air-software-updates-zu-ihren-kunden/92501. Drawn: 12.02.2017

[Han16] Hannon, E., Ramkumar, S., McKerracher, C., et al.: An integrated perspective on the future of mobility, McKinsey & Company and Bloomberg, Oct. 2016. http://www.mckinsey.com/business-functions/sustainability-and-resource-productivity/our-insights/an-integrated-perspective-on-the-future-of-mobility. Drawn: 05.02.2017

[Han17] Hannon, E., Ramkumar, S., McKerracher, C., et. al.: An integrated perspektive on the future of mobility. McKinsey & Company and Bloomberg. http://www.mckinsey.com/business-functions/sustainability-and-resource-productivity/ourinsights/an-integrated-perspective-on-the-future-of-mobility (2016, October). Drawn 05.02.2017

[Hon16] Hongwiwat, S.: Blockchain experiences, IBM presentation, 27.10.2016. http://www.slideshare.net/suwath/ibm-blockchain-experience-suwat-20161027. Drawn: 10.02.2016

[IDE17] Ideaconnection: Open innovation platform delivers autonomous bus, ideaconnection, 30.01.2017. https://www.ideaconnection.com/open-innovationsuccess/Open-Innovation-Platform-Delivers-Autonomous-Bus-00624.html. Drawn: 06.02.2017

[Jun16] Jungwirth, J.: Roadmap Mobility 2016: Der Wandel einer Schlüsselindustrie (The transformation of a key industry), 2b_Ahead Think!Tank, 30.07.2016. https://www.2bahead.com/nc/de/tv/rede/video/roadmap-mobilitaet-2026-der-wandel-einer-schluesselindustrie/. Drawn: 06.02.2017

[Jun17] Young, J., Niemeyer, S.: Künstliche Intelligenz im Tourismus – Dein Wegweiser, wie Du mit Chatbots Gäste begeisterst (Artificial Intelligence in Tourism – Your Guide to Inspire Chatbot's Guests), eBook neusta e Tourism GmbH, 12.01.2017. http://labor.neusta-etourism.de/ebook/ebook-chatbots-neusta.pdf. Drawn: 11.02.2017

[Kai15] Kain, D.: In favor of Shop-Click-Drive – the move toward digital retailing is the inevitable result of increased familiarity with self-service transactions, Auto Dealer Today, 12.05.2015. http://www.autodealermonthly.com/channel/internet-department/article/story/2015/05/in-favor-of-shop-click-drive.aspx. Drawn: 15.02.2017

[Kal16] Kalmbach, R., Hoffmann, M., Obermaier, K., et al.: Digitale Transformation stellt klassisches Neuwagengeschäft und etablierte Vertriebsstrukturen infrage (Digital transformation challenges traditional new car business and established sales structures), ATKearney car dealer study, March 2016. http://www.atkearney.de/documents/856314/7822680/BIP+Wechsel+zum+Online-Kauf+kommt+schneller+als+erwartet.pdf/b92e95cd-bb96-4448-9e11-fb1f97f5613f. Drawn: 15.02.2016

References

[Kal16] Kaltofen, T.: Blockchain im Einsatz (Blockchain in use), Computerwoche, 18.10.2016. http://www.computerwoche.de/a/blockchain-im-einsatz,3316539. Drawn: 09.02.2017

[Kel16] Keles, A.: Digital Labs – Deutsche Unternehmen trainieren für den Digitalisierungsmarathon (German companies train for the digitisation marathon), crisp research 01.04.2016. https://www.crisp-research.com/digital-labs-deutsche-unternehmen-trainieren-fur-den-digitalisierungsmarathon/. Drawn: 17.02.2017

[Kni15] Knight, W.: Rebooting the automobile, MIT Technology Review, June 23, 2015. https://www.technologyreview.com/s/538446/rebooting-the-automobile/. Drawn: 03.02.2017

[Kos14] Kossmann, M.: Mensch-Roboter-Kooperation in der Automobilindustrie – Anwendungen, Potentiale und Herausforderungen (Human-robot cooperation in the automotive industry – applications, potentials and challenges), BMW Group, 04.06.2014. http://www.teamwork-arbeitsplatzgestaltung.de/download/vortraege2014/Mensch-Roboter-Kooperation.pdf?m=1461674549. Drawn: 09.02.2017

[Lan16] Lang, K.: Alles, was digitalisiert werden kann, wird digitalisiert werden (Everything that can be digitised, will be digitised; Lecture at the BME Procurement Day, 03.02.2016. https://www.bme.de/alles-was-digitalisiert-werden-kann-wird-digitalisiert-werden-1427/. Drawn: 07.02.2017

[LaP16] LaPlante, A., Sharma, B.: Architecting data lakes – data management for advanced business use cases, O'Reilly Media, März 2016. https://www.oreilly.com/ideas/best-practices-for-data-lakes. Drawn: 15.02.2017

[Lec16] Lecklider, T.: 3D printing drives automotive innovation, Evaluation Engineering, 21.12.2016. https://www.evaluationengineering.com/3d-printing-drives-automotive-innovation. Drawn: 15.02.2017

[McK16] McKenna, D.: Making full vehicle OTA updates a reality, White Paper NXP, B.V., 2016. http://www.nxp.com/assets/documents/data/en/white-papers/Making-Full-Vehicle-OTA-Updates-Reality-WP.pdf. Drawn: 15.02.2017

[McN15] McNamara, P.: Buy a BMW in just 10 minutes, with a new online platform, CAR magazine, Bauer Consumer Media Ltd. 27.11.2015. http://www.carmagazine.co.uk/car-news/industry-news/bmw/buy-a-bmw-in-just-10-minutes-with-new-online-platform/. Drawn: 15.02.2017

[Mol16] Molitch-Hou, M.: Meet Olli: the first autonomous vehicle featuring IBM Watson, ENGINEERING.com, 16.06.2016. http://www.engineering.com/DesignerEdge/DesignerEdgeArticles/ArticleID/12421/Meet-Olli-The-First-Autonomous-Vehicle-Featuring-IBM-Watson.aspx. Drawn: 06.02.2017

[NDR17] NDR2: BMW Werbung für das 5er Fahrzeug, 1. Radiowerbung (BMW advertising for the 5er vehicle, 1. Radio advertising); NDR2, 02.02.2017, 07:58, 2. Youtube ad: https://www.youtube.com/watch?v=lyHV4SMOzn0. Drawn: 02.02.2017

[PSA16] Financial publications PSA Groupe. https://www.groupe-psa.com/en/finance/publications/presentation-des-resultats-en/. Drawn: 02.02.2017

[PSA17] PSA: CAR & WELLBEINGS – Accelerator Program, PSA Group 4 Developers #PG4D. http://developer-program.groupe-psa.com/Accelerator. Drawn: 02.02.2017

[Roe16] Roeder, D.: Wie Blockchain und Elektromobilität zusammenwachsen (How Blockchain and Electromobility are merging), t3n digital pioneers, 23.09.2016; http://t3n.de/news/blockchain-elektromobilitaet-725714/#article. Drawn: 10.02.2017

[Sat16] Satterfield, D., Roland, W., Carlson, S., etal.: IBM & Salesforce – customer experience platform, IBM Präsentation, 2016. https://www-935.ibm.com/services/multimedia/IBM_and_Salesforce_IoT_Customer_Experience_Platform.pdf. Drawn: 12.02.2017

[Sol17] Solomon, B.: Lyft rides tripled last year, but remains far behind Uber, Forbes online, 05.01.2017. http://www.forbes.com/sites/briansolomon/2017/01/05/lyft-rides-tripled-last-year-but-remains-far-behind-uber/#329d50f44f45. Drawn: 05.02.2017

[Sou16] Soule, D., Puram, A., Westerman, G., et al.: Becoming a digital organization: the journey to digital dexterity, MIT Center of Digital Business, Working Paper #301, 05.01.2016. https://papers.ssrn.com/sol3/papers2.cfm?abstract_id=2697688. Drawn: 17.02.2017

[Stö16] Stöckert, K.: Kommunikation – Herzstück erfolgreicher Digitalstrategien, Handbuch Digitalisierung – die vernetzte Gesellschaft (Communication – the heart of successful digital strategies, Handbook digitisation – the networked society), ayway media, 2016. http://handbuch-digitalisierung.de/download-handbuch-digitalisierung/. Drawn: 16.02.2016

[Tan12] Tannou, M., Westerman, G.: Case study: volvo car corporation: shifting from a B2B to a „B2B+B2C" business model, Capgemini Consulting, 22.06.2012. http://ebusiness.mit.edu/research/papers/2012.04_Tannou_Westerman_Volvo%20Cars%20Corporation_298.pdf. Drawn: 12.02.2017

[Vol16] Vollmer, A.: Platooning und autonome trucks all-electronics.de, 11.08.2016. http://www.all-electronics.de/platooning-und-autonome-trucks/. Drawn: 06.02.2017

[WELT17] Welt: CES 2017: Autonomes Fahren – Was kommt danach? (Autonomous driving – what comes thereafter?) WeltN24 09.01.2017. https://www.welt.de/motor/news/article160998502/CES-2017-Autonomes-Fahren.html. Drawn: 05.02.2017

[Wes14] Westerman, G., Bonnet, D., McAfffee: Leading digital – turning technology into business transfromation. Harvard Business Review Press, Boston (2014)

[Win16] Winterhoff, M., Keese, S., Boehler, C., et al.: Think act beyond mainstream digital factories, Roland Berger GmbH (2016). https://www.rolandberger.com/publications/publication_pdf/roland_berger_tab_digital_factories_20160217.pdf. Drawn: 07.02.2017

Chapter 10
Car Mobility 2040

The explanations in this chapter provide a glimpse into the year 2040 and, by means of a few examples, a vision of how the environment, IT, the automotive industry and mobility could develop. The futuristic outlook is intended to strengthen the attitude and the courage for the digital transformation which is running up at full steam to act innovatively and quickly in short steps. In the current situation, speed is of the essence and must come before lengthy careful weighing. It is time to "execute", as then IBM CEO Lou Gerstner at that time hammered into his squad.

In the following outlook, the exponential development of the technologies has to be taken into account, even if this is rather difficult for humans because we are used to linear thinking. The increasing number of projects and declarations of intent in the domain of mobility services, digital services in the infotainment units as well as the activities of all manufacturers in the field of electric vehicles and autonomous driving convey the impression that developments have accelerated sharply since the mid-2010s, and the bend of the exponential curve out of the gradual, steady ascent towards the very steep course is reached. Against this backdrop, the chapter makes bold predictions and bundles them in a day-in-a-life scenario. Firstly, the environment of the automotive industry in the year 2040 is described. The subsequent forecasts are based on a few studies, supplemented by assessments on the part the author, based on his many years of industrial experience.

10.1 Environment

The increasing use of intelligent IT technology, robotics and 3D printing means that in 2040, the average unemployment rate worldwide will be significantly higher than today (20%) and will continue to rise to 24% by 2050 [Win16]. Support for the unemployed is financed through the "taxation of technology usage". The population share of the elderly is growing steadily. The proportion of over 65-year-olds in Germany have increased to 31% and then makes 21.7 million out of a total of

78.2 million citizens [BMV16]. Robots are established in the household as well as in the health care sector, and purchasing drones deliver desired goods shortly after ordering to the agreed transfer point. Credit cards are abolished, and the payment is made with a personal ID chip, which can also be implanted as an option. Telemedicine replaces doctor visits, as a comprehensive sensor system continuously monitors the state of health. These data are also used to add the required vitamins and active ingredients to the food from the 3D printer.

Computers for a price of fewer than one thousand Dollars have a computing power of over one thousand human brains [Kur01]. Storage space and network width are available to a sufficient extent free of charge everywhere. E-mails have disappeared by 2040 and are replaced by real-time communication. Through novel human-machine interfaces, people control digital assistants through thoughts and gestures [Har16]. These also organise virtual meetings to communicate and to collaborate in the metaversum. Cognitive solutions train themselves and continuously learn something new. Impulses to this and also to extensions into new knowledge areas are communicated in simple language dialogue and thus "programmed". Blind persons can "see" again as reading and navigation systems transmit information via the human-machine interface. Neural implants support human organs and compensate weaknesses in eyesight, hearing and tasting [Har16].

These visionary examples as to how the world will continue to develop until 2040 also characterise the environment of the future automotive industry. After an outlook on the year 2030 has already been given in Sect. 5.3, a forecast for the year 2040 is hereinafter.

10.2 Electric Drive and Autonomous Driving

- The proportion of electric vehicles in the new car business is at least 30% [Ran16]. From the author's point of view, a rate of over 50% is to be expected based on the current manufacturer initiatives and the results of intensive battery research. Fuel cells prevail as a technology for electric drives.
- Autonomously driving vehicles account for 50–70% of the traffic [Ran16], [Mot12].
- More than 90% of the autonomous vehicles are used as robotaxis in mobility services.
- Autonomous cars travel at even speed, so traffic congestion and accidents are a thing of the past. In long-distance journeys, vehicles are grouped together as platoons to optimise space requirements and fuel consumption.
- Because of the drastically reduced accident numbers, service providers, vehicle assessors and also lawyers are losing on a significant source of revenue. The occupations are more and more replaced by innovative cognitive technologies.
- The autonomous vehicles monitor their service requirements and independently book the rarely required workshop visits. A significant part of the service is

provided by miniature servicing robots, which are permanently integrated into the vehicle components and are activated as required (cf. Sect. 2.7).
- The number of vehicles on the roads is declining significantly, as is the need for parking space, and parking lots become green areas.
- Despite falling maximum speeds of the autonomous cars, the average speed increases.

10.3 Market Shift

- Vehicle ownership is only regarded as desirable for a few years in some "emerging" countries.
- Private vehicle ownership is focussed on niche segments such as sports and vintage vehicles. The owners organise themselves in user groups with communication and exchange in virtual rooms.
- The volume of new registrations per year will drop by at least 20% due to the higher utilisation levels of the sharing models. In the mid-2010s, around 75 million new vehicles are registered per year [Ado15]. The volume will reduce to below 60 million units past 2040.
- The car trade takes place mainly in virtual showrooms online. Instead of car dealers, there will be distribution centres, from which a large number of the vehicles deliver themselves, autonomously driving, to the customers.
- For men in mid-life crisis, flying drones replace the Porsche as a symbol of prestige.
- Many everyday illnesses have been reduced since the stress associated with driving is gone. The available work capacity increases due to shorter absences.

10.4 Mobility Services and Vehicle Equipment

- In addition to major international providers of mobility services, private peer-to-peer sharing is establishing in which some from the remaining small group of private car owners share their autonomous vehicle with others when they do not need it in order to achieve a cost contribution.
- Handling of the car sharing is done automatically and cashless in the background based on blockchain technologies.
- Due to the attractive prices and the comfortable possibilities of use driven by robotaxis, carsharing becomes the defining inner-city means of transportation. As a result, the urbanisation trend is weakening as the connection of rural areas is problem-free.
- Apart from professional providers of sharing services, new business models are emerging. For example, vehicles are part of the infrastructure in residential districts, or sharing will be part of a therapy or fitness service while driving.

- The design and equipment of the autonomous vehicles differ considerably from the traditional designs with steering wheel and fixed seat alignment. The vehicles evolve to rolling club houses, restaurants, meeting and family rooms with a table in the middle [Way15]. Also equipment as cinema, doctor's practice and classroom are feasible.

10.5 Innovative Process- and Production Structures

- Vehicle production will develop in two directions: Customised models with high luxury equipment are developed in flexible production cells. Mass products for the standard mobility continue to be manufactured on synchronised flexible assembly lines.
- Spare parts are produced locally in service hubs at a rate of more than 70% in 3D printing technology, thus avoiding logistics costs and stocks.
- Business processes are automated at a rate of 80% on the basis of thinking IT solutions.
- The development of vehicles is largely automated by "robo-engineers" or "thinking" IT solutions. Tests are done virtually, and prototype construction is only required at a reduced level.
- Due to the even and continuous driving of the robotaxis, the running performance of the vehicles is increased. The load profiles of the cars also change and are to be considered when the components are designed.
- Vehicles are developed on the basis of a central IT unit. This dominates the vehicle with engine power, driving characteristics, equipment and integration. This is how the "software defined vehicle" has established itself. This also includes the fact that certain equipment features or a special performance behaviour is activated by software and paid as required.
- Connected Services do no longer distinguish between vehicle, smartphone or computer. There is a uniform user ID which allows customers to use their personal software environment in a fully synchronised manner at any time – even in rental cars or in the IT environment of a hotel.
- A superordinate intermodal traffic control is established (see Sect. 5.3.1). In continuous connection with the vehicles, the infrastructure, public transport and also the customer. The system undertakes the holistic optimisation, for example with the parameters travel time, utilisation and avoidance of traffic congestion. It can also flexibly adapt permissible travelling directions of roads. Traffic signs, traffic lights and traffic guidance systems are only interesting from a historical point of view.

This vision on the automotive industry in the year 2040 is already clearly visible in some of the above mentioned fields, for example in the areas of electric drive and autonomous driving. Therefore, many of the forecasts are likely to come true, and automotive manufacturers should align their initiatives and projects in the field of digitisation to these perspectives. The same applies to future vehicle designs, and

Fig. 10.1 Vision cars 2040 [Reh16], [Wal15], [Hol17]

especially to the redesign of the interior, which is fundamentally changing with autonomous driving. For illustrative purposes, Fig. 10.1 shows studies of some car models in 2040.

The picture summarises three different aspects of the vehicles in 2040. The upper part shows the orientation of the seats. The four users wear virtual reality glasses and are obviously roaming in different worlds. In the lower part on the left, the focus is on flexibly positionable car seats, while the design study of a futuristic Ferrari is on the lower right. It can be expected that in this vehicle segment private ownership and self-driving will continue in the future, just fun and joy are the focus here rather than mobility.

10.6 Day-in-a-Life of a "Liquid Workforce"

The outlook on the year 2040 is complemented by a day-in-a-life scenario, thus illustrating the interplay of future developments. The description of the day is about Ernest, a 30-year-old single and independent programmer in the IT industry, specialising in virtual reality simulations in the interior design area. He works as a freelance entrepreneur in various projects for different clients. Here is Ernest's daily routine:

- Wake-up at 06:28 – the time was set autonomously by the personal digital assistant based on the diary and taking into account the preferences of Ernest.

- First on the agenda is six kilometres of jogging on the treadmill. While running, the body functions are monitored and current blood values are transmitted to the virtual hospital as a basis for continuous health monitoring via the implanted "health chip". Through the use of the chip, Ernest was able to switch to the "proactive tariff" and reduce the costs of his health insurance policy by 20%.
- The chip also carries the radio-transmissible encrypted personal ID of Ernst, which is used in a variety of ways, for example, as the basis of payment services, for "walk-trough" border controls, which by now is practised in 80 countries. The ID also serves as the basis for marketing personal data to mobile service providers, insurance companies, banks and also retailers.
- After the shower, at 07:55 the robotaxi of a restaurant is at the door for the trip to a meeting. The breakfast table in the vehicle is prepared. To the desired cereals some individual additives have been added, the dosage of which has been determined by the analysis of the previously recorded health data. Freshly brewed coffee and a fresh fruit salad are also served for breakfast.
- Ernest has his breakfast during the drive. On a large screen, Ernest watches up-to-date news that is specifically configured for him based on his past interests and behaviour. Particularly interesting is the report on the first community on the Mars which also offers sites for sale to terrestrials.
- After a congestion-free journey, despite traditional rush hour, arrival at a shared office is at 08:12. The address of the office had been communicated by the personal assistant of Ernest to the control of the robotaxi, together with the breakfast wishes, when the service request was made.
- In the office, the meeting takes place with three other self-employed persons, who are jointly known as "Liquid Workforce" (cf. Sect. 3.6.2) in a project to equip a new office building with vintage furniture. The three colleagues are interior designers, fabric and wood experts. The meeting is about finalising the design for seating furniture. Prototypes were, on the basis of the results of the virtual brainstorming on the previous day, created overnight in the 3D printing process. To test the functionality and the feel of the fabric and wood samples brought along, they are meeting in person in the office. The team adapts the design in the virtual space for a better understanding of the real materials. Intermediate steps of the work are coordinated with the customer in the virtual showroom of the project. Finally, the team is satisfied and starts the 3D printing of a historic writing desk and a "stressless"-lounger.
- Meanwhile it is 11:15. The team's virtual assistant has arranged a meeting with the customer in the newly-built office building in order to finalise the furnishings on the premises – however not until 15:30.
- The team decides to spend the resulting break together and prepare a lunch with fresh ingredients in a nearby cookery studio. The digital assistant books the appointment in the studio and makes menu suggestions. The team chooses Thai curry chicken with fresh salad.
- Basic ingredients such as rice and spices are ordered automatically and delivered to the reserved kitchen area. In the meantime, the team splits up to get the fresh ingredients. Two robotaxis are waiting outside the door. Two colleagues get the

10.6 Day-in-a-Life of a "Liquid Workforce"

chicken from the poultry farm, while Ernest and his other colleague drive to the nearby "herb garden", a large garden facility on the former car park of an automobile factory, which now serves as a museum.

- At 11:45, having arrived at the garden, the pre-booked bicycles are already waiting, in order to ride to the fields for the harvest by own hands. With fresh vegetables and salad they get back to the cookery studio per bike for the sake of body fitness. Arrival 12:50.
- To abridge it at this point: cooking using the virtual instructions was a team-building and successful experience. After the meal, the team is back at the office at 15:25, where the 3D printing of the sample furniture is finished by now. With a spacious robotaxi van the team drives to the new office building.
- The team shows the customer the exhibits with the realistically simulated wood and fabric structures. To complement these impressions, Ernest launches his Virtual Reality Show on the overall furnishing of the house. The customer gains realistic impressions of the overall fit-out via hologram displays. Sound backgrounds are also played. In dialogue with the customer, changes are made to the design which appears directly in the virtual world.
- At 17:15. The overall concept was finalised and the result documented. The automatically generated drafts, parts lists and equipment data go directly to three vendors as an invitation to tender and requesting offers to become the general contractor. At 18:10 the last offer is available in the virtual project room. The customer brings the company's decision-making circle into the virtual space. Five other colleagues connect and join from different workplaces around the world to assess and decide the design and the offers. At 19:10 the order was placed.
- The project team had parted at 18:15. On his journey home, Ernest is asked to attend a virtual meeting. Because of his references, he was immediately linked into an "RFQ space" (Request for Quotation) to participate in a tender. Ernest initiates his digital assistant to compile information on the order within 12 h on which he can follow up the next morning.
- In the background, Ernest's digital assistant creates the billing for the office design services and submits it to the customer's payment machine via a blockchain-controlled process.
- Meanwhile it is 19:10 and Ernest spontaneously decides to attend the birthday party of a friend in a resort 120 km away. So Ernest is waiting for a flying drone. En route, Ernest exchanges information with his assistant in order to manage the preparations for his offer. The flight time is also enough to complete an e-learning unit on a technology concept that is relevant to the new tender.
- At about 20:00 Ernest arrives at the party. As a gift he hands over a voucher for a "vintage ride" in a classic car with manual transmission. Since the birthday child has no driver's experience, a driving lesson of 1 h with instructor is included. The exotic gift is greatly appreciated.
- At the party, Ernest also meets many people for the first time. A camera integrated in his spectacles sends shots to his digital assistant who carries out

research in the background and gives Ernest detailed information on the persons through his mini-in-ear loudspeaker.
- At around 22:30 pm Ernest decides to take an off-time and decouples from the virtual world. He would like to enjoy traditional dialogues and direct experience.
- At the end of the party, a flying drone is waiting for the trip back home – a long, exciting day.

This scenario, which in some respects certainly is somewhat overstated, illustrates the saturation of digitisation, new forms of mobility and innovative work concepts while focusing on healthy eating and harmonious work/life balance in the year 2040. At the same time, the increasing rapidity and frequency of the interactions can also be seen.

10.7 Conclusion

Adaptation of speed, agility, innovativeness as well as risk-taking are the prerequisites for successful digital transformations. In view of the continuing rapid development of digitisation, the automotive industry must, in the opinion of the author, accelerate the implementation to ensure the competitiveness of the industry and needs to remedy the hesitant approach which is still quite common. Entrepreneurship has to be put before lengthy multiple coordination. Only with these qualities can the established manufacturers, the Goliaths, manage to successfully compete against the new challengers, the Davids. This would hopefully falsify some projections which predict that the manufacturers do not have a chance in disruptive changes due the lack of just these qualities, and therefore the David always wins [Chr17], [Gla15].

References

[Ado15] Adolf, J., Rommerskirchen, S., Balzer, C., et al.: Shell PKW-Szenarien bis 2040 – Fakten, Trends und Perspektiven für Auto-Mobilität (Shell car scenarios by 2040 – facts, trends and perspectives for car mobility), ed. Shell Deutschland GmbH (2015). https://www.prognos.com/uploads/tx_atwpubdb/140900_Prognos_Shell_Studie_Pkw-Szenarien2040.pdf. Drawn: 23.02.2017

[BMV16] BMVi: Verkehr und Mobilität in Deutschland – Daten und Fakten kompakt (Transport and Mobility in Germany – facts and figures compact) – Bundesministerium für Verkehr und digitale Infrastruktur, Juli 2016.https://www.bmvi.de/SharedDocs/DE/Publikationen/G/verkehr-und-mobilitaet-in-deutschland.pdf?__blob=publicationfile. Drawn: 22.02.2017

[Chr17] Christensen, C.: Disruptive innovation – key concepts. http://www.claytonchristensen.com/key-concepts/. Drawn: 24.02.2017

[Gla15] Gladwell, M.: David and Goliath – underdogs, misfits, and the art of battling giants. Back Bay Books, New York (2015)

References

[Har16] Harari, Y.: Homo deus – a brief history of tomorrow. Pengiun Random House, London (2016)

[Hol17] Holzer, H.: Autositz der Zukunft – Sitzecke für Fahrer und Beifahrer (Car seat of the future – lounge area for driver and passengers), Motorzeitung.de, 23.02.2017. http://motorzeitung.de/news.php?newsid=414392. Drawn: 23.02.2017

[Kur01] Kurzweil, R.: Homo sapiens: Leben im 21. Jahrhundert – Was bleibt vom Menschen? (Life in the 21st century – what is left of man? 4 edn.). Ullstein Paperback (2001)

[Mot12] Motavalli, J.: Self-driving cars will take over by 2040, Forbes online, 25.09.2012. http://www.forbes.com/sites/eco-nomics/2012/09/25/self-driving-cars-will-take-over-by-2040/#1d227b2f21f2. Drawn: 22.02.2017

[Ran16] Randall, T.: Here's how electric cars will cause the next oil crisis, Bloomberg online, 25.02.2016. https://www.bloomberg.com/features/2016-ev-oil-crisis/. Drawn: 22.02.2017

[Reh16] Rehme, M.: Vernetzte Mobilität: (Ge-) Fahren in der Zukunft (Networked Mobility: Driving [and threats] in the Future) Institut für vernetzte Mobilität, 12.04.2016.http://www.new-mobility-leipzig.de/media/Programm/block2/01_Rehme_new-mobility-2016_Rehme.pdf. Drawn: 23.02.2017

[Wal15] Wallerang, L.: So sieht ein Ferrari in 25 Jahren aus (This is how a Ferrari looks in 25 years), Motorzeitung.de; 15/12/2015. http://motorzeitung.de/news.php?newsid=323759. Drawn: 23.02.2017

[Way15] Wayner, P.: Future Ride Version 2.0; CreateSpace Independent Publishing Platform, 2nd edn. 14.04.2015

[Win16] Wintermann, O., Daheim, C.: 2050: Die Zukunft der Arbeit. (The Future of Work.) Results of an international Delphi Study in the Millennium Project, Bertelsmann-Stiftung, March 2016. https://www.bertelsmann-stiftung.de/fileadmin/files/BSt/Publikationen/GrauePublikationen/BST_Delphi_Studie_2016.pdf. Drawn: 22.02.2017

Glossary

3D–Chip 3D–Chip architectures — three-dimensionally stacked chips — a promising path to increase the energy efficiency and performance of computers in the future. These architectures reduce the chip footprint, straighten the data connections and increase the bandwidth for data transfer in the chip many times over (IBM Zurich).

3D–Printing is a generative manufacturing method (unlike cutting manufacturing processes) for producing three-dimensional objects from plastic, metal or ceramics. It is therefore also referred to as an additive manufacturing process or Additive Manufacturing (AM). On the basis of a digital model of the component to be produced, the material is applied in layers of powdered or liquid material and solidified by means of curing or fusing.

Additive Manufacturing 3D–Printing.

Agile project management methods are used in inter-divisionally operating teams to achieve swift project success. Well-known examples are -> Design Thinking and -> Scrum.

API means Application Programming Interface. API is important for programmers as an interface between the device to be programmed (e.g. an operating system) and the programme. Thus it is possible to use unsophisticated commands to trigger complex functions (Computer Dictionary).

App means Application Software. An application system is a software system for performing tasks in various application areas and runs on a desktop computer, mobile device or server (Encyclopedia of Business Informatics).

Appliances are integrated turnkey systems which are optimised for a specific application. In a casing there are servers, memory, and system software including visualisation, and partially software for data management as well.

Big Data Big Data describes data volumes that can be processed only to a limited extent by current databases and data management tools due to their complexity, diversity or speed. In contrast to existing Business Intelligence (BI) and Data Warehouse Systems (DWS), Big Data applications usually work without complex data preparation. (Encyclopedia of Business Informatics).

Blockchain is a decentralised protocol through which information of any kind – e.g. financial transactions – can be transmitted and released for all parties involved. In the blockchain, information is split into blocks. Each block is connected to the preceding block by a checksum and also contains a checksum of the entire information.

Business Component Model (CBM) is a modelling method developed by IBM for systematic company and process structuring as a basis for weak-point analyses.

Business Platform is a business model that allows for simplified exchange between two or more interdependent groups, usually consumers and producers. Some examples are product, service or payment platforms.

Chatbot (derived from "chat" [talk] and "bot", shortened version of robot [work]) are software programmes which enable automated communication in software applications as a service offer for questions regarding the operation or as an information system in public transport. However, a chatbot can also be addressed directly in order to perform certain activities in interaction with the connected IT systems.

CKD (Completely Knocked Down) The abbreviation describes a product's state following complete disassembly into its individual parts. Because of customs regulations and/or high import tariffs, in particular automobiles are not dispatched as ready end products but rather as individual parts, assembled in the country of destination and prepared for distribution.

Cloud Computing includes technologies and business models to dynamically provide IT resources and to bill their use according to flexible payment models. Instead of operating IT resources, such as servers or applications, in own enterprise data centers, these are available in a flexible and demand-orientated manner in the form of a service-based business model via the Internet or an Intranet (Gabler Wirtschaftslexikon).

Cognitive Computing is an approach of computer technology which aims to make computer technology act like a human brain. A prerequisite for this type of artificial intelligence is that the system is not pre-programmed for all potential problem solutions, but rather the corresponding computer system gradually learns autonomously (onPage).

Connected Services are different service offerings around the automobile. In detail, the topics are security and remote maintenance, fleet management, mobility, navigation, infotainment, insurance and payment systems. A service bundle contains various networked vehicle services, which are based on a series of telematics functions, belong to a self-contained business segment, have access to the same profit sources, address the same target group and are based on different business models with new compensation concepts (Oliver Wyman).

Connectivity (network capability) is the ability to connect or network computers using hardware and software; it characterises a network-capable computer for instance. It also describes the quality of a connection between computers (Enzyklo.de).

Glossary

Content Management System is software for the collaborative creation, editing and organisation of content mostly on websites, but also in other forms of media (Wikipedia).

Content Provider means the provision of content for third party use and covers various applications, services and topics for purchase or free use on online platforms (content).

CRM (Customer Relationship Management) is a strategic approach which is used to fully plan, control and execute all interactive processes with customers. It includes database marketing and appropriate CRM software as a control instrument (Gabler Wirtschaftslexikon).

Crowdsourcing Is a digital form of work organisation in which companies access the knowledge, creativity, the manpower and the resources of a large mass of participants over the Internet in order to incorporate them into the operating performance (Encyclopedia of Business Informatics).

Cyber-Physical Systems (CPS) refers to the coupling of information and software components with mechanical or electronic components, which communicate with each other in real time via a communication infrastructure such as the Internet. The mechanical or electronic parts of a CPS are realised through so-called embedded systems, which can perceive their local environment by means of sensors and influence the physical environment via actuators. (Encyclopedia of Business Informatics).

Data Lakes as opposed to -> Datawarehouses, save all kinds of raw data unmodified without further preparation, including the coupling to the source data, in a flexible system.

Datastream Management Systems (DSMS) manage continuous data streams.

Datawarehouse DWH imports data from different source systems, transfers them into a target data structure and stores them in the DWH. Reports and evaluations are fed from the target data of the DWH, while the output data in the source systems are overwritten.

DevOps The term being a word combination of Development and (IT) Operations, aims to improve the collaboration between software developers and IT in order to enable fast release cycles and short deployment times.

Digital Immigrant Digital Native

Digital Native means a person who has grown up in the digital world. The counter term is the Digital Immigrant, describing someone who first became acquainted with this world in the adult age (Wikipedia).

Digital Services are services which are ordered and provided through the Internet or similar electronic media, e.g. online access to databases or programme downloads (Gabler Wirtschaftslexikon).

Digital Twin Is the digital form of a real object based on a CAD-3D model, to which all product properties, functions and process parameters have been assigned. As an intelligent 3D model, the digital twin allows a realistic simulation in a computer-supported simulation environment. (Schunk).

Real-time Monitoring means the continuous detection of a machine condition by measuring and analysing physical variables, e.g. vibrations, temperatures, position (Wikipedia).

E-Learning describes all forms of learning which use digital solutions for the presentation of study material and the dialogue between learners and teachers.

Electronic Blood A project by IBM Research Zurich, ETH Zurich and other partners to develop a microchannel system with an electrochemical flow battery that simultaneously cools and supplies energy to the 3D-Chip stack. The fluid used is also referred to as electronic blood because it both absorbs and delivers electrical energy (IBM Research Zurich).

Embedded Control Units (ECUs) control one or more systems or subsystems in a vehicle.

Embedded Software means software which is integrated in a technical device with a computer and is in charge of controlling, managing or monitoring the system without intervention by the user (Encyclopedia of Business Informatics).

ERP Enterprise Resource Planning is a business application software for the integrated planning and controlling not only of the production, but also of all the resources involved in the value creation of a company.

Foglets are microscopic nanoscale robots devised by scientists that are equipped with microelectronics, sensors and actuators and can form networks of solid structures.

FORTRAN FORmula TRANslation is a programming language that has been developed and optimised for numerical computations and is now ISO-standardised.

Gamification is an approach to transfer game principles to entrepreneurial interests and thereby to motivate personnel and spark their interest.

Gateway (protocol converter) enables communication between multiple networks, which may be based on different protocols by converting them to the appropriate format (Computer Dictionary).

Hackathon is a word combination of hack and marathon. Under this term, company events are organised on a topic, and students and interested digital natives are invited to programme an App to solve problems in the given subject area within a set time frame.

Hype Cycle is an overview of innovative technologies published by the analysis and consulting company Gartner.

IAM Identity and Access Management simplifies and automates the recording, control and management of users' electronic identities and associated access rights (by search security).

Incident Management comprises the organisational and technical process of reacting to detected or suspected safety incidents or disruptions in IT systems as well as preparatory measures and processes (Wikipedia).

Industry 4.0 Is the fourth development stage of production, after the use of water and steam power, mass production and automation, towards the Internet of Things. This opens up new possibilities to link resources, services and people in production on the basis of -> Cyber-Physical Systems in real time (WG-Standpunkt Industrie 4.0).

Infotainment Artificial word derived from information and entertainment is a media offer which both informs and entertains the recipients.

In-Memory Technology is a concept in which an execution program as well as the required data are stored in the main memory (RAM). This differentiates it from the usual data storage on physical media and allows faster execution times.

Instant Messaging are Internet services that facilitate text or character-based communication in real-time.

Internet of Things (IoT) refers to the linking of objects to the Internet, so that these objects can communicate independently via the Internet and thus perform various tasks for the owner. The scope of application ranges from general information supply and automated ordering to warning and emergency functions (Gabler Wirtschaftslexikon).

IP address is an Internet protocol-based address that is assigned to a device connected to the Internet and uniquely identifies it. This means that data packets can be transported to a physical postal address from a sender to a recipient or a group of recipients (according to Wikipedia).

IT Container consist of a complete runtime environment, an application including all dependencies, libraries and configuration functions. Containers ensure that software runs reliably after being moved from one environment to another, for example, from the developer's laptop to a test environment, or from the test environment to production (Rubens, computerwoche 2015).

Collaboration Tools also called Groupware, are software programs which support communication and collaboration processes in teams and organisations (Encyclopedia of Business Informatics).

Machine Learning is an umbrella term for the "artificial" generation of knowledge from experience: an artificial system learns from examples and can generalise these after the end of the learning phase (Wikipedia).

Massive Open Online Course (MOOC) are free, open online learning offers with very large numbers of participants. With the interactive format, the participants develop their learning material themselves out of the specifications.

Master Data Management (MDM) covers all strategic, organisational, methodological and technological activities related to the master data of a company. Its task is to ensure the consistent, complete, up-to-date, correct and high-quality master data to support the performance processes of a company (Encyclopedia of Business Informatics).

Micro Services are an architectural model of information technology, in which complex application software is composed of small, independent processes that communicate with one another using language-independent programming interfaces (Wikipedia).

Mobile Development Platform MDP allows the rapid development of -> Apps for smartphones, tablets, desktops and TV sets.

Moore's Law An observation formulated by Gordon Moore in 1965 according to which the number of circuit components on an integrated circuit doubles about every 2 years. From Moore's Law it cannot be concluded that with the number of transistors on a computer chip the processing power of the computer increased linearly as well (Wikipedia).

Nanotubes are extremely small hollow bodies with a diameter of less than 100 nanometers (0.0001 mm).

Nanotechnology is a broad spectrum of new cross-sectional technologies with materials, components and systems whose functions and applications are based on the special properties of nanoscaled (≤ 100 nm) dimensions (Fraunhofer Nanotech).

Neuromorphous Chips consist of conventional silicon-based components, however imitate the structure of nerve cells and brain. They work with learning neuronal networks and are particularly suitable for pattern recognition, but still under development (IBM).

Neuronal network means in the neurosciences a number of interconnected neurons, which form a functional connection as part of a nervous system (Wikipedia).

OEM Original Equipment Manufacturer stands for the producer of the original equipment. An OEM partner uses software or hardware under his license in his own products or in the form of packages (for instance CD burner and burning software) (Thewald).

Open Innovation means the active strategic development of the collective knowledge base, creativity and innovation potentials outside the own company (Community of Knowledge).

OpenStack Technology This is a comprehensive software portfolio for building open Cloud solutions, developed by the OpenStack Foundation and available as an Open Source solution.

Platform as a Service (PaaS) is a service that provides a programming model and developer tools to create and run Cloud-based applications (computerwoche).

RFID means "Radio Frequency Identification" and refers to technologies for object identification via radio waves (Encyclopedia of Business Informatics).

Robocabs denotes autonomous-driving taxis.

Scrum is a term from the Rugby sports where it describes an "arranged crowd", in order to restart the game after minor fouls. The Scrum process is carried out in iterations, whereby a target project goal is divided into partial steps, which are then worked on step by step in a given time frame in creation loops, so-called sprints.

Shared Service Centre is an organisational model for in-house services. Similar services of the company headquarters are linked to those of the individual divisions, business units or departments and are grouped together in an organisational unit (Centre). The individual business units, disciplines or departments can then access this unit jointly and as required (Shared) in order to receive the appropriate service (businesss-wissen.de).

Single Sign-On (SSO) is an authentication process for a user's session. Thus a user specifies a name and a password to access multiple applications. There are no additional input prompts for an identification. (TechTarget).

SOA Service-Oriented Architecture: SOA tries to adjust the software directly to the business processes of a company. For this purpose, the system is divided into so-called Services. Services are small, loosely linked and stand-alone software components. Combining these services creates an application system which should remain easily adaptable and modifiable (Software Engineering Faculty University of Hanover).

Social Media is a collective term for internet-based media offers, which are based on social interaction and the technical possibilities of the so-called Web 2.0. The focus is on communication and the exchange of user-generated content. Webblogs, forums, social networks, wikis and podcasts are used as technologies (Gabler Wirtschaftslexikon).

Social Navigation refers to concepts in which users can orientate themselves by the behaviour and suggestions of other users when navigating the World Wide Web. Navigation notes can either be exchanged in direct dialogue or arise indirectly through the traces of past navigation activities or artifacts left in the information space (Baier, Weinreich, Wollenweber).

Software Defined Storage (SDS) is a key element in building a service-oriented infrastructure. It allows to procure, add and provide storage resources simply depending on demand (Computerwoche).

SWOT analysis is an abbreviation of Analysis of strengths, weakness, opportunities and threats; The analysis allows a positioning of one's own activities versus the competition (Gabler Wirtschaftslexikon).

Total Cost of Ownership (TCO) Is the sum of all costs for the purchase of an asset (e.g. a computer system), its use and, where applicable, disposal costs. Total Costs of Ownership are a design aspect during the product development phase; the intention is to try to understand and influence the reasons for the purchase decision of the customer (Gabler Wirtschaftslexikon).

Vulnerability Management addresses the security-relevant weak points in IT systems. Vulnerability management is designed to develop processes and techniques that enable a security configuration to be introduced and managed in the company (IT knowledge) in order to increase IT security.

Wearable also called Wearable Computer, is a computer system that is attached to the user's body during use. Wearable computing differs from the use of other mobile computer systems in that the main activity of the user is not the use of the computer itself, yet a computer-assisted activity in the real world (Wikipedia).

Web 2.0 Is an evolutionary stage with regard to the offer and use of the World Wide Web, where the focus is no longer the mere distribution of information by website operators, but the participation of users on the web and the generation of further additional usage (Encyclopedia of Business Informatics).

Printed by Printforce, the Netherlands